Introduction to Enzyme and Coenzyme Chemistry

Second Edition

TIM BUGG

Professor of Biological Chemistry, Department of Chemistry,
University of Warwick, UK

Blackwell
Publishing

© 1997, 2004 by Blackwell Publishing Ltd

Editorial offices:
Blackwell Publishing Ltd, 9600 Garsington Road, Oxford OX4 2DQ, UK
 Tel: +44 (0)1865 776868
Blackwell Publishing Inc., 350 Main Street, Malden, MA 02148-5020, USA
 Tel: +1 781 388 8250
Blackwell Publishing Asia Pty Ltd, 550 Swanston Street, Carlton, Victoria 3053, Australia
 Tel: +61 (0)3 8359 1011

First published 1997 by Blackwell Science
Second edition published 2004 by Blackwell Publishing

Library of Congress Cataloging-in-Publication Data

Bugg, Tim.
  Introduction to enzyme and coenzyme chemistry / Tim Bugg.–2nd ed.
    p. cm.
  Includes bibliographical references and index.
  ISBN 1-4051-1452-5 (pbk. : alk. paper)
  1. Enzymes. 2. Coenzymes. I. Title.

  QP601.B955 2004
  572′.7–dc22
  2003025117

ISBN 1-4051-1452-5

A catalogue record for this title is available from the British Library

Set in 10/13 pt Times
by Kolam Information Services Pvt. Ltd, Pondicherry, India
Printed and bound in Great Britain
by Ashford Colour Press, Gosport

The publisher's policy is to use permanent paper from mills that operate a sustainable forestry policy, and which has been manufactured from pulp processed using acid-free and elementary chlorine-free practices. Furthermore, the publisher ensures that the text paper and cover board used have met acceptable environmental accreditation standards.

For further information on Blackwell Publishing, visit our website:
www.blackwellpublishing.com

Contents

Preface

Since the publication of the first edition in 1998, the field of chemical biology has, I would say, become more a part of the core research and teaching of Chemistry Departments around the world. As the genes and proteins involved in important biological problems are elucidated, so they become accessible for study at the molecular level by chemists. Enzymology is a core discipline within chemical biology, since enzymes are the biological catalysts that make things happen within cells: they translate gene sequence into biological function.

The main feature of the first edition that I wanted to improve was the figures. My original intention was to have a case study in each chapter, where I would illustrate the chemical mechanism, with curly arrows, and on the facing page would show the three-dimensional structure of the enzyme active site, and/ or tertiary structure of the enzyme. One of the fascinations of enzymology is the interplay between structure and function, and I wanted to try to convey this to students. In the first edition I was only able to include a set of colour plates in the middle of the book. In the second edition, I have prepared a series of two-colour pictures using Rasmol, to accompany the text. I hope that these pictures convey the desired ideas.

I have also updated the first edition with recent observations and references from the literature, and have added a few new topics. I have written a new chapter entitled Radicals in Enzyme Catalysis (Chapter 11), which includes the discovery of protein radicals and the discovery of SAM-dependent radical reactions. I have also mentioned the recent proposals for protein dynamics (Chapter 3) and proton tunnelling (Chapter 4) in enzyme catalysis, in a way that I hope will be accessible to undergraduate students.

I would like to thank my colleagues, researchers, and students at Warwick for their encouragement, and hope that the book is useful to chemical biologists everywhere.

Tim Bugg
University of Warwick

Representation of Protein Three-Dimensional Structures

In the second edition I have used the program Rasmol to draw representations of protein three-dimensional structures. Rasmol was developed by Roger Sayle (GlaxoSmithKline Pharmaceuticals), is freely available from the internet (http://www.umass.edu/microbio/rasmol), and can be downloaded with instructions for its use. There are several packages available for representation of protein structures, but Rasmol is straightforward to learn and freely available.

In order to view a protein structure, you must first download the PDB file from the Brookhaven Protein Database, which contains all the data for published X-ray crystal structures and NMR structures of proteins and nucleic acids. I have included the PDB filename for each of the pictures that I have drawn, in the figure legend. I recommend that you try downloading a few of these, and viewing each one on a computer screen, as you can turn the structure around and hence get a really good feel for the three-dimensional structure of the protein. You can download the PDB file from http://www.rcsb.org/pdb.

Once you have downloaded the PDB file, then you run the Rasmol program, and open the PDB structure file to view the structure. You can view the protein backbone in several different ways: as individual atoms (wireframe), protein backbone (backbone), strands or cartoons. In most of the pictures in this edition I have drawn the protein backbone in cartoon format. I have then selected certain catalytic amino acid residues, and highlighted them in red, and selected any bound substrate analogues or coenzymes, and highlighted them in black. In preparing figures for the book I used only black and red, but on a computer screen you can use a wide range of colours and you can prepare your own multi-colour pictures!

Further reading

R.A. Sayle & E.J. Milner-White (1995) RasMol: Biomolecular graphics for all. *Trends Biochem. Sci.*, **20**, 374–6.

1 From Jack Beans to Designer Genes

1.1 Introduction

Enzymes are giant macromolecules which catalyse biochemical reactions. They are remarkable in many ways. Their three-dimensional structures are highly complex, yet they are formed by spontaneous folding of a linear polypeptide chain. Their catalytic properties are far more impressive than synthetic catalysts which operate under more extreme conditions. Each enzyme catalyses a single chemical reaction on a particular chemical substrate with very high enantio-selectivity and enantiospecificity at rates which approach 'catalytic perfection'. Living cells are capable of carrying out a huge repertoire of enzyme-catalysed chemical reactions, some of which have little or no precedent in organic chemistry. In this book I shall seek to explain from the perspective of organic chemistry what enzymes are, how they work, and how they catalyse many of the major classes of enzymatic reactions.

1.2 The discovery of enzymes

Historically, biological catalysis has been used by mankind for thousands of years, ever since fermentation was discovered as a process for brewing and bread-making in ancient Egypt. It was not until the 19th century AD, however, that scientists addressed the question of whether the entity responsible for processes such as fermentation was a living species or a chemical substance. In 1897 Eduard Buchner published the observation that cell-free extracts of yeast containing no living cells were able to carry out the fermentation of sugar to alcohol and carbon dioxide. He proposed that a species called 'zymase' found in yeast cells was responsible for fermentation. The biochemical pathway involved in fermentation was subsequently elucidated by Embden and Meyerhof – the first pathway to be discovered.

The exquisite selectivity of enzyme catalysis was recognised as early as 1894 by Emil Fischer, who demonstrated that the enzyme which hydrolyses sucrose, which he called 'invertin', acts only upon α-D-glucosides, whereas a second enzyme 'emulsin' acts only upon β-D-glucosides. He deduced that these two enzymes must consist of 'asymmetrically built molecules', and that 'the enzyme and glucoside must fit each other like a lock and key to be able to exert a chemical influence upon each other'. Fischer's 'lock and key' hypothesis remained a powerful metaphor of enzyme action for many years. The crystallisation in

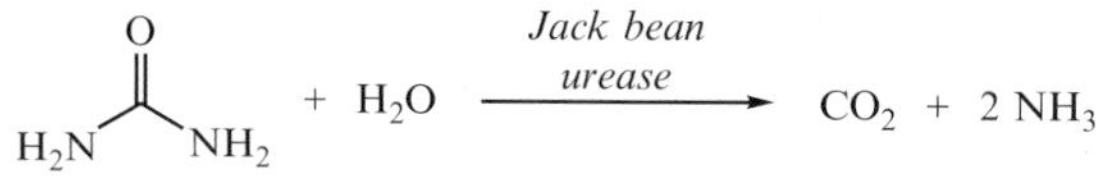

Figure 1.1 Reaction catalysed by the enzyme urease.

1926 of the enzyme urease from Jack beans by Sumner proved beyond doubt that biological catalysis was carried out by a chemical substance (Figure 1.1).

The recognition that biological catalysis is mediated by enzymes heralded the growth of biochemistry as a subject, and the elucidation of the metabolic pathways catalysed by enzymes. Each reaction taking place on a biochemical pathway is catalysed by a specific enzyme. Without enzyme catalysis the uncatalysed chemical process would be too slow to sustain life. Enzymes catalyse reactions involved in all facets of cellular life: metabolism (the production of cellular building blocks and energy from food sources); biosynthesis (how cells make new molecules); detoxification (the breakdown of toxic foreign chemicals); and information storage (the processing of deoxyribonucleic acid (DNA)).

In any given cell there are present several thousand different enzymes, each catalysing its specific reaction. How does a cell know when it needs a particular enzyme? The production of enzymes, as we shall see in Chapter 2, is controlled by a cell's DNA, both in terms of the specific structure of a particular enzyme and the amount that is produced. Thus different cells in the same organism have the ability to produce different types of enzymes and to produce them in differing amounts according to the cell's requirements.

Since the majority of the biochemical reactions involved in cellular life are common to all organisms, a given enzyme will usually be found in many or all organisms, albeit in different forms and amounts. By looking closely at the structures of enzymes from different organisms which catalyse the same reaction, we can in many cases see similarities between them. These similarities are due to the evolution and differentiation of species by natural selection. So by examining closely the similarities and differences of an enzyme from different species we can trace the course of molecular evolution, as well as learning about the structure and function of the enzyme itself.

Recent developments in biochemistry, molecular biology and X-ray crystallography now allow a far more detailed insight into how enzymes work at a molecular level. We now have the ability to determine the amino acid sequence of enzymes with relative ease, whilst the technology for solving the three-dimensional structure of enzymes is developing apace. We also have the ability to analyse their three-dimensional structures using molecular modelling and then to change the enzyme structure rationally using site-directed mutagenesis. We are now starting to enter the realms of enzyme engineering where, by rational design, we can modify the genes encoding specific enzymes, creating the 'designer genes' of the title. These modified enzymes could in future perhaps be used to catalyse new types of chemical reactions, or via gene therapy used to correct genetic defects in cellular enzymes which would otherwise lead to human diseases.

1.3   The discovery of coenzymes

At the same time as the discovery of enzymes in the late 19th and early 20th centuries, a class of biologically important small molecules was being discovered which had remarkable properties to cure certain dietary disorders. These molecules were christened the vitamins, a corruption of the phrase 'vital amines' used to describe their dietary importance (several of the first-discovered vitamins were amines, but this is not true of all the vitamins). The vitamins were later found to have very important cellular roles, shown in Table 1.1.

The first demonstration of the importance of vitamins in the human diet took place in 1753. A Scottish naval physician, James Lind, found that the disease scurvy, prevalent amongst mariners at that time, could be avoided by deliberately including green vegetables or citrus fruits in the sailors' diets. This discovery was used by Captain James Cook to maintain the good health of his crew during his voyages of exploration in 1768–76. The active ingredient was elucidated much later as vitamin C, ascorbic acid.

The first vitamin to be identified as a chemical substance was thiamine, lack of which causes the limb paralysis beriberi. This nutritional deficiency was first identified in the Japanese Navy in the late 19th century. The incidence of beriberi in sailors was connected with their diet of polished rice by Admiral Takaki, who eliminated the ailment in 1885 by improvement of the sailors' diet. Subsequent investigations by Eijkman identified a substance present in rice husks able to cure beriberi. This vitamin was subsequently shown to be an essential 'cofactor' in cellular decarboxylation reactions, as we shall see in Chapter 7.

Over a number of years the family of vitamins shown in Table 1.1 was identified and their chemical structures elucidated. Some, like vitamin C, have simple structures, whilst others, like vitamin B_{12}, have very complex structures. It has taken much longer to elucidate the molecular details of their biochemical mode of action. Many of the vitamins are in fact coenzymes: small organic cofactors which are used by certain types of enzyme in order to carry out particular classes of reaction. Table 1.1 gives a list of the coenzymes that we are going to encounter in this book.

1.4   The commercial importance of enzymes in biosynthesis and biotechnology

Many plants and micro-organisms contain natural products that possess potent biological activities. The isolation of these natural products has led to the

Table 1.1 The vitamins.

Vitamin	Chemical name	Deficiency disease	Biochemical function	Coenzyme chemistry
A	Retinol	Night blindness	Visual pigments	—
B_1	Thiamine	Beriberi	Coenzyme (TPP)	Decarboxylation of α-keto acids
B_2	Riboflavin	Skin lesions	Coenzyme (FAD, FMN)	$1e^-/2e^-$ redox chemistry
Niacin	Nicotinamide	Pellagra	Coenzyme (NAD)	Redox chemistry
B_6	Pyridoxal	Convulsions	Coenzyme (PLP)	Reactions of α-amino acids
B_{12}	Cobalamine	Pernicious anaemia	Coenzyme	Radical re-arrangements
C	Ascorbic acid	Scurvy	Coenzyme, anti-oxidant	Redox agent (collagen formation)
D	Calciferols	Rickets	Calcium homeostasis	—
E	Tocopherols	Newborn haemolytic anaemia	Anti-oxidant	—
H	Biotin	Skin lesions	Coenzyme	Carboxylation
K	Phylloquinone	Bleeding disorders	Coenzyme, anti-oxidant	Carboxylation of glutamyl peptides
	Folic acid	Megaloblastic anaemia	Coenzyme (tetrahydrofolate)	1-carbon transfers
	Pantothenic acid	Burning foot syndrome	Coenzyme (CoA, phosphopantotheine)	Acyl transfer

CoA, coenzyme A; FAD, flavin adenine dinucleotide; FMN, flavin mononucleotide; NAD, nicotinamide adenine dinucleotide; PLP, pyridoxal-5′-phosphate; TPP, thiamine pyrophosphate.

discovery of many biologically active compounds such as quinine, morphine and penicillin (Figure 1.2) which have been fundamental to the development of modern medicine.

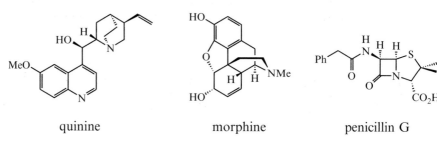

quinine   morphine   penicillin G

Figure 1.2 Structures of quinine, morphine and penicillin G.

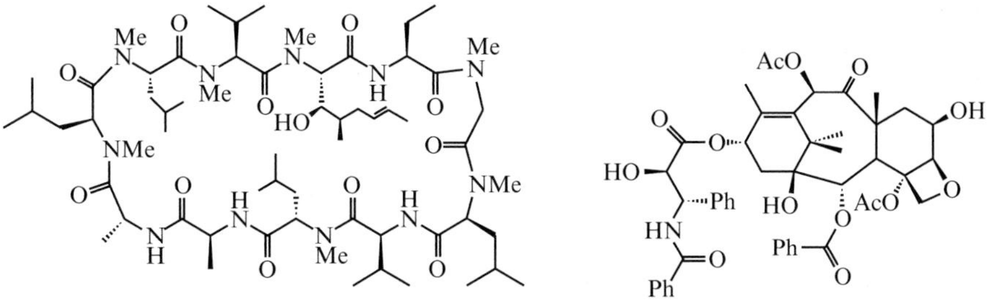

cyclosporin A   taxol

Figure 1.3 Structures of cyclosporin A and taxol.

The process of natural product discovery continues today, with the recent identification of important compounds such as cyclosporin A, a potent immunosuppressant which has dramatically reduced the rejection rate in organ transplant operations; and taxol, an extremely potent anti-cancer drug isolated from yew bark (Figure 1.3).

Many of these natural products are structurally so complex that it is not feasible to synthesise them in the laboratory at an affordable price. Nature, however, is able to biosynthesise these molecules with apparent ease using enzyme-catalysed biosynthetic pathways. Hence, there is considerable interest in elucidating the biosynthetic pathways for important natural products and using the enzymes to produce natural products *in vitro*. One example of this is the industrial production of semi-synthetic penicillins using a naturally occurring enzyme, penicillin acylase (Figure 1.4). Penicillin G, which is obtained from growing *Penicillium* mould, has certain clinical disadvantages; enzymatic deacylation and chemical re-acylation give a whole range of 'semi-synthetic' penicillins which are clinically more useful.

The use of enzyme catalysis for commercial applications is an exciting area of the biotechnology industry. One important application that we shall encounter is the use of enzymes in asymmetric organic synthesis. Since enzymes are highly efficient catalysts that work under mild conditions and are enantiospecific, they can in many cases be used on a practical scale to resolve racemic mixtures of chemicals into their optically active components. This is becoming increasingly important in drug synthesis, since one enantiomer of a drug usually

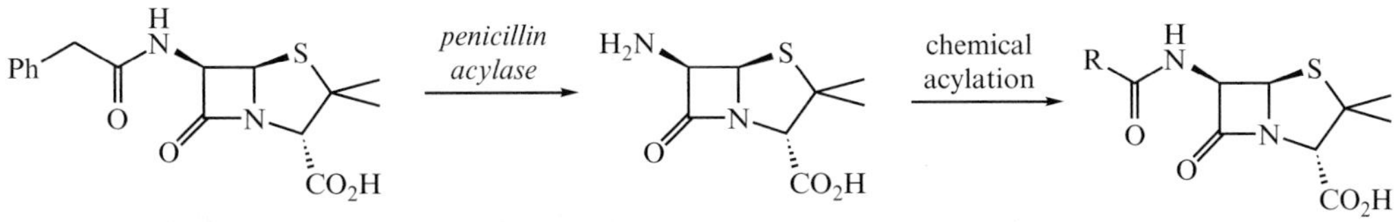

Figure 1.4 Industrial production of a semi-synthetic penicillin using penicillin acylase.

has very different biological properties from the other. The unwanted enantiomer might have detrimental side-effects, as in the case of thalidomide, where
one enantiomer of the drug was useful in relieving morning sickness in pregnant
women, but the other enantiomer caused serious deformities in the newborn
child when the racemic drug was administered.

1.5   The importance of enzymes as targets for drug discovery

If there is an *essential* enzyme found uniquely in a certain class of organism or
cell type, then a selective *inhibitor* of that enzyme could be used for selective
toxicity against that organism or cell type. Similarly, if there is a significant
difference between a particular enzyme found in bacteria as compared with the
same enzyme in humans, then a selective inhibitor could be developed for
the bacterial enzyme. If this inhibitor did not inhibit the human enzyme, then
it could be used as an antibacterial agent. Thus, *enzyme inhibition is a basis for
drug discovery*.

This principle has been used for the development of a range of pharmaceut-
ical and agrochemical agents (Table 1.2). We shall see examples of important
enzyme targets later in the book. In many cases resistance to these agents has
emerged due to mutation in the structures of the enzyme targets. This has
provided a further incentive to study the three-dimensional structures of

Table 1.2 Commercial applications of enzyme inhibitors.

Anti-bacterial agents	**Penicillins** and **cephalosporins** inactivate the *transpeptidase* enzyme which normally makes cross-links in the bacterial cell wall (peptidoglycan), leading to weakened cell walls and eventual cell lysis. **Streptomycin** and **kanamycin** inhibit protein synthesis on bacterial ribosomes, whereas mammalian ribosomes are less affected.
Anti-fungal agents	**Ketoconazole** inhibits *lanosterol 14α-demethylase*, an enzyme involved in the biosynthesis of an essential steroid component of fungal cell membranes. **Nikkomycin** inhibits *chitin synthase*, an enzyme involved in making the chitin cell walls of fungi.
Anti-viral agents	**AZT** inhibits the *reverse transcriptase* enzyme required by the HIV virus in order to replicate its own DNA.
Insecticides	Organophosphorus compounds such as **dimethoate** derive their lethal activity from the inhibition of the insect enzyme *acetylcholinesterase* involved in the transmission of nerve impulses.
Herbicides	**Glyphosate** inhibits the enzyme *EPSP synthase* which is involved in the biosynthesis of the essential amino acids phenylalanine, tyrosine and tryptophan (see Section 8.5).

AZT, 3′-azido,3′-deoxythymidine; EPSP, 5-enolpyruvyl-shikimate-3-phosphate.

enzyme targets, and has led to the development of powerful molecular modelling software for analysis of enzyme structure and *de novo design* of enzyme inhibitors.

The next two chapters are 'theory' chapters on enzyme structure and enzyme catalysis, followed by a 'practical' chapter on methods used to study enzymatic reactions. Chapters 5–11 cover each of the major classes of enzymatic reactions, noting each of the coenzymes used for enzymatic reactions. Finally, there is a brief introduction in Chapter 12 to other types of biological catalysis. In cases where discussion is brief the interested reader will find references to further reading at the end of each chapter.

Further reading

Historical development of enzymology

T.D.H. Bugg (2001) The development of mechanistic enzymology in the 20th century. *Nat. Prod. Reports*, **18**, 465–93.

Enzymes in biosynthesis and biotechnology

J. Mann (1987) *Secondary Metabolism*, 2nd edition. Clarendon Press, Oxford.
C.H. Wong & G.M. Whitesides (1994) *Enzymes in Synthetic Organic Chemistry*. Pergamon, Oxford.

Medicinal chemistry

G.L. Patrick (2001) *An Introduction to Medicinal Chemistry*, 2nd edition. OUP, Oxford.
R.B. Silverman (2001) *The Organic Chemistry of Drug Design and Drug Action*. Academic Press, San Diego.

2   All Enzymes are Proteins

2.1   Introduction

Enzymes are giant molecules. Their molecular weight varies from 5000 to
5 000 000 Da, with typical values in the range 20 000–100 000 Da. At first
sight this size suggests a bewildering complexity of structure, yet we shall see
that enzymes are structurally assembled in a small number of steps in a fairly
simple way.

Enzymes belong to a larger biochemical family of macromolecules known as
proteins. The common feature of proteins is that they are polypeptides: their
structure is made up of a linear sequence of α-amino acid building blocks joined
together by amide linkages. This linear polypeptide chain then 'folds' to give a
unique three-dimensional structure.

2.2   The structures of the L-α-amino acids

Proteins are composed of a family of 20 α-amino acid structural units whose
general structure is shown in Figure 2.1. α-Amino acids are chiral molecules:
that is, their mirror image is not superimposable upon the original molecule.

Each α-amino acid can be found as either the L- or D-isomer depending on
the configuration at the α-carbon atom (except for glycine where R=H). All
proteins are composed only of L-amino acids, consequently enzymes are inherently chiral molecules – an important point. D-amino acids are rare in biological
systems, although they are found in a number of natural products and notably
in the peptidoglycan layer of bacterial cell walls (see Chapter 9).

The α-amino acids used to make up proteins number only 20, whose structures are shown in Figure 2.2. The differences between these 20 lie in the nature
of the side chain R. The simplest amino acids are glycine (abbreviated Gly or
simply G), which has no side chain, and alanine (Ala or A), whose side chain is a
methyl group. A number of side chains are hydrophobic ('water-hating') in
character, for example the thioether of methionine (Met); the branched aliphatic
side chains of valine (Val), leucine (Leu) and isoleucine (Ile); and the aromatic

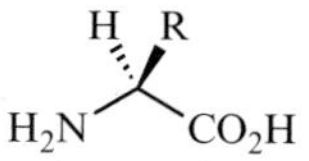

general structure of
an L-α-amino acid

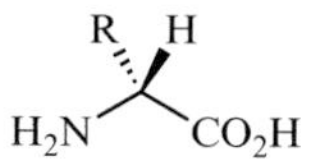

general structure of
a D-α-amino acid

Figure 2.1 General structure of L- and
D-amino acids.

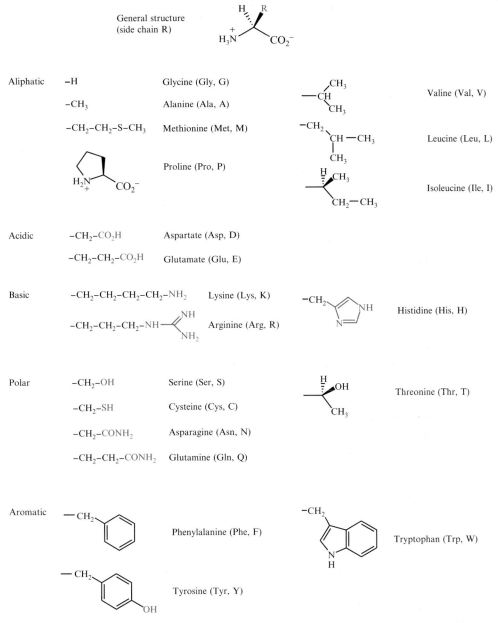

Figure 2.2 The side chains of the 20 α-amino acids found in proteins. Whole amino acid structure shown for proline. Functionally important groups highlighted in red.

side chains of phenylalanine (Phe) and tryptophan (Trp). The remainder of the amino acid side chains are hydrophilic ('water-loving') in character. Aspartic acid (Asp) and glutamic acid (Glu) contain carboxylic acid side chains, and their corresponding primary amides are found as asparagine (Asn) and glutamine

(Gln). There are three basic side chains consisting of the ε-amino group of lysine (Lys), the guanidine group of arginine (Arg), and the imidazole ring of histidine (His). The polar nucleophilic side chains that will assume a key role in enzyme catalysis are the primary hydroxyl of serine (Ser), the secondary hydroxyl of threonine (Thr), the phenolic hydroxyl group of tyrosine (Tyr) and the thiol group of cysteine (Cys).

The nature of the side chain confers certain physical and chemical properties upon the corresponding amino acid, and upon the polypeptide chain in which it is located. The amino acid side chains are therefore of considerable structural importance and, as we shall see in Chapter 3, they play key roles in the catalytic function of enzymes.

2.3   The primary structure of polypeptides

To form the polypeptide chain found in proteins each amino acid is linked to the next via an amide bond, forming a linear sequence of 100–1000 amino acids – this is the primary structure of the protein. A portion of the amino-terminal (or N-terminal) end of a polypeptide is shown in Figure 2.3, together with the abbreviated representations for this peptide sequence.

The sequence of amino acids in the polypeptide chain is all-important. It contains all the information to confer both the three-dimensional structure of proteins in general and the catalytic activity of enzymes in particular. How is this amino acid sequence controlled? It is specified by the nucleotide sequence of the corresponding *gene*, the piece of DNA (deoxyribonucleic acid) which encodes for that particular protein in that particular organism. To give an idea of how this is achieved, I will give a simplified account of how the polypeptide sequence is derived from the gene sequence. For a more detailed description the reader is referred to biochemical textbooks.

Genes are composed of four deoxyribonucleotides (or 'bases'): deoxyadenine (dA), deoxycytidine (dC), deoxyguanine (dG) and deoxythymidine (dT),

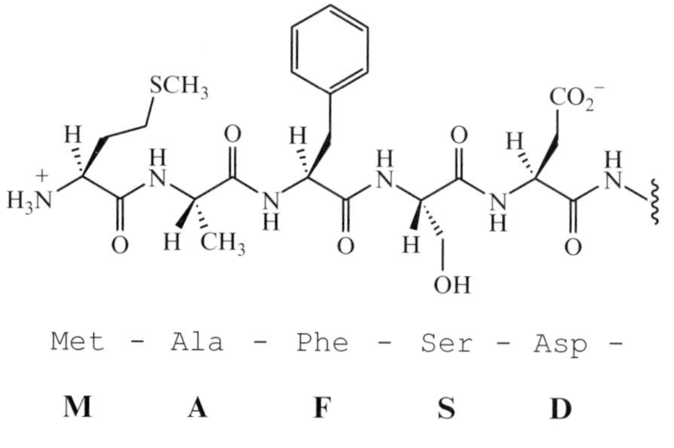

Met – Ala – Phe – Ser – Asp –

M   **A**   **F**   **S**   **D**

Figure 2.3 Structure of the N-terminal portion of a polypeptide chain.

arranged in a specific linear sequence. To give some idea of size, a typical gene might consist of a sequence of 1000 nucleotide bases encoding the information for the synthesis of a protein of approximately 330 amino acids, whose molecular weight would be 35–40 kDa.

How is the sequence encoded? First the deoxyribonucleotide sequence of the DNA strand is transcribed into messenger ribonucleic acid (mRNA) containing the corresponding ribonucleotides adenine (A), cytidine (C), guanine (G) and uridine (U, corresponding to dT). The RNA strand is then translated into protein by the biosynthetic machinery known as ribosomes, as shown in Figure 2.4. The RNA sequence is translated into protein in sets of three nucleotide bases, one set of three bases being known as a 'triplet codon'. Each codon encodes a single amino acid. The code defining which amino acid is derived from which triplet codon is the 'universal genetic code', shown in Figure 2.5. This universal code is followed by the protein biosynthetic machinery of all organisms.

As an example we shall consider in Figure 2.6 the N-terminal peptide sequence Met–Ala–Phe–Ser–Asp illustrated in Figure 2.3. The first amino acid at the N-terminus of each protein is always methionine, whose triplet codon is AUG. The next triplet GCC encodes alanine; UUC encodes phenyl-alanine; UCC encodes serine; and GAC encodes aspartate. Translation then continues in triplets until one of three 'stop' codons is reached; at this point protein translation stops. Note that for most amino acids there is more than one possible codon: thus if UUC is changed to UUU, phenylalanine is still encoded, but if changed to UCC then serine is encoded as above.

In this way the nucleotide sequence of the gene is translated into the amino acid sequence of the encoded protein. An important practical consequence is that the amino acid sequence of an enzyme can be determined by nucleotide sequencing of the corresponding gene, which is now the most convenient way to determine a protein sequence.

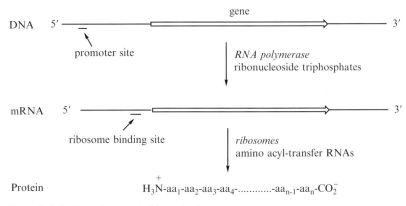

Figure 2.4 Pathway for protein biosynthesis.

AAA	Lys	ACA	Thr	AGA	Arg	AUA	Ile
AAG	Lys	ACG	Thr	AGG	Arg	AUG	Met
AAC	Asn	ACC	Thr	AGC	Ser	AUC	Ile
AAU	Asn	ACU	Thr	AGU	Ser	AUU	Ile
CAA	Gln	CCA	Pro	CGA	Arg	CUA	Leu
CAG	Gln	CCG	Pro	CGG	Arg	CUG	Leu
CAC	His	CCC	Pro	CGC	Arg	CUC	Leu
CAU	His	CCU	Pro	CGU	Arg	CUU	Leu
GAA	Glu	GCA	Ala	GGA	Gly	GUA	Val
GAG	Glu	GCG	Ala	GGG	Gly	GUG	Val
GAC	Asp	GCC	Ala	GGC	Gly	GUC	Val
GAU	Asp	GCU	Ala	GGU	Gly	GUU	Val
UAA	Stop	UCA	Ser	UGA	Stop	UUA	Leu
UAG	Stop	UCG	Ser	UGG	Trp	UUG	Leu
UAC	Tyr	UCC	Ser	UGC	Cys	UUC	Phe
UAU	Tyr	UCU	Ser	UGU	Cys	UUU	Phe

Figure 2.5 The universal genetic code.

```
                     start
                     codon
mRNA....GGATCAUGGCCUUCUCCGACUACCGGA....
                AUG GCC UUC UCC GAC ...
                Met Ala Phe Ser Asp
```

Figure 2.6 Translation of mRNA into protein.

2.4   Alignment of amino acid sequences

Most biochemical reactions are found in more than one organism, in some cases in all living cells. If the enzymes which catalyse the same reaction in different organisms are purified and their amino acid sequences are determined, then we often see similarity between the two sequences. The degree of similarity is usually highest in enzymes from organisms which have evolved recently on an evolutionary timescale. The implication of such an observation is that the two enzymes have evolved divergently from a common ancestor.

Over a long period of time, changes in the DNA sequence of a gene can occur by random mutation or by several types of rare mistakes in DNA replication. Many of these mutations will lead to a change in the encoded protein sequence in such a way that the mutant protein is inactive. These mutations are likely to be lethal to the cell and are hence not passed down to the next generation. However, mutations which result in minor modifications to non-essential residues in an enzyme will have little effect on the activity of the enzyme, and will therefore be passed onto the next generation.

So if we look at an alignment of amino acid sequences of 'related' enzymes from different organisms, we would expect that catalytically important amino

```
Alignment of N-terminal 15 amino acids of four sequences in 3-letter codes:

                         1             5               10              15
E. coli MhpB             Met His Ala Tyr Leu His Cys Leu Ser His Ser Pro Leu Val Gly
A. eutrophus MpcI        Met Pro Ile Gln Leu Glu Cys Leu Ser His Thr Pro Leu His Gly
P. paucimobilis LigB Met Ala Arg Val Thr Thr Gly Ile Thr Ser Ser His Ile Pro Ala Leu Gly
E. coli HpcB             Met Gly Lys Leu Ala Leu Ala Ala Lys Ile Thr His Val Pro Ser Met Tyr
                                                                 +    *           *


Alignment of N-terminal 60 amino acids of two sequences in 1-letter codes:

                  1         11        21        31        41        51
E. coli MhpB      MHAYLHCLSH SPLVGYVDPA QEVLDEVNGV IASARERIAA FSPELVVLFA PDHYNGFFYD
A. eutrophus MpcI MPIQLECLSH TPLHGYVDPA PEVVAEVERV QAAARDRVRA FDPELVVVFA PDHFNGFFYD
                  * **** +** ****** **+ ** * *+**+*+ * * *****+** ***+******

* = identically conserved residue   + = functionally conserved residue
```

Figure 2.7 Amino acid sequence alignment.

acid residues would be invariant or 'conserved' in all species. In this way sequence alignments can provide clues for identifying important amino acid residues in the absence of an X-ray crystal structure. For example, in Figure 2.7 there is an alignment of the N-terminal portion of the amino acid sequence of a dioxygenase enzyme MhpB from *Escherichia coli* with 'related' dioxygenase enzymes from *Alcaligenes eutrophus* (MpcI) and *Pseudomonas* (LigB) and another *E. coli* enzyme HpcB. Clearly there are a small number of conserved residues (indicated by a *) which are very important for activity, and a further set of residues for which similar amino acid side chains are found (e.g. hydroxyl-containing serine and threonine, indicated with a +).

Furthermore, sequence similarity is sometimes observed between different enzymes which catalyse similar reactions or use the same cofactor, giving rise to 'sequence motifs' found in a family of enzymes. We shall meet some examples of sequence motifs later in this book.

2.5   Secondary structures found in proteins

When the linear polypeptide sequence of the protein is formed inside cells by ribosomes, a remarkable thing happens: the polypeptide chain spontaneously folds to form the three-dimensional structure of the protein. All the more remarkable is that from a sequence of 100–1000 amino acids a *unique* stable three-dimensional structure is formed. It has been calculated that if the protein folding process were to sample each of the available conformations then it would take longer than the entire history of the universe – yet, in practice, it takes a few seconds! The mystery of protein folding is currently a topic of intense research, and the interested reader is referred to specialist articles on this topic. Factors that seem to be important in the folding process are:

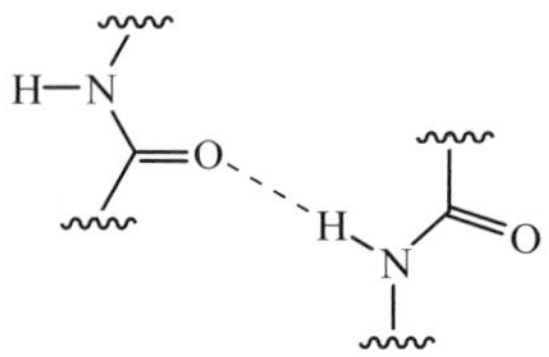

Figure 2.8 A hydrogen bond.

(1) packing of hydrophobic amino acid side chains and exclusion of solvent water;
(2) formation of specific non-covalent interactions;
(3) formation of secondary structures.

Secondary structure is the term given to local regions (10–20 amino acids) of stable, ordered three-dimensional structures held together by hydrogen-bonding, that is non-covalent bonding between acidic hydrogens (O−H, N−H) and lone pairs as shown in Figure 2.8.

There are at least three stable forms of secondary structure commonly observed in proteins: the α-helix, the β-sheet and the β-turn. The α-helix is a helical structure formed by a single polypeptide chain in which hydrogen bonds are formed between the carbonyl oxygen of one amide linkage and the N−H of the amide linkage four residues ahead in the chain, as shown in Figure 2.9.

In this structure each of the amide linkages forms two specific hydrogen bonds, making it a very stable structural unit. All of the amino acid side chains point outwards from the pitch of the helix, consequently amino acid side chains that are four residues apart in the primary sequence will end up close in space. Interactions between such side chains can lead to further favourable inter-actions within the helix, or with other secondary structures. A typical α-helix is shown in Figure 2.10a, showing the positions of the side chains of the amino acid residues. In Figure 2.10b, the same helix is drawn in 'ribbon' form, a convenient representation that is used for drawing protein structures.

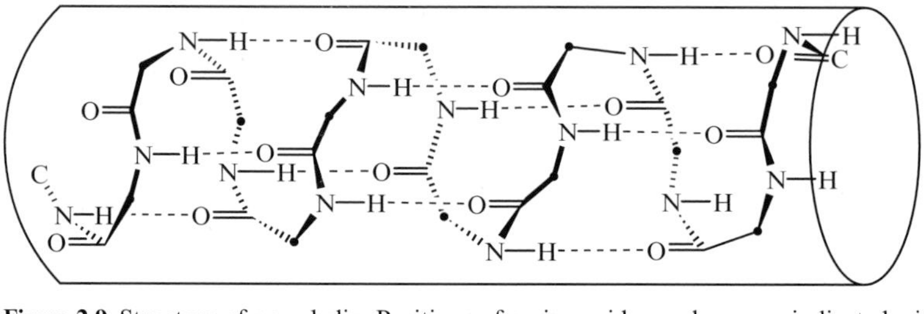

Figure 2.9 Structure of an α-helix. Positions of amino acid α-carbons are indicated with dots.

(a)   (b)

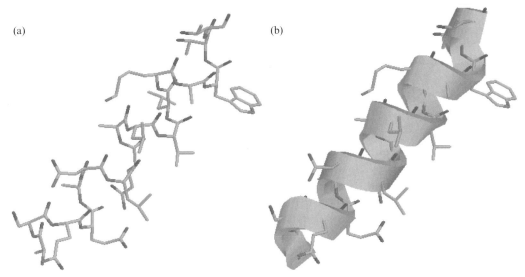

Figure 2.10 Structure of an α-helix, (a) showing positions of the polypeptide chain and side chains and (b) showing the same structure in ribbon format.

The β-sheet is a structure formed by two or more linear polypeptide strands, held together by a series of interstrand hydrogen bonds. There are two types of β-sheet structures: parallel β-sheets, in which the peptide strands both proceed in the same amino-to-carboxyl direction; and anti-parallel, in which the peptide strands proceed in opposite directions. Both types are illustrated in Figure 2.11. Figure 2.12a shows an example of two anti-parallel β-sheets in a protein structure, with Figure 2.12b showing the same β-sheets in 'ribbon' form.

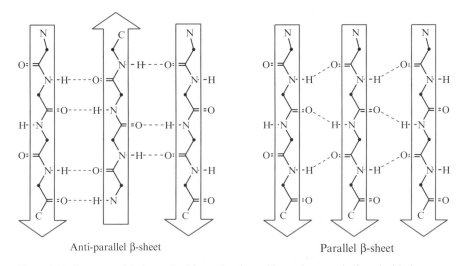

Anti-parallel β-sheet   Parallel β-sheet

Figure 2.11 Structure of β-sheets. Positions of amino acid α-carbons are indicated with dots.

(a)

(b)

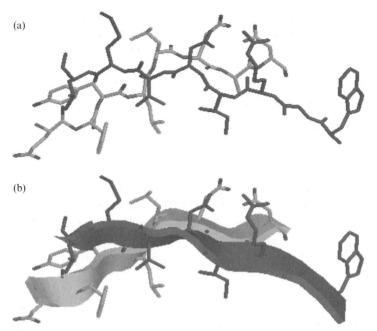

Figure 2.12 Structure of two anti-parallel β-sheets, (a) showing positions of the polypeptide chain and side chains and (b) showing the same structure in ribbon format.

The β-turn is a structure often formed at the end of a β-sheet which leads to a 180° turn in the direction of the peptide chain. An example of a β-turn is shown in Figure 2.13, where the role of hydrogen bonding in stabilising such structures can be seen.

2.6   The folded tertiary structure of proteins

The three-dimensional structure of protein sub-units, known as the tertiary structure, arises from packing together elements of secondary structure to form a stable global conformation, which in the case of enzymes is catalytically active. The packing of secondary structural units usually involves burying

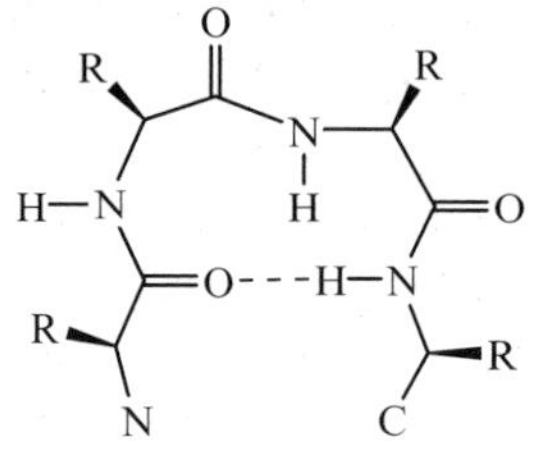

Figure 2.13 Structure of a β-turn.

hydrophobic amino acid side chains on the inside of the protein and positioning hydrophilic amino acid side chains on the surface.

Although in theory the number of possible protein tertiary structures is virtually infinite, in practice proteins are often made up of common structural motifs, from which the protein structure can be categorised. Common families of protein structure are:

(1) α-helical proteins;
(2) α/β structures;
(3) anti-parallel β structures.

Members of each class are illustrated below, with α-helices and β-sheets represented in ribbon form. The α-helical proteins are made up only of α-helices which pack onto one another to form the tertiary structure. Many of the haem-containing cytochromes which act as electron carriers (see Chapter 6) are four-helix 'bundles', illustrated in Figure 2.14 in the case of cytochrome b_{562}. The family of globin oxygen carriers, including haemoglobin, consist of a more complex α-helical tertiary structure. The α/β structures consist of regular arrays of β-sheet–α-helix–*parallel* β-sheet structures. The redox flavoprotein flavodoxin contains five such parallel β-sheets, forming a twisted β-sheet surface interwoven with α-helices, as shown in Figure 2.15. Anti-parallel β structures consist of regular arrays of β-sheet–β-turn–*anti-parallel* β-sheet. For example, the metallo-enzyme superoxide dismutase contains a small barrel of anti-parallel β-sheets, as shown in Figure 2.16.

Frequently, proteins consist of a number of 'domains', each of which contains a region of secondary structure. Sometimes a particular domain has a specific function, such as binding a substrate or cofactor. Larger proteins often consist of more than one tertiary structure, which fit together to form the active 'quaternary' structure. In some cases a number of identical sub-units can bind together to form a homodimer (two identical sub-units), trimer or tetramer, or in other cases non-identical sub-units fit together to form highly complex quaternary structures. One familiar example is the mammalian oxygen transport protein haemoglobin, which consists of a tetramer of identical 16-kDa sub-units.

How are protein tertiary structures determined experimentally? The most common method for solving three-dimensional structures of proteins is to use X-ray crystallography, which involves crystallisation of the protein, and analysis of the diffraction pattern obtained from X-ray irradiation of the crystal. The first protein structure to be solved by this method was lysozyme in 1965, since which time several hundred crystal structures have been solved. Recent advances in nuclear magnetic resonance (NMR) spectroscopy have reached the point where the three-dimensional structures of small proteins (<15 kDa) in solution can be solved using multi-dimensional NMR techniques.

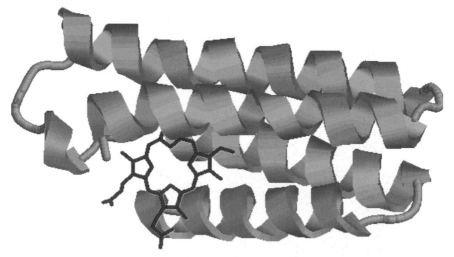

Figure 2.14 Structure of cytochrome b$_{562}$ (PDB file 256B), a four-helix bundle protein. Haem cofactor shown in red.

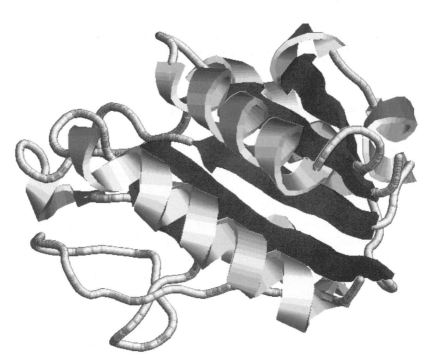

Figure 2.15 Structure of flavodoxin (PDB file 1AHN), a redox carrier protein containing five parallel β-sheets, each connected by an intervening α-helix. Parallel β-sheets shown in red.

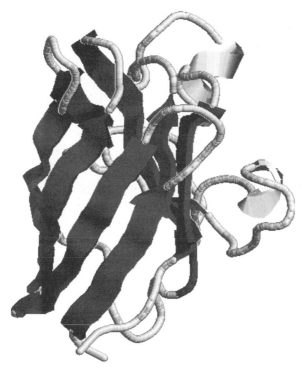

Figure 2.16 Structure of superoxide dismutase (PDB file 1CB4), a β-barrel protein containing eight anti-parallel β-sheets. Anti-parallel β-sheets shown in red.

2.7   Enzyme structure and function

All enzymes are proteins, but not all proteins are enzymes, the difference being that enzymes possess catalytic activity. The part of the enzyme tertiary structure which is responsible for the catalytic activity is called the 'active site' of the enzyme, and often makes up only 10–20% of the total volume of the enzyme. This is where the enzyme chemistry takes place.

The active site is usually a hydrophilic cleft or cavity containing an array of amino acid side chains which bind the substrate and carry out the enzymatic reaction, as shown in Figure 2.17a. In some cases the enzyme active site also binds one or more cofactors which assist in the catalysis of particular types of enzymatic reactions, as shown in Figure 2.17b.

How does the enzyme bind the substrate? One of the hallmarks of enzyme catalysis is its high substrate selectivity, which is due to a series of highly specific non-covalent enzyme–substrate binding interactions. Since the active site is chiral, it is naturally able to bind one enantiomer of the substrate over the other, just as a hand fits a glove. There are four types of enzyme–substrate interactions used by enzymes, as follows:

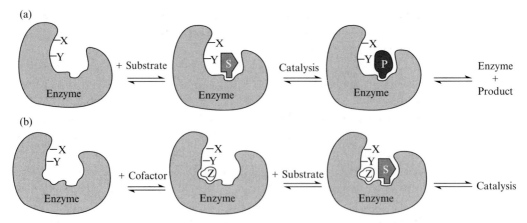

Figure 2.17 Schematic figure of (a) enzyme plus substrate and (b) enzyme plus substrate plus cofactor.

(1) *Electrostatic interactions.* Substrates containing ionisable functional groups which are charged in aqueous solution at or near pH 7 are often bound via electrostatic interactions to oppositely charged amino acid side chains at the enzyme active site. Thus, for example, carboxylic acids (pK$_a$ 4–5) are found as the negatively charged carboxylate anion at pH 7, and are often bound to positively charged side chains such as the protonated ε-amino side chain of a lysine or the protonated guanidine side chain of arginine, shown in Figure 2.18.

   Similarly, positively charged substrate groups can be bound electrostatically to negatively charged amino acid side chains of aspartate and glutamate. Energetically speaking, the binding energy of a typical electrostatic interaction is in the range 25–50 kJ mol^{-1}, the strength of the electrostatic interaction varying with $1/r^2$, where r is the distance between the two charges.

(2) *Hydrogen bonding.* Hydrogen bonds can be formed between a hydrogen-bond donor containing a lone pair of electrons and a hydrogen-bond acceptor containing an acidic hydrogen. These interactions are widely used for binding polar substrate functional groups. The strength of hydrogen bonds depends upon the chemical nature and the geometrical alignment of the interacting groups. Studies of enzymes in which hydrogen-bonding groups have been specifically mutated has revealed that hydrogen

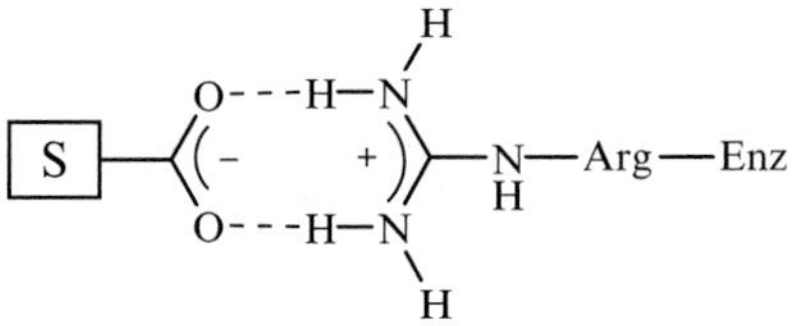

Figure 2.18 Electrostatic enzyme–substrate interaction.

bonds between uncharged donors/acceptors are of energy $2.0-7.5\,kJ\,mol^{-1}$, whilst hydrogen bonds between charged donors/acceptors are much stronger, in the range $12.5-25\,kJ\,mol^{-1}$.

(3) *Non-polar (Van der Waals) interactions.* Van der Waals interactions arise from interatomic contacts between the substrate and the active site. Since the shape of the active site is usually highly complementary to the shape of the substrate, the sum of the enzyme–substrate Van der Waals interactions can be quite substantial $(50-100\,kJ\,mol^{-1})$, even though each individual interaction is quite weak $(6-8\,kJ\,mol^{-1})$. Since the strength of these interactions varies with $1/r^6$ they are only significant at short range $(2-4\,\text{Å})$, so a very good 'fit' of the substrate into the active site is required in order to realise binding energy in this way.

(4) *Hydrophobic interactions.* If the substrate contains a hydrophobic group or surface, then favourable binding interactions can be realised if this is bound in a hydrophobic part of the enzyme active site. These hydrophobic interactions can be visualised in terms of the tendency for hydrophobic organic molecules to aggregate and extract into a non-polar solvent rather than remain in aqueous solution. These processes of aggregation and extraction are energetically favourable due to the maximisation of inter-water hydrogen-bonding networks which are otherwise disrupted by the hydrophobic molecule, as shown in Figure 2.19.

There are many examples of hydrophobic 'pockets' or surfaces in enzyme active sites which interact favourably with hydrophobic groups or surfaces in the substrate and hence exclude water from the two hydrophobic surfaces. As mentioned above, these hydrophobic interactions may be very important for

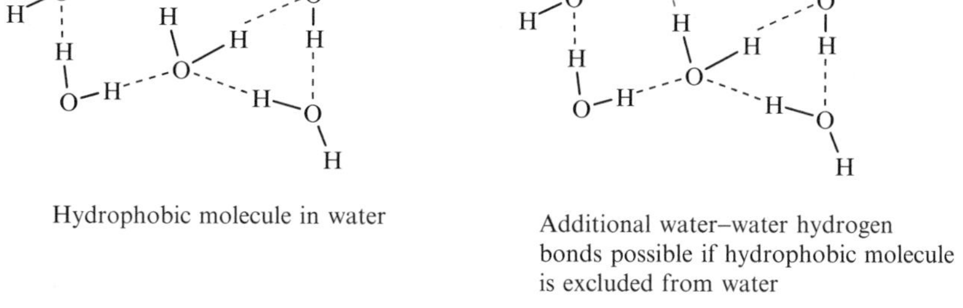

Hydrophobic molecule in water

Additional water–water hydrogen bonds possible if hydrophobic molecule is excluded from water

Figure 2.19 Hydrophobic interaction.

maintaining protein tertiary structure and, as we shall see below, they are central to the behaviour of biological membranes.

Having bound the substrate, the enzyme then proceeds to catalyse its specific chemical reaction using active site catalytic groups, and finally releases its product back into solution. Enzyme catalysis will be discussed in the next chapter. However, before finishing the discussion of enzyme structure three special classes of enzyme structural types will be introduced.

2.8   Metallo-enzymes

Although the polypeptide backbone of proteins is made up only of the 20 common L-amino acids, many proteins bind one or more metal ions. Enzymes which bind metal ions are known as metallo-enzymes: in these enzymes the metal cofactor is usually found at the active site of the enzyme, where it may have either a structural or a catalytic role.

A brief summary of the more common metal ions is given in Table 2.1. Magnesium ions are probably the most common metal ion cofactor: they are found in many enzymes which utilise phosphate or pyrophosphate substrates, since magnesium ions effectively chelate polyphosphates (see Figure 2.20).

Zinc ions are used structurally to maintain tertiary structure, for example in the 'zinc finger' DNA-binding proteins by co-ordination with the thiolate side chains of four cysteine residues, as shown in Figure 2.21a. In contrast, zinc is also used in a number of enzymes as a Lewis acid to co-ordinate carbonyl groups present in the substrate and hence activate them towards nucleophilic attack, as shown in Figure 2.21b.

Table 2.1 Metallo-enzymes.

Metal	Types of enzyme	Role of metal	Redox active?
Mg	Kinases, phosphatases, phosphodiesterases	Binding of phosphates/polyphosphates	×
Zn	Metalloproteases, dehydrogenases	Lewis acid carbonyl activation	×
Fe	Oxygenases (P_{450}, non-haem)	Binding and activation of oxygen	✓
	[FeS] Clusters	Electron transport, hydratases	
Cu	Oxygenases	Activation of oxygen	✓
Mn	Hydrolases, hydratases	Lewis acid?	✓
Co	Vitamin B_{12} coenzyme	Homolysis of Co–carbon bond	✓
Mo	Nitrogenase	Component of Mo/Fe cluster	✓

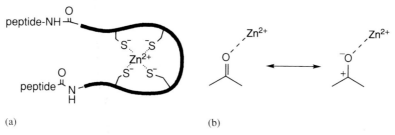

Figure 2.20 Binding of polyphosphate by magnesium.

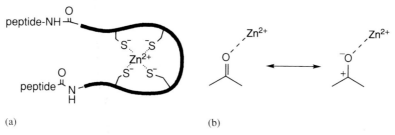

(a)   (b)

Figure 2.21 Uses of zinc (a) in a structural role and (b) as a Lewis acid.

The other common role of metal ions is as redox reagents. Since none of the 20 common amino acids are able to perform any useful catalytic redox chemistry, it is not surprising that many redox enzymes employ redox-active metal ions. We shall meet a number of examples of these redox-active metallo-enzymes in Chapter 6. For a more detailed discussion of the role of metal ions in biological systems the reader is referred to several excellent texts in bio-inorganic chemistry.

2.9   Membrane-associated enzymes

Although the majority of enzymes are freely soluble in water and exist in the aqueous cytoplasm of living cells, there is a substantial class of enzymes that are associated with the biological membranes which encompass all cells. Biological membranes are made up of a lipid bilayer composed of phospholipid molecules containing a polar head group and a hydrophobic fatty acid tail. The phospholipid molecules assemble spontaneously to form a stable bilayer in which the hydrophilic head groups are exposed to solvent water and the hydrophobic tails are packed together in a hydrophobic interior.

Enzymes that are associated with biological membranes fall into two classes, as illustrated in Figure 2.22:

(1) extrinsic membrane proteins which are bound loosely to the surface of the membrane, often by a non-specific hydrophobic interaction, or in some cases by a non-peptide membrane 'anchor' which is covalently attached to the protein;

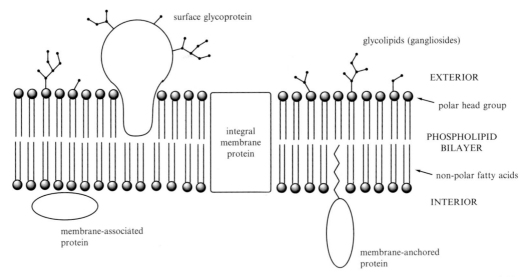

Figure 2.22 Membrane proteins. Black dots represent monosaccharide units attached to glycolipids and glycoproteins.

(2) intrinsic or integral membrane proteins which are buried in the membrane bilayer.

Why should some enzymes be membrane associated? Many biological processes involve passage of either a molecule or a 'signal' across biological membranes, and these processes are often mediated by membrane proteins. These membrane processes have very important cellular functions such as cell–cell signalling, response to external stimuli, transport of essential nutrients and export of cellular products. In many cases these membrane proteins have an associated catalytic activity and are therefore enzymes.

Intrinsic membrane proteins which completely span the membrane bilayer often possess multiple transmembrane α-helices containing exclusively hydrophobic or non-polar amino acid side chains which interact favourably with the hydrophobic environment of the lipid bilayer. The structure of bacteriorhodopsin, a light-harvesting protein found in photosynthetic bacteria, is shown in Figure 2.23.

2.10   Glycoproteins

A significant number of proteins found in animal and plant cells contain an additional structural feature attached covalently to the polypeptide backbone of the protein: they are glycosylated by attachment of carbohydrates. The attached carbohydates can be monosaccharides such as glucose, or complex oligosaccharides. The glycoproteins are usually membrane proteins residing in

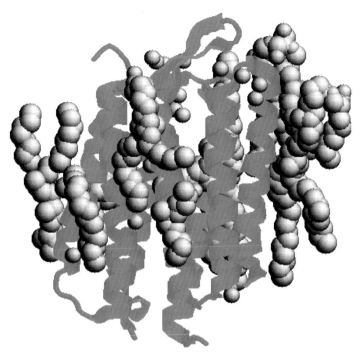

Figure 2.23 Structure of bacteriorhodopsin (PDB file 1C3W), a seven-transmembrane helix membrane protein, solved in the presence of phospholipid. Protein structure shown in red; phospholipid and water molecules shown in monochrome (spacefill).

the cytoplasmic membrane of the cell, in which the sugar residues attached to the protein are located on the exterior of the cell membrane. Since these glycoproteins are exposed to the external environment of the cell, they are often important for cell–cell recognition processes. In this respect they act as a kind of 'bar-code' for the type of cell on which they are residing. This function has been exploited in a sinister fashion, as a means of recognition and entry into mammalian cells, by viruses such as influenza virus and human immunodeficiency virus (HIV).

The carbohydrate residues are attached in one of two ways shown in Figure 2.24: either to the hydroxyl group of a serine or threonine residue (O-linked glycosylation); or to the primary amide nitrogen of an asparagine residue (N-linked glycosylation).

The level of glycosylation can be very substantial: in some cases up to 50% of the molecular weight of a glycoprotein can be made up of the attached carbohydrate residues. The pattern of glycosylation can also be highly complex, for example highly branched mannose-containing oligosaccharides are often found. The sugar attachments are generally not involved in the active site catalysis, but are usually required for full activity of the protein.

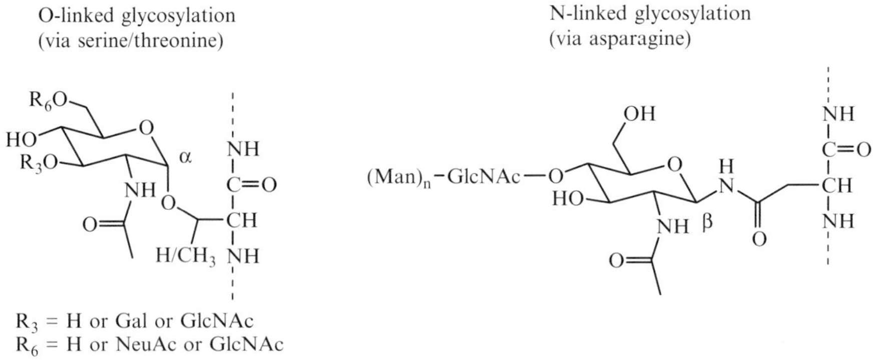

O-linked glycosylation
(via serine/threonine)

N-linked glycosylation
(via asparagine)

R_3 = H or Gal or GlcNAc
R_6 = H or NeuAc or GlcNAc

Figure 2.24 O- and N-linked glycosylation. Gal, galactose; GlcNAc, N-acetylglucosamine; Man, mannose; NeuAc, N-acetylneuraminic acid.

Problems

(1) Which of the amino acid side chains found in proteins would be (a) positively charged or (b) negatively charged at pH 4, 7 and 10, respectively?

(2) The amide bonds found in polypeptides all adopt a *trans*-conformation in which the N−H bond is transcoplanar with the C=O. Why? Certain peptides containing proline have been found to contain *cis*-amide bonds involving the amine group of proline. Explain.

(3) The following segment of RNA sequence is found in the middle of a gene, but the correct reading frame is not known. What amino acid sequences would be encoded from each of the three reading frames? Comment on which is the most likely encoded amino acid sequence.
5′-ACGGCUGAAAACUUCGCACCAAGUCGAUAG-3′

(4) You have just succeeded in purifying a new enzyme, and you have obtained an N-terminal sequence for the protein, which reads Met–Ala–Leu–Ser–His–Asp–Trp–Phe–Arg–Val. How many possible nucleotide sequences might encode this amino acid sequence? If you want to design a 12-base oligonucleotide 'primer' with a high chance of matching the nucleotide sequence of the gene as well as possible, what primer sequence would you suggest?

(5) α-Helices in proteins have a 'pitch' of approximately 3.6 amino acid residues. In order to visualise the side chain–side chain interactions in α-helices, the structure of the helix is often represented as a 'helical wheel'. This representation is constructed by viewing along the length of the helix from the N-terminal end, with the amino acid side chains protruding from the central barrel of the helix, as shown below.

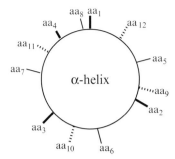

Draw helical wheels for the following synthetic peptides, which were designed to form α-helices with specific functions. Suggest what that function might be.

(a) Gly–Glu–Leu–Glu–Glu–Leu–Leu–Lys–Lys–Leu–Lys–Glu–Leu–Leu–Lys–Gly

(b) Leu–Ala–Lys–Leu–Leu– Lys–Ala–Leu–Ala–Lys–Leu– Leu–Lys–Lys

Inspired by the above examples, suggest a synthetic peptide which would fold into an α-helix containing aspartate, histidine and serine side chains in a line along one face of the helix.

Further reading

Protein structure

C. Branden & J. Tooze (1991) *Introduction to Protein Structure*. Garland, New York.

C. Chothia (1984) Principles that determine the structure of proteins. *Annu. Rev. Biochem.*, **53**, 537–72.

T.E. Creighton (1993) *Proteins – Structures and Molecular Properties*, 2nd edn. Freeman, New York.

G.E. Schulz & R.H. Schirmer (1979) *Principles of Protein Structure*. Springer-Verlag, New York.

Protein folding

R. Jaenicke (1991) Protein folding: local structures, domains, subunits and assemblies. *Biochemistry*, **30**, 3147–61.

C.R. Matthews (1993) Pathways of protein folding. *Annu. Rev. Biochem.*, **62**, 653–83.

M.G. Rossmann & P. Argos (1981) Protein folding. *Annu. Rev. Biochem.*, **50**, 497–532.

Protein evolution

M. Bajaj & T. Blundell (1984) Evolution and the tertiary structure of proteins. *Annu. Rev. Biophys. Bioeng.*, **13**, 453–92.

R. F. Doolittle (1979) Protein evolution. In: *The Proteins* (eds H. Neurath & R.L. Hill), Vol. 4, pp. 1–118. Academic Press, New York.

Metalloproteins

I. Bertini, H.B. Gray, S.J. Lippard & J.S. Valentine (1994) *Bio-inorganic Chemistry.* University Science Books, Mill Valley, California.

Biological membranes

J.B.C. Findlay & W.H. Evans (1987) *Biological Membranes – A Practical Approach.* IRL Press, Oxford.
R.G. Gennis (1989) *Biomembranes: Molecular Structure and Function.* Springer-Verlag, New York.

Glycoproteins

T.W. Rademacher, R.B. Parekh & R.A. Dwek (1988) Glycobiology. *Annu. Rev. Biochem.*, **57**, 785–838.
Y.C. Lee & R.T. Lee (1995) Carbohydrate-protein interactions: the basis of glycobiology. *Acc. Chem. Res.*, **28**, 321–7.

3 Enzymes are Wonderful Catalysts

3.1 Introduction

The function of enzymes is to catalyse biochemical reactions. Each enzyme has evolved over millions of years to catalyse one particular reaction, so it is perhaps not surprising to find that they are extremely good catalysts when compared with man-made catalysts.

The hallmarks of enzyme catalysis are: speed, selectivity and specificity. Enzymes are capable of catalysing reactions at rates well in excess of a million-fold faster than the uncatalysed reaction, typical ratios of k_{cat}/k_{uncat} being 10^6–10^{14}. Figure 3.1 shows an illustration of the speed of enzyme-catalysed glycoside hydrolysis. The rate of acid-catalysed glycoside catalysis is accelerated 10^3-fold by intramolecular acid catalysis, but enzyme-catalysed glycoside hydrolysis is 10^4-fold faster still – some 10^7 faster than the uncatalysed reaction carried out at pH 1.

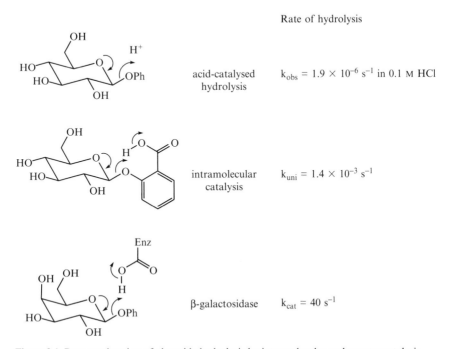

Figure 3.1 Rate acceleration of glycoside hydrolysis by intramolecular and enzyme catalysis.

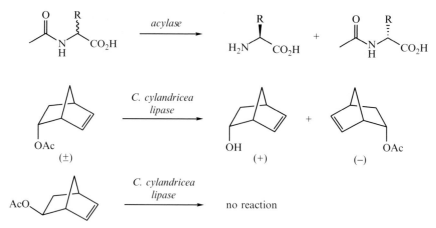

Figure 3.2 Stereoselectivity in enzymatic hydrolysis reactions.

Enzymes are highly selective in the reactions that they catalyse. Since they bind their substrates via a series of selective enzyme–substrate binding inter-actions at a chiral active site, they are able to distinguish the most subtle changes in substrate structure, and are able to distinguish between regioisomers and between enantiomers, as shown in Figure 3.2. Finally, enzymes carry out their reactions with near faultless precision: they are able to select a unique site of action within the substrate, and carry out the enzymatic reaction stereospe-cifically, as illustrated in Figure 3.3.

In this chapter we shall examine the factors that contribute to the remarkable rate acceleration achieved in enzyme-catalysed reactions. Examples of enzyme stereospecificity will be discussed in Chapter 4. It is worth at this point distin-guishing between *selectivity*, which is the ability of the *enzyme* to select a certain substrate or functional group out of many; and *specificity*, which is a property of

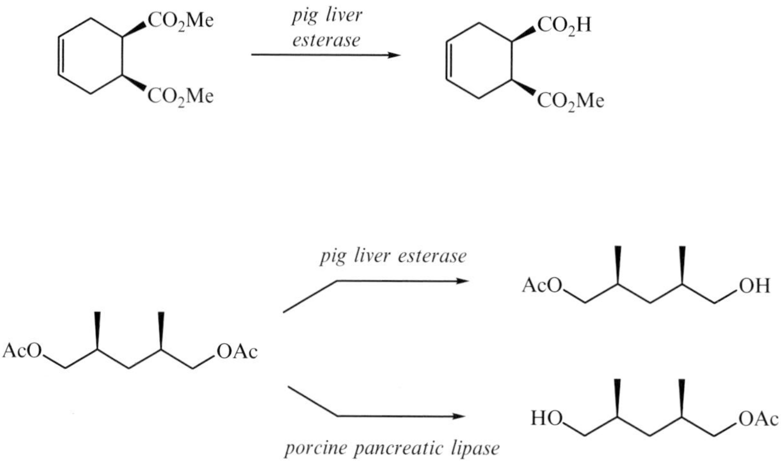

Figure 3.3 Stereoselectivity in enzymatic hydrolysis reactions.

the *reaction* catalysed by the enzyme, being the production of a single regio- and stereo-isomer of the product. Both are properties which are highly prized in synthetic reactions used in organic chemistry: enzymes are able to do both.

3.2   A thermodynamic model of catalysis

A catalyst may be defined as a species which accelerates the rate of a chemical reaction whilst itself remaining unchanged at the end of the catalytic reaction. In thermodynamic terms, catalysis of a chemical reaction is achieved by reducing the *activation energy* for that reaction, the activation energy being the difference in free energy between the reagent(s) and the transition state for the reaction. This reduction in activation energy can be achieved either by stabilisation (and hence reduction in free energy) of the transition state by the catalyst, or by the catalyst finding some other lower energy pathway for the reaction.

Figure 3.4 illustrates the free energy profile of a typical acid-catalysed chemical reaction which converts a substrate, S, to a product, P. In this case an intermediate chemical species SH^+ is formed upon protonation of S. If the conversion of SH^+ to PH^+ is 'easier' than the conversion of S to P, then the activation energy for the reaction will be reduced and hence the reaction will go faster. It is important at this point to define the difference between an intermediate and a transition state: an intermediate is a stable (or semi-stable) chemical species formed during the reaction and is therefore a *local energy minimum*, whereas a transition state is by definition a *local energy maximum*.

The rate of a chemical reaction is related to the activation energy of the reaction by the following equation:

$$k = A.e^{(-E_{act}/RT)}$$

Therefore, the rate acceleration provided by the catalysis can simply be calculated:

$$k_{cat}/k_{uncat} = e^{(E_{uncat}-E_{cat}/RT)}$$

If, for example, a catalyst can provide $10\,kJ\,mol^{-1}$ of transition stabilisation energy for a reaction at $25°C$ a 55-fold rate acceleration will result, whereas a $20\,kJ\,mol^{-1}$ stabilisation will give a 3000-fold acceleration and a $40\,kJ\,mol^{-1}$ stabilisation a 10^7-fold acceleration! A consequence of the exponential relationship between activation energy and reaction rate is that a little extra transition state stabilisation goes a long way!

An enzyme-catalysed reaction can be analysed thermodynamically in the same way as the acid-catalysed example, but is slightly more complicated. As explained in Chapter 2, enzymes function by binding their substrate reversibly at their active site, and then proceeding to catalyse the biochemical reaction

(a)

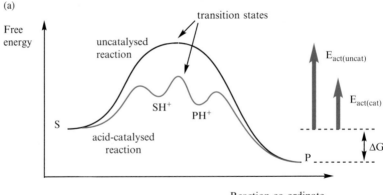

(b)

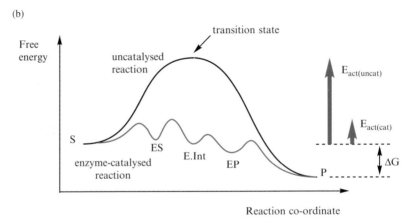

Figure 3.4 Free energy profiles for (a) an acid-catalysed reaction and (b) an enzyme-catalysed reaction which converts substrate S to product P.

using the active site amino acid side chains. Often enzyme-catalysed reactions are multi-step sequences involving one or more intermediates, as illustrated in Figure 3.4. An enzyme–substrate intermediate, ES, is formed upon binding of the substrate, which is then converted to the enzyme–product complex, EP, either directly or via one or more further intermediates.

In both catalysed reactions shown in Figure 3.4 the overriding consideration as far as rate acceleration is concerned is transition state stabilisation. Just as in non-enzymatic reactions there is acid–base and nucleophilic catalysis taking place at enzyme active sites. However, the secret to the extraordinary power of enzyme catalysis lies in the fact that the reaction is taking place as the substrate is bound to the enzyme active site. So, what was in the non-enzymatic case an intermolecular reaction has effectively become an intramolecular reaction. The rate enhancements obtained from these types of *proximity effects* can be illustrated by intramolecular reactions in organic chemistry, which is where we shall begin the discussion.

3.3   Proximity effects

There are many examples of organic reactions that are intramolecular: that is, they involve two or more functional groups within the same molecule, rather than functional groups in different molecules. Intramolecular reactions generally proceed much more rapidly and under much milder reaction conditions than their intermolecular counterparts, which makes sense since the two reacting groups are already 'in close proximity' to one another. But how can can we explain these effects?

A useful concept in quantitating proximity effects is that of effective concentration. In order to define the effective concentration of a participating group (nucleophile, base, etc.), we compare the rate of the intramolecular reaction with the rate of the corresponding intermolecular reaction where the reagent and the participating group are present in separate molecules. The effective concentration of the participating group is defined as the concentration of reagent present in the intermolecular reaction required to give the same rate as the intramolecular reaction.

I will illustrate this using data for the rates of hydrolysis of a series of phenyl esters in aqueous solution at pH 7, given in Figure 3.5. The reference reaction in this case is the hydrolysis of phenyl acetate catalysed by sodium acetate at the same pH. Introduction of a carboxylate group into the same molecule as the ester leads to an enhancement of the rate of ester hydrolysis, which for phenyl succinate (see Figure 3.5 (**3**)) is 23 000-fold faster than phenyl acetate (see Figure 3.5 (**1**)). This remarkable rate acceleration is because the neighbouring carboxylate group can attack the ester to form a cyclic anhydride intermediate, shown in Figure 3.6. This intermediate is more reactive than the original ester group and so hydrolyses rapidly.

Note that the rate acceleration is largest when a five-membered anhydride is formed, since five-membered ring formation is kinetically favoured over six-membered ring formation, which in turn is greatly favoured over three-, four- and seven-membered ring formation. The effective concentration can be worked out by comparing the rates of these intramolecular reactions with the rates of the intermolecular reaction between phenyl acetate and sodium acetate in water. For phenyl succinate an effective concentration of 4 000 M is found, so the hydrolysis of phenyl succinate proceeds much faster than if phenyl acetate was surrounded completely by acetate ions! Here we start to see the catalytic potential of proximity effects.

In the same series of phenyl esters, if the possible ring size of five is maintained, but a *cis-* double bond is placed in between the reacting groups,

	Ester	k_{rel}	Effective concentration (mol l^{-1})	Reference Reaction
(1)		1		
(2)		50	9	
(3)		23,000	4,000	
(4)		150	25	
(5)		1,100,000	2×10^5	
(6)		5×10^{10}	10^{10}	
(7)		1	–	
(8)		1,000,000	490	
(9)		2,500,000	1,300	

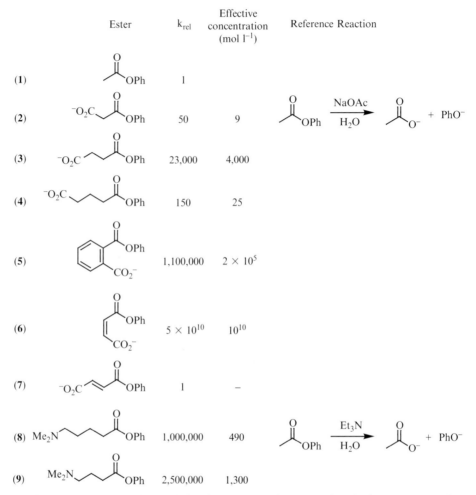

Figure 3.5 Intramolecular catalysis of ester hydrolysis. Et_3N, triethylamine; NaOAc, sodium acetate.

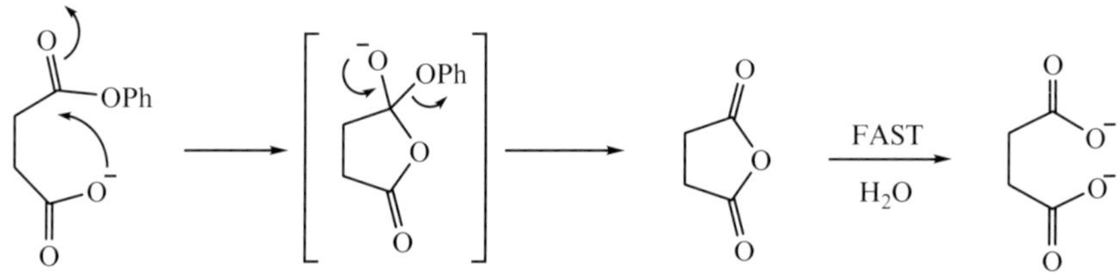

Figure 3.6 Mechanism for intramolecular hydrolysis of phenyl succinate (**3**).

(3)   unreactive
conformation

reactive
conformation

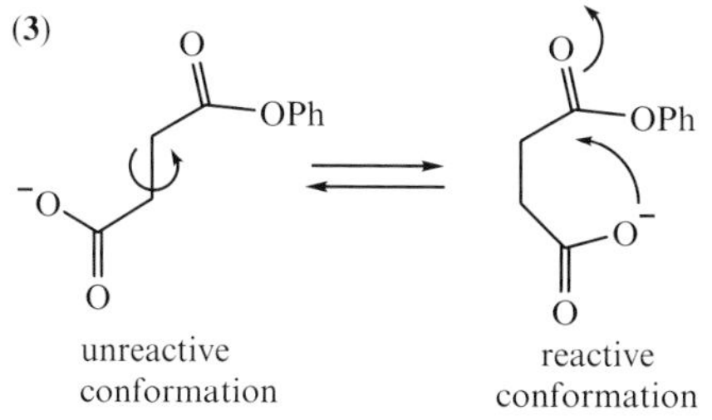

(6)   held in reactive
conformation

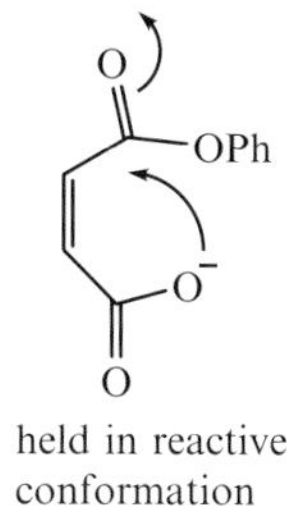

Figure 3.7 Intramolecular hydrolysis of phenyl succinate (**3**) versus phenyl maleate (**6**).

the observed rates of hydrolysis are even faster. Phenyl phthalate (see Figure 3.5 (**5**)) has an effective concentration of acetate ions of 2×10^5 M, whilst phenyl maleate (see Figure 3.5 (**6**)) has an astonishing effective concentration of 10^{10} M! Yet the same molecule containing a *trans*- double bond has no rate acceleration at all. So it is clear that by holding the reactive groups rigidly in close proximity to one another remarkable rate acceleration can be achieved. Why is the hydrolysis of phenyl maleate, in which a five-membered anhydride is formed, so much faster than the hydrolysis of phenyl succinate, in which an apparently similar five-membered anhydride is formed? The answer is that in phenyl maleate the reactive groups are held in the right orientation to react, as shown in Figure 3.7, so the *probability* of the desired reaction is increased.

In thermodynamic terms, the restriction of the double bond in the case of phenyl maleate has removed rotational degrees of freedom, so that in going to the transition state for the intramolecular reaction fewer degrees of freedom are lost, which means that the reaction is *entropically* more favourable. If you think of entropy as a measure of order in the system, then in the case of phenyl maleate the molecule is already ordered in the right way with respect to the reacting groups. Thus, there is a large kinetic advantage in intramolecular chemical reactions due to the ordering of reactive groups.

The same effect operates in enzyme active sites, and is a major factor in enzyme catalysis. The binding of substrates and cofactors at an enzyme active site of defined three-dimensional structure brings the reagents into close proximity to one another and to the enzyme active site functional groups. This increases the probability of correct positioning for reaction to take place, so it speeds up the reaction. An important factor in this analysis is that the enzyme structure is already held rigidly (or fairly rigidly at least) in the correct conformation for binding and catalysis. Recent studies have argued that the thermodynamic origin of this kinetic advantage in enzyme catalysis is primarily enthalpic, rather than entropic. Nevertheless, the catalytic power of proximity effects, or 'pre-organisation', has been demonstrated by the synthesis of host–guest systems which mimic enzymes by binding substrates non-covalently (see Chapter 12).

3.4   The importance of transition state stabilisation

All catalysts operate by reducing the activation energy of the reaction, by stabilising the transition state for the reaction. Enzymes do the same, but the situation is somewhat more complicated since there are usually several transition states in an enzymatic reaction. We have already seen how an enzyme binds its substrate reversibly at the enzyme active site. One might imagine that if an enzyme were to bind its substrate very tightly that this would lead to transition state stabilisation. This, however, is not the case: in fact, the enzyme does not want to bind its substrate(s) too tightly!

Shown in Figure 3.8 are free energy curves for a hypothetical enzyme-catalysed reaction proceeding via a single rate determining transition state. Suppose that we can somehow alter the enzyme, E, so that it binds the substrate, S, or the transition state, TS, more tightly. In each case the starting free energy (of E + S) is the same. In the presence of high substrate concentrations the enzyme will in practice be fully saturated with substrate, so the activation energy for the reaction will be governed by the energy difference between the ES complex and the transition state. In Figure 3.8b the enzyme is able to bind both the substrate and the transition state more tightly (and hence lower their free energy equally). This, however, leads to no change in the activation energy, and hence no rate acceleration. In Figure 3.8c the enzyme

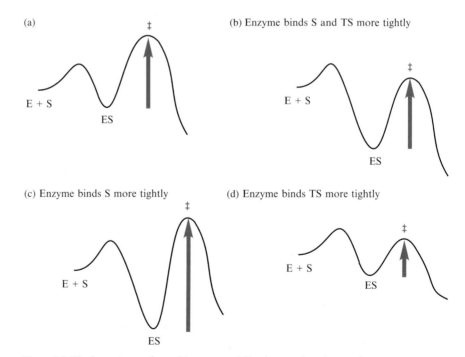

Figure 3.8 The importance of transition state stabilisation – a thought experiment.

binds only the substrate more tightly: this generates a 'thermodynamic pit' for the ES complex and hence increases the activation energy, so the reaction is slower! However, if, as in Figure 3.8d, the enzyme can bind selectively the transition state, then it can reduce the activation energy and hence speed up the reaction.

The conclusion of this thought experiment is that in order to achieve optimal catalysis, enzymes should selectively bind the transition state, rather than the substrate. Hence, it is not advantageous for enzymes to bind their substrates too tightly. This is evident when one looks at substrate binding constants for enzymes (these will be discussed in more detail in Section 4.3). Typical K_M values for enzymes are in the mM–μM range (10^{-3}–10^{-6} M), whereas dissociation constants for binding proteins and antibodies whose function is to bind small molecules tightly are in the range nM–pM (10^{-9}–10^{-12} M). We shall meet several examples of specific transition state-stabilising interactions later in the chapter.

3.5   Acid/base catalysis in enzymatic reactions

Acid and base catalysis is involved in all enzymatic processes involving proton transfer, so in practice there are very few enzymes that do not have acidic or basic catalytic groups at their active sites. However, unlike organic reactions which can be carried out under a very wide range of pH conditions to suit the reaction, enzymes have a strict limitation that they must operate at physiological pH, in the range 5–9. Given this restriction, and the fairly small range of amino acid side chains available for participation in acid/base chemistry (shown in Figure 3.9), a remarkably diverse range of acid/base chemistry is achieved.

General acid catalysis takes place when the substrate is protonated by a catalytic residue which in turn gives up a proton, as shown in Figure 3.10. The active site acidic group must, therefore, be protonated at physiological pH but its pK_a must be just above (i.e. in the range 7–10). If the pK_a of a side chain was in excess of 10 then it would become thermodynamically unfavourable to transfer a proton.

General base catalysis takes place either when the substrate is deprotonated, or when water is deprotonated prior to attack on the substrate, as shown in Figure 3.11.

Enzyme active site bases must therefore be deprotonated at physiological pH but have pK_a values just below. Typical pK_a ranges for amino acid side chains in enzyme active sites are shown in Figure 3.12. They can be measured by analysis of enzymatic reaction rate versus pH, as described in Section 4.7.

Although the pK_a values given in Figure 3.12 are the typical values found in proteins, in some cases the pK_a values of active site acidic and basic groups can be strongly influenced by their micro-environment. Thus, for example, the

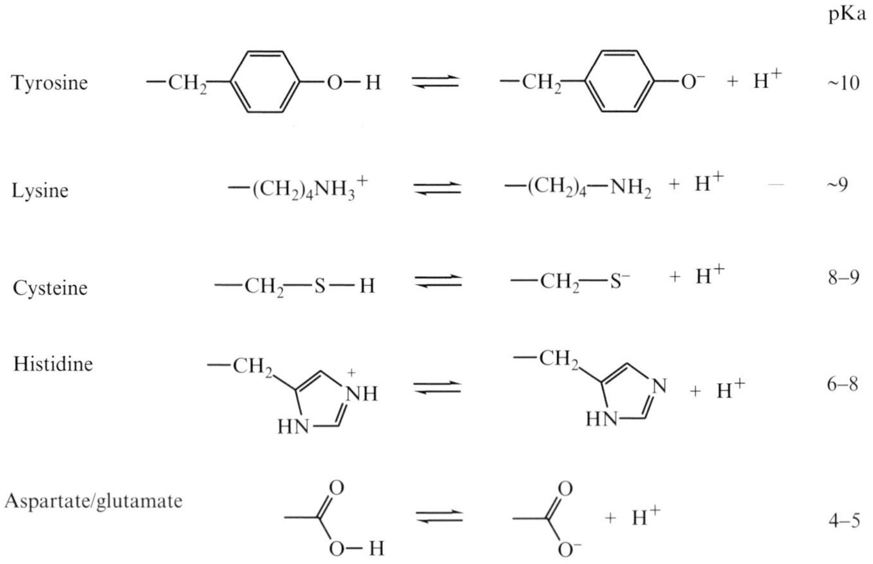

pKa

Tyrosine   ~10

Lysine   ~9

Cysteine   8–9

Histidine   6–8

Aspartate/glutamate   4–5

Figure 3.9 Amino acid side chains used for acid/base catalysis.

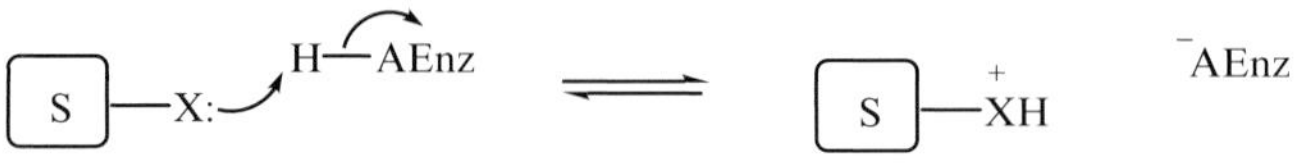

Figure 3.10 General acid catalysis.

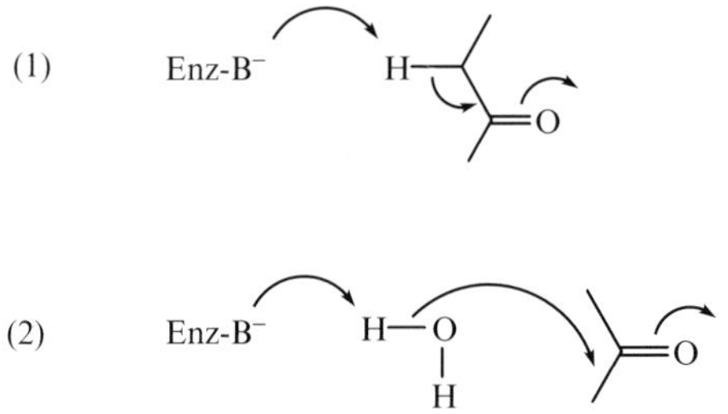

(1)   Enz-B⁻

(2)   Enz-B⁻

Figure 3.11 General base catalysis.

enzyme acetoacetate decarboxylase contains an active site lysine residue which forms an imine linkage with its substrate – its pK_a value was found to be 5.9, much less than the expected value of about 9. When an active site peptide was obtained containing the catalytic lysine, it was found to be adjacent to another lysine residue. So it is likely that the proximity of another positively charged residue would make the protonated form thermodynamically less favourable, and hence reduce the pK_a. The same effect can be observed in the pK_a values of ethylenediamine (see Figure 3.13), where the pK_a for the monoprotonated form is 10.7 as usual, but the pK_a for the doubly protonated form is 7.5.

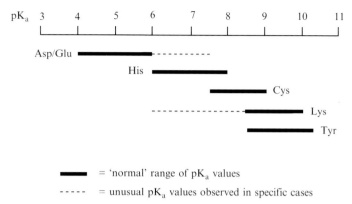

Figure 3.12 Range of pK$_a$ values observed for amino acid side chains in enzyme active sites.

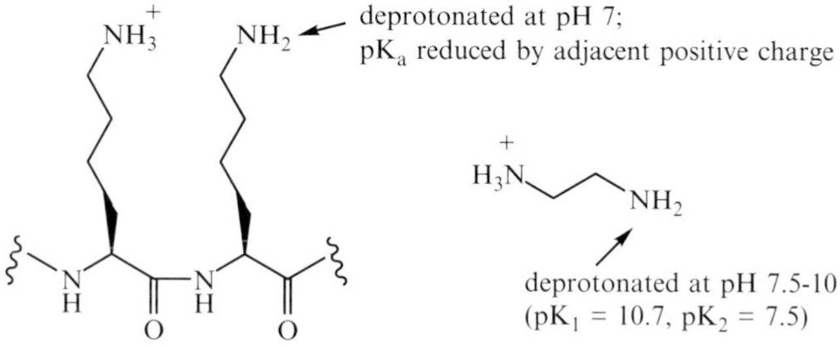

deprotonated at pH 7;
pK$_a$ reduced by adjacent positive charge

deprotonated at pH 7.5–10
(pK$_1$ = 10.7, pK$_2$ = 7.5)

Figure 3.13 Abnormally low lysine pK$_a$ in acetoacetate decarboxylase.

Similarly, if a charged group was involved in a salt bridge with an oppositely charged residue, its pK$_a$ would be altered, or if it was in a hydrophobic region of the active site, which would destabilise the charged form of the group. So, for example, we shall later on see protonated aspartic acid and glutamic acid residues acting as acidic groups in some cases (i.e. with elevated pK$_a$ values of 7.0 or above). Finally, it is worth noting that histidine, whose side chain contains an imidazole ring of pK$_a$ 6–8, can act either as an acidic or a basic residue, depending on its particular local pK$_a$, making it a versatile reagent for enzymatic acid/base chemistry.

Acid/base catalysis by enzymes is made that much more effective by the optimal positioning of the active site acid/base groups in close proximity to the substrate, generating a high effective concentration of the enzyme reagent. This can be illustrated in the case of glycoside hydrolysis using the data from Figure 3.1. The mechanism of the non-enzymatic reaction involves protonation of the glycosidic group by external acid to form a good leaving group, followed by formation of an oxonium intermediate. Glycoside hydrolysis can be accelerated dramatically by positioning an acidic group in close proximity to the glycosidic leaving group, as shown in Figure 3.1. Enzymes which catalyse glycoside hydrolysis (see Section 5.7) also employ an acidic catalytic group to protonate the glycosidic leaving group – but the enzymatic reaction is some 30 000-fold faster even than the intramolecular reaction, suggesting that the enzyme is able to further stabilise the transition state for this reaction.

Enzymes also have the ability to carry out bifunctional catalysis: protonation of the substrate at the same time as deprotonation in another part of the molecule. An example of bifunctional catalysis is the enzyme ketosteroid isomerase, whose active site (see Figure 3.14) contains two catalytic residues: Asp-38 which acts as a catalytic base; and Tyr-14 which acts as an acidic group. The mechanism (see Figure 3.15) involves the formation of a dienol intermediate via a concerted step involving simultaneous deprotonation of the substrate by Asp-14 and protonation of the substrate carbonyl by Tyr-14. In the second step the tyrosinate group acts as a base, and the substrate is re-protonated by the protonated Asp-38.

Bifunctional catalysis is thought to make possible the deprotonation of substrates with apparently high pK_a values. Thus, in the above example deprotonation adjacent to a ketone in solution to form an enolate species would involve removal of a proton of pK_a 18–20, which would be impractical at pH 7. However, simultaneous protonation to form an enol intermediate makes the reaction thermodynamically much more favourable.

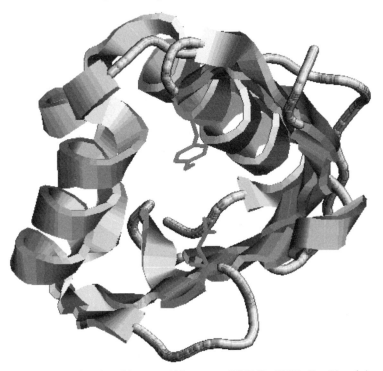

Figure 3.14 Active site of ketosteroid isomerase (PDB file 1E3V). Tyr-14 and Asp-38 are shown in red.

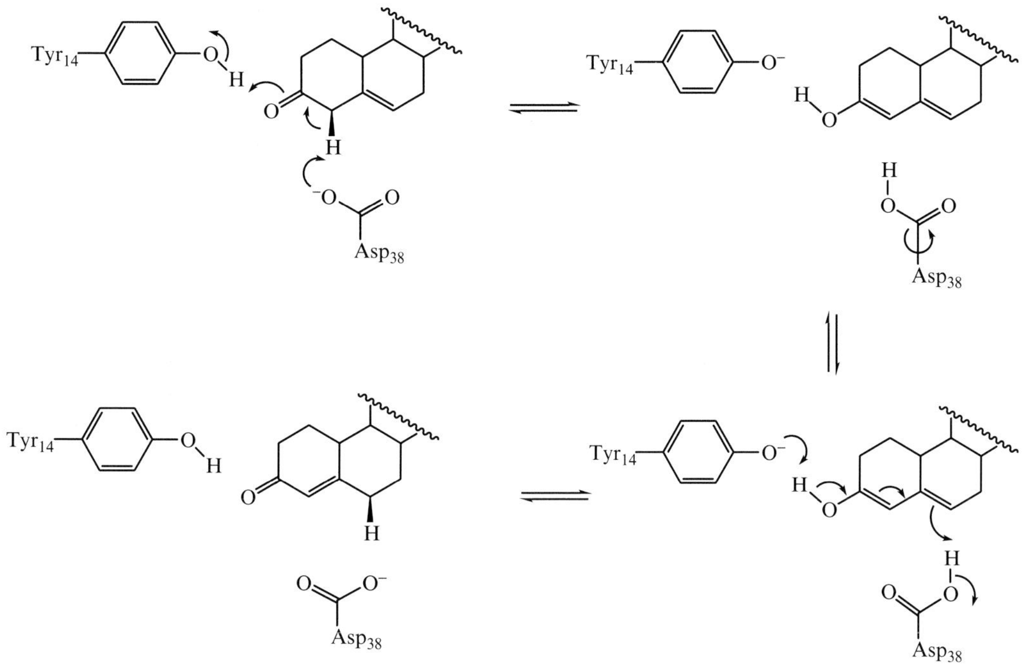

Figure 3.15 Mechanism for ketosteroid isomerase.

Finally, enzymes which bind metal cofactors such as Zn^{2+} and Mg^{2+} can utilise their properties as Lewis acids, i.e. electron pair acceptors. An example is the enzyme thermolysin, whose mechanism is illustrated in Figure 3.16. In this enzyme, Glu-143 acts as an active site base to deprotonate water for attack on the amide carbonyl, which is at the same time polarised by co-ordination by an active site Zn^{2+} ion. The protonated glutamic acid probably then acts as an acidic group for the protonation of the departing amine.

3.6   Nucleophilic catalysis in enzymatic reactions

Nucleophilic (or covalent) catalysis is a type of catalysis seen relatively rarely in organic reactions, but which is used quite often by enzymes. It involves nucleophilic attack of an active site group on the substrate, forming a covalent bond between the enzyme and the substrate, and hence a covalent intermediate in the reaction mechanism. This is a particularly effective strategy for enzyme-catalysed reactions for two reasons. First, as we have seen before, an enzyme is able to position an active site nucleophile in close proximity and correctly aligned to attack its substrate, generating a very high effective concentration of nucleophile.

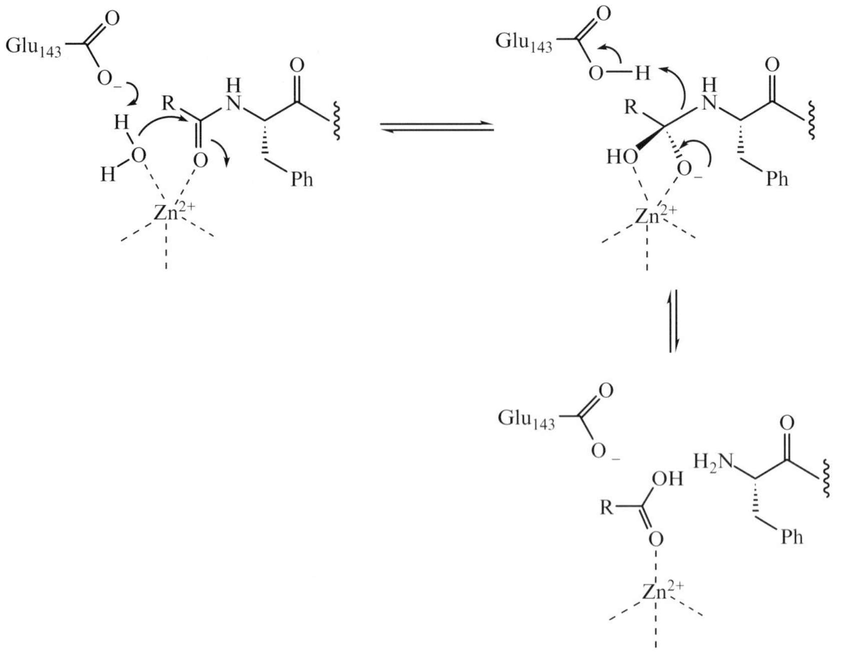

Figure 3.16 Mechanism for thermolysin. R, peptide chain.

Second, since enzyme active sites are often largely excluded from water molecules, an enzyme active site nucleophile is likely to be 'desolvated'. Thus, a charged nucleophile in aqueous solution would be surrounded by several layers of water molecules, which greatly reduce the polarity and effectiveness of the nucleophile. However, a desolvated nucleophile at a water-excluded active site will be a much more potent nucleophile than its counterpart in solution. This effect can be illustrated in organic reactions carried out in dipolar aprotic solvents such as dimethylsulphoxide (DMSO) or dimethylformamide (DMF), in which nucleophiles are not hydrogen bonded as they would be in aqueous solution. A consequence of this desolvation effect is that nucleophilic displacement reactions occur much more readily in these solvents.

Enzymes have a range of potential nucleophiles available to them, which are shown in Table 3.1.

Probably the best nucleophile available to enzymes is the thiol side chain of cysteine, which we shall see operating in proteases and acyl transfer enzymes. The ε-amino group of lysine is used in a number of cases to form imine linkages with ketone groups in substrates, as in the example of acetoacetate decarboxylase, shown in Figure 3.17.

This enzyme catalyses the decarboxylation of acetoacetate to acetone. Two lines of evidence were used to show that an imine linkage is formed between the ketone of acetoacetate and the ε-amino group of an active site lysine. First,

Table 3.1 Amino acid side chains used for nucleophilic catalysis in enzymatic reactions.

Amino acid	Side chain	Examples
Serine	$-CH_2-OH$	Serine proteases, esterases, lipases
Threonine	$-CH(CH_3)-OH$	Phosphotransferases
Cysteine	$-CH_2-SH$	Cysteine proteases, acyl transferases
Aspartate, glutamate	$-(CH_2)_n-CO_2H$	Epoxide hydrolase, haloalkane dehalogenase
Lysine	$-(CH_2)_4-NH_2$	Acetoacetate decarboxylase, class I aldolases
Histidine	imidazole$-NH$	Phosphotransferases
Tyrosine	$Ar-OH$	DNA topoisomerases

Ar, aromatic group.

treatment of enzyme with substrate and sodium borohydride leads to irreversible enzyme inactivation, via *in situ* reduction of the enzyme-bound imine intermediate by borohydride. Treatment of enzyme with borohydride alone gives no inactivation. Second, incubation of acetoacetate labelled with ^{18}O at the ketone position leads to the rapid exchange of ^{18}O label out of this position, consistent with reversible imine formation. As mentioned above, the pK_a of this lysine group is abnormally low at 5.9, which is sufficiently low for it to act as a nucleophile at pH 7.

The other nitrogen nucleophile available to enzymes is the versatile imidazole ring of histidine. This group is more often used for acid/base chemistry, but is occasionally used as a nucleophile in, for example, phosphotransfer reactions. Finally, enzymes have oxygen nucleophiles available in the form of

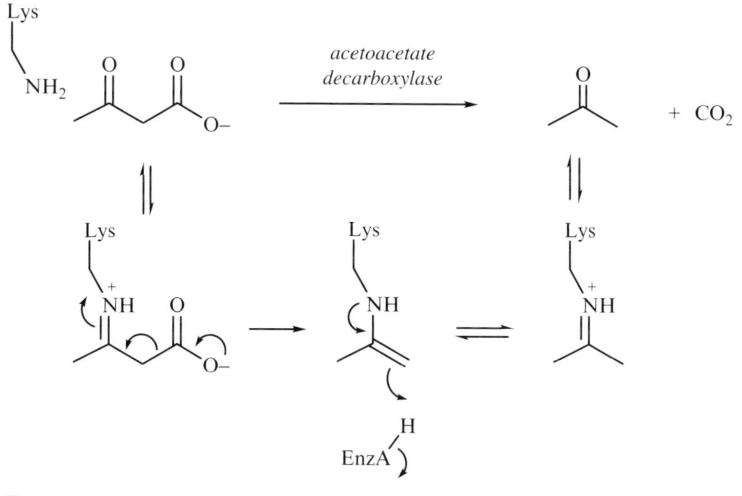

Figure 3.17 Mechanism for acetoacetate decarboxylase.

the hydroxyl groups of serine, threonine and tyrosine, and the carboxylate groups of aspartate and glutamate. There are examples of each of these groups being used for nucleophilic catalysis, especially serine, which we shall see in Chapter 5 used for the serine proteases.

An example of the use of aspartate as a nucleophile is the enzyme haloalkane dehalogenase from *Xanthobacter autotrophicus*, which is involved in the dechlorination of organochlorine chemicals found in industrial waste. This 35-kDa protein contains seven strands of β-sheet arranged centrally with intervening α-helices. This type of αβ structure is found in many hydrolase enzymes such as the serine proteases discussed in Chapter 5. The catalytic residues Asp-124 and His-289 are situated on loops at the ends of β-sheets. The active site cavity of volume 37 Å^3 is lined with hydrophobic residues, which can form favourable hydrophobic interactions with its non-polar substrates. Shown in Figure 3.18 is a view of the protein structure, highlighting the active site catalytic residues. The catalytic mechanism proceeds by displacement of chloride by Asp-128 residue, to give a covalent ester intermediate, followed by base-catalysed hydrolysis involving His-289, as shown in Figure 3.19.

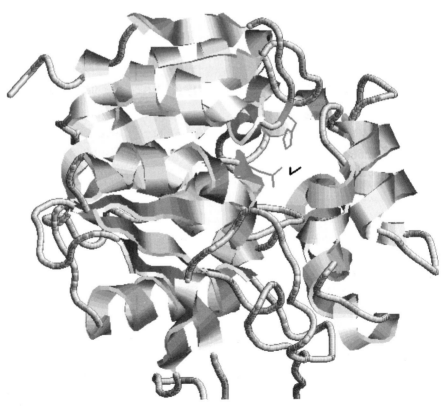

Figure 3.18 Structure of haloalkane dehalogenase (PDB file 2DHD). Asp-124 and His-289 are shown in red; the product ethanol is shown in black.

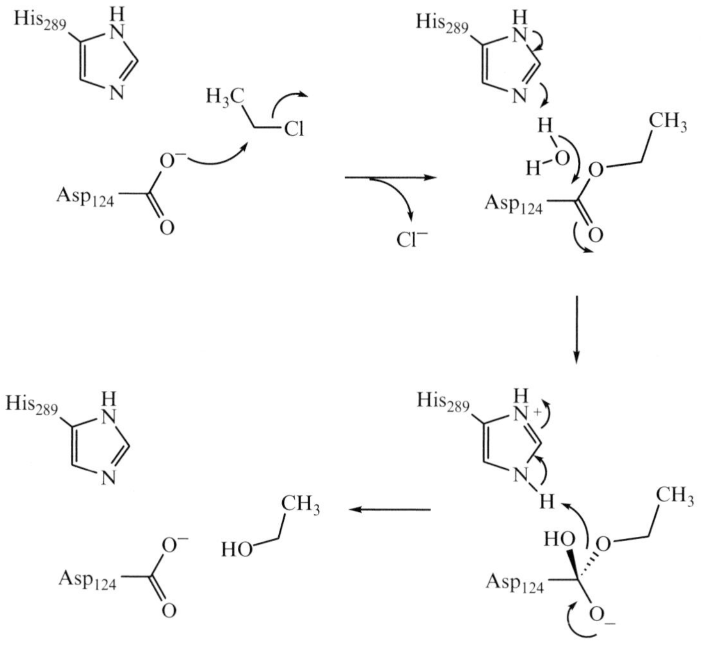

Figure 3.19 Mechanism for haloalkane dehalogenase.

3.7   The use of strain energy in enzyme catalysis

The concept of 'strain' is one that is rather difficult to explain, since it occurs very rarely in organic reactions, and there are only a few examples of enzymatic reactions in which there is evidence that it operates. Remember that the over-riding factor in achieving rate acceleration in enzyme-catalysed reactions is the difference in free energy between the ES complex and the transition state of the enzymatic reaction. If the enzyme can somehow bind the substrate in a strained conformation which is *closer to the transition state* than the ground state conformation, then the difference in energy between the bound conformation and the transition state will be reduced, and the reaction will be accelerated (see Figure 3.20).

How can an enzyme bind its substrate in a strained conformation? Is that not energetically unfavourable? The answer to these questions is that if the substrate is of a reasonable size, the enzyme can form a number of enzyme–substrate binding interactions, and the total enzyme–substrate binding energy can be quite substantial. In some cases, in order to benefit from the most favourable overall binding interactions, the substrate must adopt an unfavour-able conformation in a *part* of the molecule. That part of the substrate may cunningly happen to be where the reaction is going to take place! In thermo-dynamic terms, the enzyme uses its favourable binding energy in the rest of the

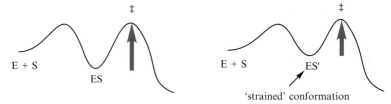

Figure 3.20 Rate acceleration from a strained ES′ complex.

substrate to compensate for the adoption of an unfavourable conformation in the strained part of the molecule.

To illustrate this concept, we shall look at the example of carboxypeptidase A, a Zn^{2+}-containing protease similar to thermolysin. When the X-ray crystal structure of carboxypeptidase A was solved, it was found that in order to bind the peptide substrate with the most favourable enzyme–substrate interactions, a 'twist' needed to be introduced into the scissile amide bond. In this conformation the carbonyl group of the amide being hydrolysed was bound slightly out of plane of the amide N−H bond, assisted by co-ordination to the active site zinc ion. This has the effect of reducing the overlap of the nitrogen lone pair of electrons with the carbonyl π-bond, which requires the amide bond to be planar. This makes the carbonyl much more reactive, more like a ketone group than an amide, so it is much more susceptible to nucleophilic attack, in this case by the carboxylate side chain of Glu-270. Figure 3.21 shows that in this strained conformation the carbonyl oxygen has already moved some distance towards where it will be in the transition state, so the energy difference between the bound conformation and the transition state is reduced, hence we see rate acceleration.

Thus, if an enzyme is able to bind its substrate in a less favourable but more reactive conformation, then it is able to realise additional rate acceleration in this way. Analysis of this type requires detailed insight from X-ray crystallography, so it is not surprising that there are only a few well-documented examples of this phenomenon. One other is that of lysozyme, which we shall see in Chapter 5. However, it is possible that this strategy may be used in many enzyme-catalysed reactions.

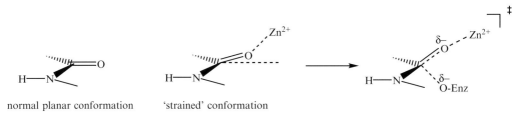

Figure 3.21 Strained substrate conformation in carboxypeptidase A.

3.8   Catalytic perfection

How fast is it possible for enzyme-catalysed reactions to proceed? Is there a limit to the rate acceleration achievable by enzymes? The answer is yes, and a small number of enzymes have achieved it. For extremely efficient enzymes, the rate of reaction becomes limited by the rate at which a substrate can diffuse onto its active site and diffuse away into solution – the so-called *diffusion limit*. This diffusion limit for collision of enzyme and substrate corresponds to a bimolecular rate constant of approximately 10^8 M^{-1} s^{-1}. We can compare this value with the bimolecular rate constant for reaction of free enzyme with free substrate, which is the catalytic efficiency k_{cat}/K_M (see Section 4.3).

Many enzymes have catalytic efficiencies of 10^6 to 10^7 M^{-1} s^{-1}, but a small number have k_{cat}/K_M values which are at the diffusion limit – these are listed in Table 3.2. One of these is the enzyme acetylcholinesterase, which is involved in the propagation of nerve impulses at synaptic junctions: a process for which the utmost speed is necessary. For these enzymes the rate-determining step has become the diffusion of substrates onto the active site. As fast as a substrate diffuses onto the active site it is processed by the enzyme before the next molecule of substrate appears. These are truly wonderful catalysts.

3.9   The involvement of protein dynamics in enzyme catalysis

Our understanding of such high rates of catalysis in enzymes is still incomplete. Chemical models for enzyme catalysts, and catalytic antibodies which function via transition state stabilisation (see Chapter 12), are orders of magnitude less active catalysts than the above examples. Therefore, there are likely to be other factors that enzymes use to achieve high rates of catalysis.

Table 3.2 Catalytic efficiencies of some diffusion-limited enzymes.

Enzyme	Reaction type	$k_{cat}/K_M (M^{-1} s^{-1})$
Superoxide dismutase	Redox dismutation	3.0×10^9
Fumarase	Hydration	2.0×10^9
Cytochrome c peroxidase	Redox peroxidase	6.0×10^8
Triose phosphate isomerase	Keto/enol isomerase	3.8×10^8
Acetylcholinesterase	Ester hydrolysis	1.4×10^8
Ketosteroid isomerase	Keto/enol isomerase	1.3×10^8
β-Lactamase	β-lactam hydrolysis	1.0×10^8

Examination of protein structure in solution by nuclear magnetic resonance (NMR) spectroscopy has revealed that there is a significant amount of internal motion in a protein, on a timescale of 1–10 ns. Such internal motion could transmit kinetic energy from a distant part of the protein to the active site in order to assist in catalysis. It has been proposed that dynamic fluctuations in the protein structure are used by enzymes to organise the enzyme–substrate complex into a reactive conformation.

The role of protein dynamics in enzyme catalysis is therefore a topic of considerable interest. Selection of enzyme catalysts by 'directed evolution' has revealed, in a number of cases, that amino acid residues distant from active sites can have a dramatic effect on enzyme activity. Replacement of Gly-120 to valine in dihydrofolate reductase (see Figure 3.22) disrupts the internal motion of a protein loop, and so interferes with the conversion of the ternary complex to the reactive conformation, hence this step becomes partially rate-limiting, reducing the rate of hydride transfer 500-fold. Thus, protein dynamics may have an important role to play in enzyme catalysis.

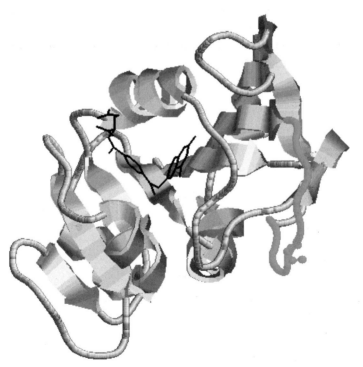

Figure 3.22 Structure of dihydrofolate reductase (PDB file 1DDS), showing Gly-120 and the loop whose internal motion is implicated in catalysis. Gly-120 and the FG loop involving Gly-120 are shown in red; the bound inhibitor (methotrexate) at the active site is shown in black.

Problems

(1) Rationalise the rate accelerations observed for the phenyl esters (8) and (9) in Figure 3.5 which contain tertiary amine groups. What type of catalysis is operating in this case?

(2) Explain why the hydrolysis of the substituted benzoic acid (A) at pH 4 is 1000-fold faster than the hydrolysis of the corresponding methyl ester (B) under the same conditions.

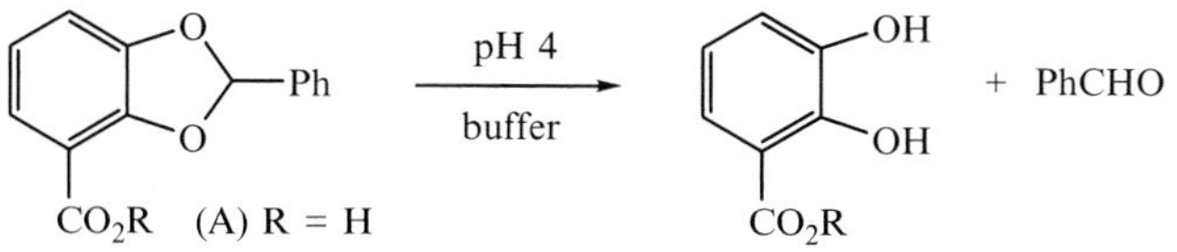

CO_2R (A) R = H
       (B) R = Me

(3) (a) Incubation of compound (C) under alkaline conditions leads to a unimolecular reaction of rate constant 7.3×10^{-2} s^{-1} to give a bicyclic product – suggest a structure. Incubation of compound (D) with phenoxide (PhO^-) leads to a bimolecular reaction of rate constant 10^{-6} $M^{-1} s^{-1}$. Work out the effective concentration of phenoxide in (C) and comment on the rate acceleration observed.

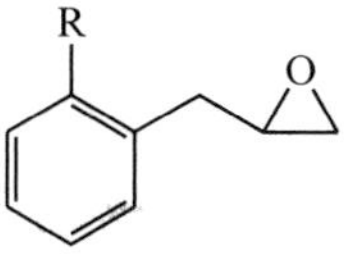

(C) R = OH
(D) R = H

(b) Epoxide hydrolase catalyses the hydrolysis of a wide range of epoxide substrates. Its active site contains an aspartate residue which is essential for catalytic activity. Given that the enzyme is most active at pH 8–9, propose two possible mechanisms for the enzymatic reaction, and suggest how you might distinguish them experimentally. Comparing the k_{cat} of 1.1 s^{-1} for epoxide hydrolase with the rate constant for reaction of (C) above, suggest how the enzyme might achieve its additional rate acceleration.

(4) Incubation of haloalkane dehalogenase (see Figure 3.19) with substrate under multiple turnover conditions in $H_2^{18}O$ leads to incorporation of one atom of ^{18}O into the alcohol product. However, incubation of a large quantity of enzyme with less than one equivalent of substrate in $H_2^{18}O$ revealed that the product contained no ^{18}O. Explain this observation.

(Note: Similar results were obtained with this type of experiment for the epoxide hydrolase enzyme in Problem 3).

Further reading

Intramolecular catalysis

N.S. Isaacs (1987) *Physical Organic chemistry*. Longman Scientific, Harlow.
A.J. Kirby (1980) Effective molarities for intramolecular reactions. *Adv. Phys. Org. Chem.*, **17**, 183–278.
F.M. Menger (1985) On the source of intramolecular and enzymatic reactivity. *Acc. Chem. Res.*, **18**, 128–34.

Enzyme catalysis

T.C. Bruice (2002) A view at the millennium: the efficiency of enzyme catalysis. *Acc. Chem. Res.*, **35**, 139–48.
T.C. Bruice & S.J. Benkovic (2000) Chemical basis for enzyme catalysis. *Biochemistry*, **39**, 6267–74.
A. Fersht (1985) *Enzyme Structure and Mechanism*, 2nd Ed. Freeman, New York.
W.P. Jencks (1969) *Catalysis in Chemistry and Enzymology*. McGraw-Hill, New York.
W.P. Jencks (1975) Binding energy, specificity and enzyme catalysis – the Circe effect. *Adv. Enzymol.*, **43**, 219–410.
R. Wolfenden & M.J. Snider (2001) The depth of chemical time and the power of enzymes as catalysts. *Acc. Chem. Res.*, **34**, 938–45.

Haloalkane dehalogenase

S.M. Franken, H.J. Rozeboom, K.H. Kalk & B.W. Dijkstra (1991) Crystal structure of haloalkane dehalogenase: an enzyme to detoxify halogenated alkanes. *EMBO J.*, **10**, 1297–302.
F. Pries, J. Kingma, M. Pentenga, G. van Pouderoyen, C.M. Jeronimus-Stratingh, A.P. Bruins & D.B. Janssen (1994) Site-directed mutagenesis and oxygen isotope incorporation studies of the nucleophilic aspartate of haloalkane dehalogenase. *Biochemistry*, **33**, 1242–7.

Dihydrofolate reductase

G.P. Miller & S.J. Benkovic (1998) Stretching exercises – flexibility in dihydrofolate reductase catalysis. *Chemistry & Biology*, **5**, R105–13.

4   Methods for Studying Enzymatic Reactions

4.1   Introduction

Having established the general principles of enzyme structure and enzyme catalysis, the remaining chapters will deal with each major class of enzymes and their associated coenzymes, and a range of enzyme mechanisms will be discussed. In this chapter we will meet the kind of experimental methods that are used to study enzymes and to elucidate the mechanisms that will be given later. There will be only a brief discussion of the biochemical techniques involved in enzyme purification and characterisation, since such methods are described in much more detail in many biochemistry texts. The chapter will focus on those experimental techniques that provide insight into the enzymatic reaction and active site chemistry.

4.2   Enzyme purification

If we want to study a particular enzymatic reaction, the first thing we need to do is to find a source of the enzyme and purify it. In order to test the activity of the enzyme we must first of all have an *assay*: a quantitative method for measuring the conversion of substrate into product (Figure 4.1). In some cases conversion of substrate to product can be monitored directly by ultraviolet (UV) spectroscopy, if the substrate or product has a distinctive UV absorbance. Failing this, a chromatographic method can be used to separate substrate from product and hence monitor conversion. In order to quantify a chromatographic assay a radioactive label is usually required in the substrate, so that after separation from substrate the amount of product can be quantitated by scintillation counting. Such an assay is highly specific and highly sensitive, but unfortunately is rather tedious for kinetic work.

A more convenient assay for kinetic purposes is to monitor consumption of a stoichiometric cofactor or cosubstrate, for example the cofactor nicotinamide adenine dinucleotide (NADH) by UV absorption at 340 nm, or consumption of oxygen by an oxygenase enzyme using an oxygen electrode. In other cases a coupled assay is used, in which the product of the reaction is immediately consumed by a second enzyme (or set of enzymes) which can be conveniently monitored.

Once a reliable assay has been developed, it can be used to identify a rich source of the enzyme, which might be a plant, an animal tissue, or

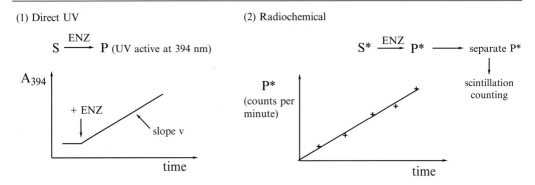

(1) Direct UV

$$S \xrightarrow{\text{ENZ}} P \text{ (UV active at 394 nm)}$$

A_{394}

+ ENZ

slope v

time

(2) Radiochemical

$$S^* \xrightarrow{\text{ENZ}} P^* \longrightarrow \text{separate } P^*$$

scintillation
counting

P^*
(counts per
minute)

time

(3) Indirect UV

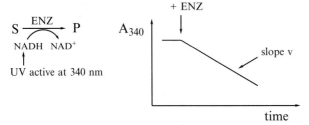

$$S \xrightarrow{\text{ENZ}} P$$
NADH NAD$^+$

UV active at 340 nm

+ ENZ

A_{340}

slope v

time

(4) Coupled UV assay

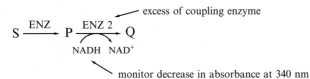

excess of coupling enzyme

$$S \xrightarrow{\text{ENZ}} P \xrightarrow{\text{ENZ 2}} Q$$
NADH NAD$^+$

monitor decrease in absorbance at 340 nm

Figure 4.1 Types of enzyme assays. A_{340}, ultraviolet (UV) absorbance at 340 nm; A_{394}, UV absorbance at 394 nm; ENZ, enzyme; NADH, nicotinamide adenine dinucleotide, P, product; Q, product of coupling enzyme; S, substrate.

a micro-organism. Enzymes are generally produced in the cytoplasm contained within the cells of the producing organism, so in order to isolate the enzyme we must break open the cells to release the enzymes inside (Figure 4.2). If it is a bacterial source, the bacteria can be grown in culture media and the cells harvested by centrifugation. The bacterial cell walls are then broken by treat-

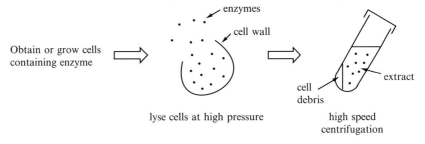

enzymes

cell wall

Obtain or grow cells
containing enzyme

cell
debris

extract

lyse cells at high pressure

high speed
centrifugation

Figure 4.2 Preparation of an enzyme extract.

ment in a high pressure cell. Animal cells can be readily broken by homogenisation, but plant cells sometimes require rapid freeze/thaw methods in order to break their tough cell walls.

Using the enzyme assay mentioned above, the enzyme activity can then be purified from the crude extract using precipitation methods such as ammonium sulphate precipitation, and chromatographic methods such as ion exchange chromatography, gel filtration chromatography, hydrophobic interaction chromatography, etc. Purification can be monitored at each stage by measuring the enzyme activity, in units per ml, where 1 unit conventionally means the activity required to convert 1 μmole of substrate per minute. Protein concentration can also be measured using colorimetric assays, in mg protein per ml. The ratio of enzyme activity to protein concentration (i.e. units per mg of protein) is known as the *specific activity* of the enzyme, and is a measure of the purity of the enzyme. As the enzyme is purified the specific activity of the enzyme should increase until the protein is homogeneous and pure, which can be demonstrated using sodium dodecyl sulphate (SDS)-polyacrylamide gel electrophoresis. A purification scheme for a hydratase enzyme from the author's laboratory is shown in Table 4.1. The purification of the enzyme can be seen from the increase in specific activity at each stage of the purification. An SDS-polyacrylamide gel containing samples of protein at each stage of the purification is shown in Figure 4.3. You can see that in the crude extract there are hundreds of protein bands, but that as the purification proceeds the 28-kDa protein becomes more and more predominant in the gel.

Why do we need pure enzyme? If we can see enzyme activity in the original extract, why not use that? The problem with using unpurified enzyme for kinetic or mechanistic studies is that there may be interference from other enzymes in the extract that use the same substrate or cofactor. There may also be enzymes that give rise to UV absorbance changes which might interfere with a UV-based assay. If the enzyme can be purified then the *turnover number* of the enzyme can be measured, which is the number of μmoles of substrate converted per μmole of enzyme per second. The turnover number can be simply

Table 4.1 Enzyme purification table (see Figure 4.3). DEAE, diethylaminoethyl.

	Volume (ml)	Enzyme activity (units ml^{-1})	Protein concentration (mg ml^{-1})	Specific activity (units mg^{-1})	Purification (-fold)
Crude extract	14	13.1	62.0	0.212	1.0
DEAE sephadex pool	19	11.6	17.0	0.684	3.2
Phenyl agarose pool	11	11.8	0.085	140	662
MonoQ anion-exchange pool	6.0	29.2	0.037	787	3710

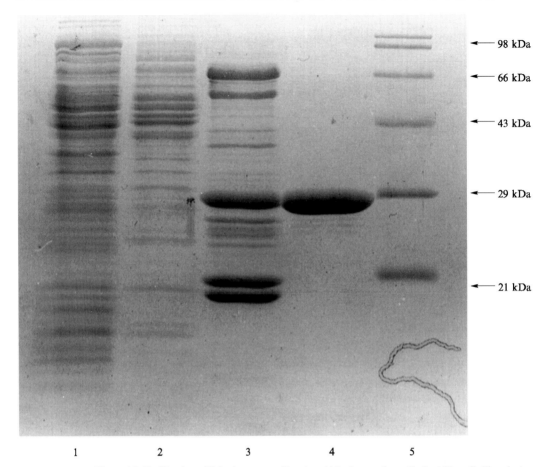

Figure 4.3 Purification of 2-hydroxypentadienoic acid hydratase from *Escherichia coli*. The photo shows an SDS-polyacrylamide gel of samples taken from the purification of this enzyme. Lane 1, *E. coli* crude extract; lane 2, DEAE sephadex pool; lane 3, phenyl agarose pool; lane 4, monoQ anion-exchange pool; lane 5, molecular weight standards (2×98 kDa, 66 kDa, 43 kDa, 29 kDa, 21 kDa). (see also Table 4.1).

calculated from the specific activity of the pure enzyme (in units per mg of protein) and the molecular weight of the enzyme (see Problem 1). The isolation of pure enzyme also allows active site studies to be carried out on the homogeneous protein, and crystallisation of the enzyme for X-ray crystallographic analysis.

4.3   Enzyme kinetics

It is possible to learn a great deal about how an enzyme works from a detailed kinetic study of the enzymatic reaction. For the purposes of this chapter I will give only a brief discussion of a simple model of enzyme kinetics; a full discussion of enzyme kinetics is given in texts such as Segel (see Further reading).

The Michaelis–Menten model for enzyme kinetics assumes that the following steps are involved in the enzymatic reaction: reversible formation of the enzyme–substrate (ES) complex, followed by conversion to product:

Michaelis–Menten Model

$$E + S \; \underset{k_{-1}}{\overset{k_1}{\rightleftharpoons}} \; ES \; \xrightarrow{k_2} \; E + P$$

There are several assumptions implicit in this model: that the enzyme binds only a single substrate; that there is only one kinetically significant step between the ES complex and product formation; and that product formation is irreversible. Despite the fact that these assumptions are not strictly correct for most enzymes, this proves to be a useful model for a very wide range of enzymes. Derivation of a rate equation uses a kinetic criterion known as the *steady state approximation*: that the enzymatic reaction will quickly adopt a situation of steady state in which the concentration of the intermediate species, ES, remains constant. Under these conditions the rate of formation of ES is equal to the rate of consumption of ES. The other criterion that is used is that the total amount of enzyme in the system E_0, made up free enzyme, E, and enzyme–substrate complex, ES, is constant. The derivation is as follows:

Derivation of Michaelis–Menten equation

Steady state approximation: Rate of formation of ES = Rate of breakdown of ES

$$k_1 [E] [S] = k_2 [ES] + k_{-1} [ES]$$

Total amount of enzyme $[E]_0 = [E] + [ES]$

$$\Rightarrow k_1 [E]_0 [S] - k_1 [ES] [S] = k_2 [ES] + k_{-1} [ES]$$

$$\Rightarrow [ES] = \frac{k_1 [E]_0 [S]}{k_1 [S] + k_2 + k_{-1}}$$

$$\text{Rate of production of P} = k_2 [ES] = \frac{k_2 [E]_0 [S]}{\left(\frac{k_{-1} + k_2}{k_1}\right) + [S]} = \frac{k_{cat} [E]_0 [S]}{K_M + [S]}$$

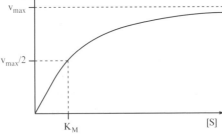

$$\text{Observed rate } v = \frac{v_{max} [S]}{K_M + [S]}$$

$$\text{where } v_{max} = k_{cat} [E_0]$$

The two kinetic constants in the Michaelis–Menten rate equation have special significance. The k_{cat} parameter is the *turnover number* mentioned above: it is a unimolecular rate constant whose units are s^{-1} (or min^{-1} if it is a very slow enzyme!), and it represents the number of μmoles of substrate converted per μmole of enzyme per second. Alternatively, in molecular terms it represents the number of molecules turned over by one molecule of enzyme per second, which gives a good feel for how quickly the enzyme is operating. Typically values are in the range $0.1–100\ s^{-1}$.

The K_M parameter is known as the Michaelis constant for the enzyme: its units are $mol\ l^{-1}$ or M. In practice, the K_M is the concentration of substrate at which half-maximal rate is observed. It can be taken as a rough indication of how tightly the enzyme binds its substrate, so a substrate bound weakly by an enzyme will have a large K_M value, and a substrate bound tightly will have a small K_M. However, it must be stressed that K_M is *not* a true dissociation constant for the substrate, since it also depends on the rate constant k_2. Values of K_M are typically in the range $1\ \mu M–1\ mM$.

Values of k_{cat} and K_M can be measured for a particular enzyme by measuring the rate of the enzymatic reaction at a range of different substrate concentrations. At high substrate concentrations ($[S] \gg K_M$) the rate equation reduces to $v = k_{cat}\ [E_0]$, so a maximum rate is observed however high the substrate concentration. Under these conditions the enzyme is fully saturated with substrate, and no free enzyme is present. So as soon as an enzyme molecule releases a molecule of product it immediately picks up another molecule of substrate. In other words the enzyme is working flat out: the observed rate of reaction is limited only by the rate of catalysis.

Under low substrate concentrations the rate equation reduces to $v = (k_{cat}/K_M)[E][S]$, so the observed rate is proportional to substrate concentration, and the reaction has effectively become a bimolecular reaction between free enzyme, E, and free substrate, S. Under these conditions the majority of enzyme is free enzyme, and the observed rate of reaction depends on how efficiently the enzyme can bind the substrate at that concentration. The bimolecular rate constant under these conditions k_{cat}/K_M is known as the *catalytic efficiency* of the enzyme, since it represents how efficiently free enzyme will react with free substrate.

A schematic representation of the energetic profiles at high and low substrate concentrations is given in Figure 4.4. At high substrate concentrations the enzyme is fully saturated with substrate, so the activation energy for the enzymatic reaction is the free energy difference between the ES complex and the transition state. At $[S] = K_M$ the enzyme is half-saturated with substrate. At low substrate concentrations the majority of the enzyme is free of substrate, so the activation energy for the reaction is the free energy difference between free enzyme + substrate and the transition state.

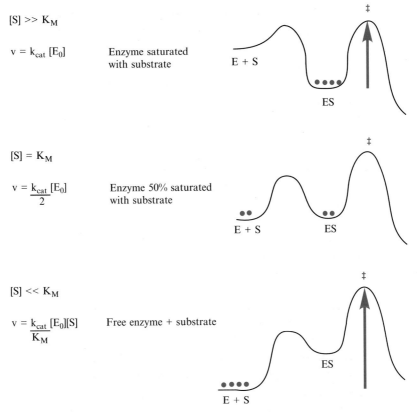

$[S] \gg K_M$

$v = k_{cat}[E_0]$ Enzyme saturated
                        with substrate

$[S] = K_M$

$v = \dfrac{k_{cat}[E_0]}{2}$ Enzyme 50% saturated
                        with substrate

$[S] \ll K_M$

$v = \dfrac{k_{cat}[E_0][S]}{K_M}$ Free enzyme + substrate

Figure 4.4 Significance of K_M, k_{cat}. Plots are of free energy versus reaction co-ordinate. Red dots indicate the population of enzyme molecules present.

How do we actually determine k_{cat} and K_M? The value of k_{cat} can be roughly visualised from the plot of v versus [S] by estimating the rate at high substrate concentrations. The K_M value corresponds to the substrate concentration at which half-maximal rate is observed. However, a more accurate way to determine K_M and k_{cat} from the data is to use either of the Lineweaver–Burk or Eadie–Hofstee plots, shown in Figure 4.5, both of which give straight lines from which the kinetic constants can be determined as indicated.

Enzyme inhibition takes place if a molecule other than the substrate binds at the active site and prevents the enzymatic reaction from taking place. Broadly speaking, there are two types of enzyme inhibition that are commonly observed: reversible and irreversible inhibition. Reversible inhibition occurs if an inhibitor is bound non-covalently at the active site, preventing one of the substrates from binding. The most common type of reversible inhibition observed for a single substrate enzyme reaction is competitive inhibition, where the inhibitor binds at the same site as the substrate, S, and therefore competes for the same binding site. The kinetic scheme and associated kinetic behaviour

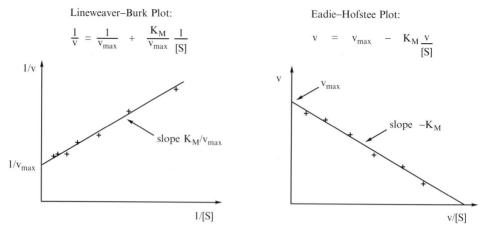

Lineweaver–Burk Plot:

$$\frac{1}{v} = \frac{1}{v_{max}} + \frac{K_M}{v_{max}}\frac{1}{[S]}$$

Eadie–Hofstee Plot:

$$v = v_{max} - K_M\frac{v}{[S]}$$

Figure 4.5 Graphical methods for determination of K_M and k_{cat}. v, reaction velocity; [S], substrate concentration.

observed for competitive inhibition is shown in Figure 4.6. If the rate of the enzyme reaction is measured at varying substrate concentrations but fixed inhibitor concentrations, apparent K_M values ($(K_M)_{app}$) can be measured at varying inhibitor concentrations. If plotted on a Lineweaver–Burk plot a series of straight lines are obtained, intersecting on the y-axis. Thus the v_{max} is unaffected by competitive inhibition, since at high substrate concentrations the substrate can competitively displace the inhibitor.

Non-competitive reversible inhibition is observed when an inhibitor, I, binds to another part of the enzyme active site leading to a non-productive EIS complex. This type of inhibition, illustrated in Figure 4.6, typically occurs in multi-substrate reactions when the inhibitor, I, binds to the binding site of the co-substrate. For a full discussion of these and other types of inhibition in multi-substrate reactions reference to texts on enzyme kinetics is recommended.

Irreversible inhibition occurs when an inhibitor first binds at the active site, then reacts with an active site group to form a covalent bond (E−I). The active site is then irreversibly blocked by the inhibitor and is permanently inactivated. Irreversible inhibitors usually contain electrophilic functional groups such as halogen substituents or epoxides. The kinetic characteristic associated with irreversible inhibition is that it is time-dependent. This is because as time goes by more and more enzyme will be blocked irreversibly by conversion of the reversible EI complex to E−I. Since [S] ≫ [E] then in practice this is a unimolecular reaction, so the observed kinetic behaviour follows unimolecular reaction kinetics. Thus, if enzyme activity (i.e. v_{max}) is plotted versus time, an exponential decrease of activity is observed versus time (Figure 4.7).

(a) Competitive inhibition

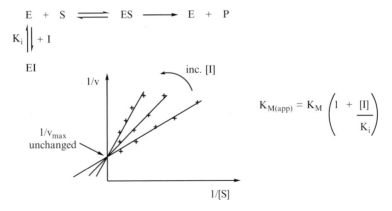

$$K_{M(app)} = K_M \left(1 + \frac{[I]}{K_i} \right)$$

(b) Non-competitive inhibition

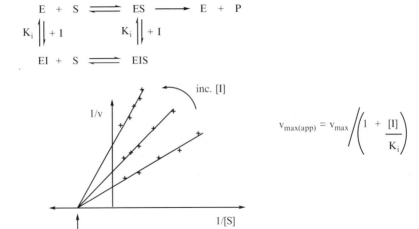

$$v_{max(app)} = v_{max} \left/ \left(1 + \frac{[I]}{K_i} \right) \right.$$

Figure 4.6 Reversible inhibition.

All of the above kinetic data can be obtained by steady state kinetics, which can be observed conveniently over a 1–10-minute assay period. However, if one is able to examine an enzymatic reaction before the attainment of steady state, then individual enzymatic rate constants can be measured directly using pre-steady-state kinetics. If the turnover number for an enzyme is $10\,s^{-1}$ then under saturating conditions a molecule of substrate will be converted to product in $0.1\,s$. Therefore, in order to examine a single catalytic cycle one must examine the enzymatic reaction in the range 0–100 ms. This can be done using a stopped flow apparatus, shown in simplified form in Figure 4.8.

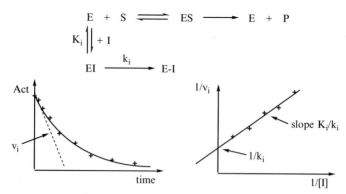

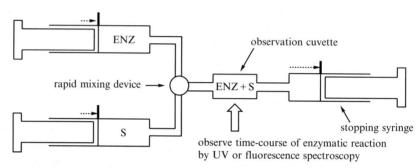

Figure 4.7 Irreversible inhibition.

Figure 4.8 Stopped flow apparatus.

This apparatus consists of two syringes containing: (i) substrate and (ii) a stoichiometric amount (i.e. 1–100 nmol) of enzyme, connected to a rapid mixing device. The syringes are driven so as to fill the observation cuvette with freshly mixed enzyme and substrate. At this point, the reaction time of the mixture is given by the distance from the mixer and the linear flow rate. The syringes are physically stopped, and the enzyme/substrate mixture is allowed to react along the enzymatic reaction pathway. Changes in absorbance or fluorescence in the observation cuvette are measured on a 10–1000-ms timescale, giving a complete time course of a single catalytic cycle of the enzymatic reaction. Several substrate concentrations are used to enable individual rate constants to be measured. In this way a multi-step enzymatic reaction can in principle be broken down into its individual steps, the slowest of which (the rate-determining step) should correspond to the steady state k_{cat} value measured by steady state kinetics.

4.4   The stereochemical course of an enzymatic reaction

The vast majority of biological molecules are chiral, that is a molecule whose mirror image is non-superimposable upon the original. In the case of carbon-based compounds, if a carbon atom is surrounded by four different groups, then it will be chiral. One simple example shown in Figure 4.9 is that of the amino acid L-alanine, which is a substrate for several pyridoxal 5′-phosphate enzymes described in Chapter 9. Chiral centres can be designated as *R* or *S* depending on the relative orientation of the four groups, according to the Cahn–Ingold–Prelog rule (see Appendix 1). Thus L-alanine is more formally written as 2*S*-alanine.

Enzyme-catalysed reactions are in general both stereoselective and stereo-specific. Stereoselectivity is the ability to select a single enantiomer of the substrate in the presence of other isomers. Stereospecificity is the ability to catalyse the production of a single enantiomer of the product via a specific reaction pathway. Stereospecificity in enzyme catalysis arises from the fact that catalysis is taking place in an enzyme active site in which the bound substrate is held in a defined orientation relative to the active site groups (unnatural substrates which can adopt more than one orientation in an enzyme active site may be processed with less stereospecificity). How do we elucidate the stereochemical course of an enzymatic reaction?

Generation of a chiral product

In many cases enzymes are able to generate a product containing one or more chiral centres from achiral substrates – a process known as *asymmetric induction*. For example, the aldolase enzyme fructose-1,6-bisphosphate aldolase (see Chapter 7) catalyses the aldol condensation of achiral substrates dihydroxy-acetone phosphate and acetaldehyde to generate a chiral aldol product, as shown in Figure 4.10.

If a chiral product is formed, then its absolute configuration can be deter-mined in the following ways:

(1) by chemical conversion to a compound of established absolute configura-tion, and measurement of optical activity;

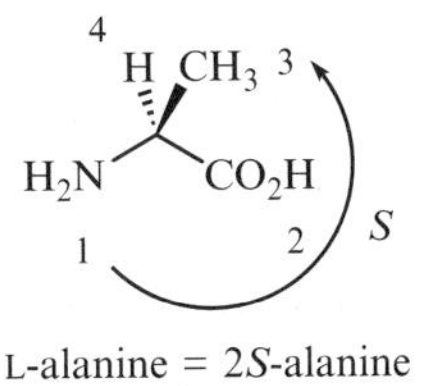

L-alanine = 2*S*-alanine

Figure 4.9 A chiral molecule.

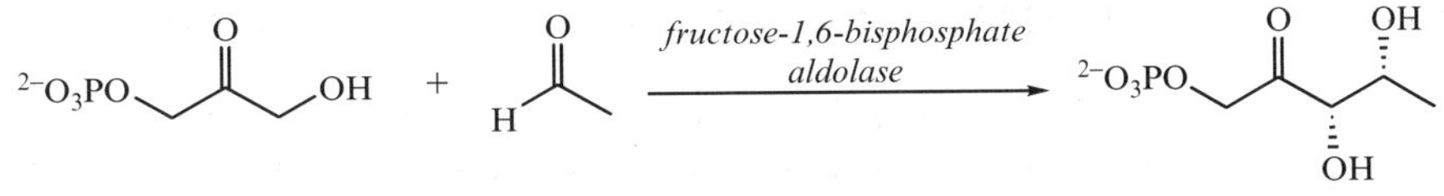

Figure 4.10 Production of a chiral product.

(2) by treatment with another known chiral reagent to make a diastereomeric derivative whose configuration can be determined by methods such as nuclear magnetic resonance (NMR) spectroscopy or X-ray crystallography;

(3) by treatment with an enzyme whose reaction proceeds with known stereo-specificity.

Prochiral selectivity

Carbon centres which are surrounded by XXYZ groups are known as *prochiral* centres. For example, the C-1 carbon atom of ethanol is prochiral since it is attached to two hydrogens, one methyl group, and one hydroxyl group. This is not a chiral centre, since there are two hydrogens attached, yet these two protons can be distinguished by the enzyme alcohol dehydrogenase, which oxidises ethanol to acetaldehyde as shown in Figure 4.11.

Prochiral hydrogens can be designated using a variation of the Cahn–Ingold–Prelog rule. The convention is that the hydrogen to be assigned is replaced by a deuterium atom (making the centre chiral), and the chirality of the resulting centre determined using the Cahn–Ingold–Prelog rule (see Appendix 1). If the resulting chiral centre has configuration R, then the hydrogen atom replaced by deuterium is labelled *proR*. Conversely, if the deuterium-containing centre is S, then the hydrogen atom is labelled *proS*. In the case of alcohol dehydrogenase, the enzyme removes stereospecifically the *proR* hydrogen. This was demonstrated by synthesising authentic samples of $(1R\text{-}^2H)$- and $(1S\text{-}^2H)$-ethanol. Each sample was separately incubated with the enzyme and the deuterium content of the product analysed in each case. In the case of the $1R$ substrate deuterium was removed by the enzyme, whereas with the $1S$ substrate deuterium was retained in the product, as shown in Figure 4.12.

How does this enzyme achieve this remarkable selectivity? If you imagine that the ethanol molecule is fixed in the plane of the page with the methyl group pointing left and the hydroxyl group pointing right, as in Figure 4.12, then one of the two hydrogens is pointing up out of the page, and the other is pointing down into the page. Since we can visualise this situation in three dimensions, we can easily distinguish between these two hydrogens. Thus, two hydrogen atoms attached to a prochiral centre can be distinguished *if they are held in a fixed orientation in a chiral environment*. Enzyme active sites satisfy both these criteria, since they are able to bind molecules in a defined orientation using

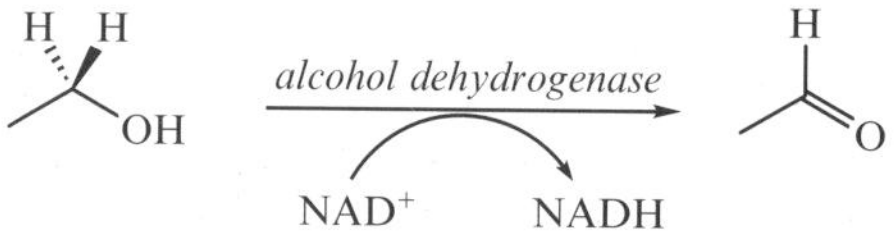

Figure 4.11 Alcohol dehydrogenase reaction.

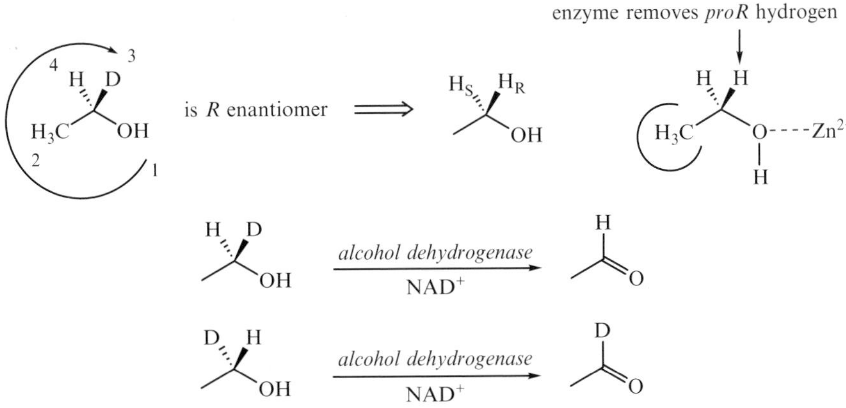

enzyme removes *proR* hydrogen

is *R* enantiomer $\Longrightarrow$

Figure 4.12 Stereochemistry of alcohol dehydrogenase.

specific enzyme–substrate binding interactions, and of course the enzyme active site is chiral.

Examination of the X-ray crystal structure of alcohol dehydrogenase reveals that the C-1 oxygen substituent is bound by an active site Zn^{2+} cofactor, and the methyl group is bound in such a way that the *proR* hydrogen is pointing directly at the NAD^+ cofactor, as shown in Figure 4.13. The prochiral selectivity can therefore easily be explained by the orientation adopted by the substrate in the enzyme active site.

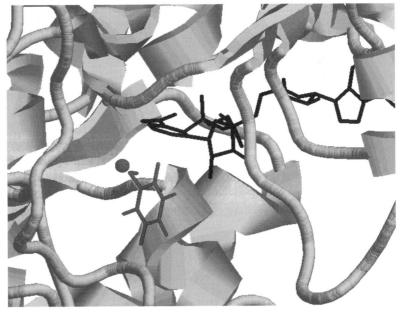

Figure 4.13 Active site of alcohol dehydrogenase (PDB file 1HLD), showing the positioning of a bound pentafluorobenzyl alcohol substrate. Zn^{2+} cofactor and bound substrate shown in red, NAD^+ cofactor shown in black.

In general the stereochemical course of an enzymatic reaction is usually determined by replacement of particular atoms of the substrate by isotopes of carbon, hydrogen, oxygen and nitrogen, whose fate at the end of the reaction can then be monitored. In particular, enzymatic reactions involving prochiral centres can only be studied by replacement of one of the prochiral substituents by another isotope, thus generating a chiral centre. The isotopes available are listed in Table 4.2, together with the methods available for their analysis. In some cases such as ^1H, ^{13}C and ^{15}N these nuclei possess nuclear spin, which allows analysis by NMR spectroscopy. In other cases, such as ^3H and ^{14}C, the nuclei are radioactive and their presence can be detected by scintillation counting. Isotopes such as ^{18}O which are neither radioactive nor possess nuclear spin must be detected either by mass spectrometry or by their effect on neighbouring nuclei: in the case of ^{18}O their attachment to a neighbouring ^{13}C shifts the ^{13}C NMR signal upfield by 0.01–0.05 ppm.

Enzymatic reactions which involve the substitution of one group for another group in a defined relative orientation can either take place with retention or inversion of stereochemistry. Where such reactions occur at a prochiral centre, the stereochemical course can be examined by synthesis of a stereospecifically labelled substrate. For example, the haem enzyme P450cam

Table 4.2 Isotopes available for stereochemical elucidation.

	Isotope	Natural abundance (%)	Method of analysis
Hydrogen	^1H	99.985	NMR (I = 1/2)
	^2H	0.015	NMR (I = 1) or shift in ^{13}C NMR
	^3H	—	Scintillation counting
Carbon	^{12}C	98.9	
	^{13}C	1.1	NMR (I = 1/2)
	^{14}C	—	Scintillation counting
Nitrogen	^{14}N	99.63	
	^{15}N	0.37	NMR (I = 1/2)
Oxygen	^{16}O	99.8	
	^{17}O	0.037	NMR (I = 5/2)
	^{18}O	0.20	Mass spectrometry or shift in ^{13}C NMR

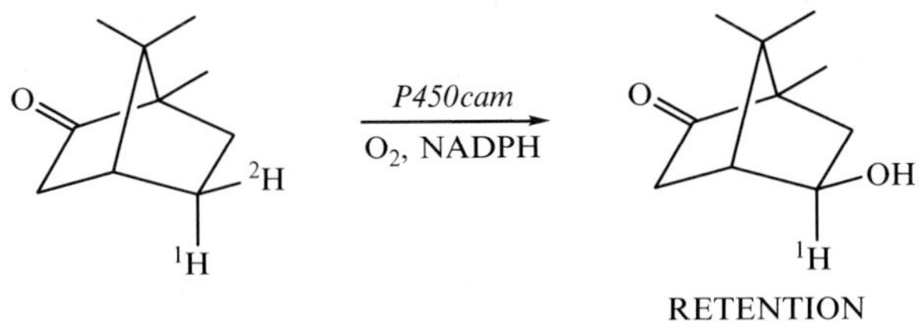

Figure 4.14 Stereochemistry of P450cam-catalysed reaction.

which catalyses the hydroxylation of camphor was shown to proceed with retention of stereochemistry by use of the labelled substrate illustrated in Figure 4.14.

Interconversions of methylene ($=CH_2$ or $-CH_2-$) groups to methyl ($-CH_3$) groups require a special type of stereochemical analysis, since the resulting methyl group contains three apparently identical hydrogen atoms (i.e. an XXXY system). However, it is possible to analyse these methylene-to-methyl interconversions using all three of the isotopes of hydrogen – ^1H, ^2H and ^3H – in the form of a chiral methyl group. It is important to note that in this analysis the ^1H and ^2H substituents are present in 100% abundance, whereas only a small proportion of molecules contain ^3H (since ^3H is only available and only safe to handle in relatively low abundance). Therefore the analysis of chiral methyl groups must focus on those molecules containing ^3H, by detecting the presence or absence of ^3H label.

Chiral methyl groups can be generated from enzymatic reactions by preparing the methylene substrate in a stereospecifically labelled form using two of the isotopes of hydrogen, and carrying out the enzymatic reaction in the presence of the third isotope. If the product can be degraded to chiral acetic acid, then the configuration of the chiral methyl group can be determined using a method developed independently by Cornforth and Arigoni, shown in Figure 4.15.

The method of analysis involves conversion to chiral acetyl coenzyme A (CoA) (see Section 5.4), followed by incubation with malate synthase (see Section 7.3), which removes one of the hydrogens on the methyl group, and combines with glyoxalate to form malic acid. The malate synthase reaction has a preference for removal of ^1H rather than ^2H or ^3H (i.e. a kinetic isotope effect of $k_H/k_T = 2.7$), so in the majority of molecules ^1H is removed. The reaction with glyoxalate then occurs with inversion of configuration. Therefore, the 2S enantiomer of acetyl CoA is converted into the 2S,3R enantiomer of malate containing ^2H and ^3H stereospecifically at C-3, as illustrated in Figure 4.15. Treatment with the enzyme fumarase then results in a stereospecific *anti*-elimination of water, the enzyme removing only the *proR* hydrogen at C-3. The configuration of the major product at C-3 can, therefore, be deduced by

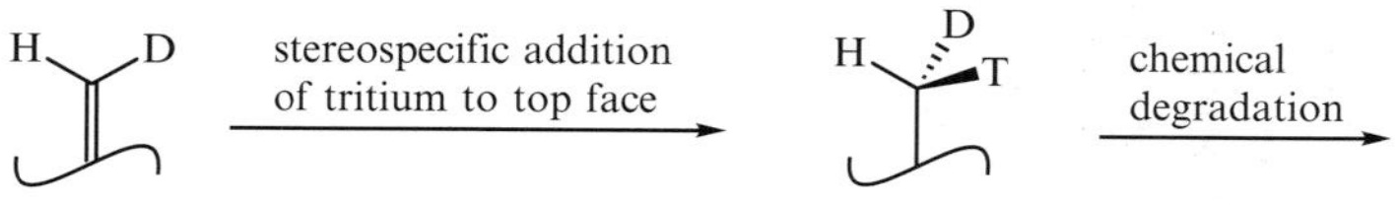

Analysis of chiral acetic acid:

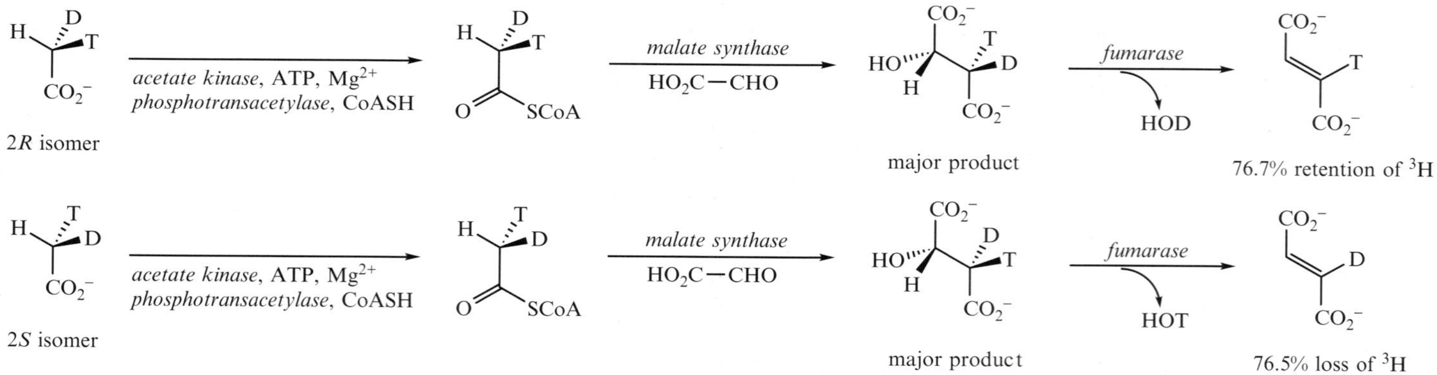

Figure 4.15 Chiral methyl groups. ATP, adenosine triphosphate; CoASH, coenzyme A.

monitoring the fate of the ^{3}H, either to water or to tritiated fumaric acid. Since the stereochemistry of the fumarase and malate synthase reactions is known, the configuration of the chiral methyl group can be deduced. By this method a number of such methylene-to-methyl interconversions have been analysed.

A similar stereochemical strategy is used to analyse the stereochemistry of phosphoryl transfer reactions, since phosphates also contain three apparently identical oxygen substituents. Three isotopes of oxygen are also available: ^{16}O, ^{17}O and ^{18}O. Using skilful synthetic chemistry approaches, phosphate ester substrates can be prepared containing all three isotopes of oxygen. Incubation of the chiral phosphate ester substrate with the corresponding phosphotransferase enzyme generates a chiral phosphate ester product, as shown in Figure 4.16. The configuration of the chiral product reveals whether the enzymatic reaction proceeds with retention or inversion of configuration. Analysis of the configuration of the chiral phosphate ester product is complicated, but in essence involves chemical or enzymatic conversion to a diastereomeric derivative, followed by ^{31}P NMR spectroscopic analysis.

The stereochemistry of reactions releasing inorganic phosphate presents an even more difficult problem, since there are four apparently identical oxygens to be distinguished, but only three isotopes of oxygen. This has been solved by incorporating one atom of sulphur as a substituent, since thiophosphate ester substrates are accepted by these enzymes (Figure 4.17). Again the configuration of the [^{16}O, ^{17}O, ^{18}O]-thiophosphate product can be deduced by conversion to a diastereomeric derivative followed by NMR spectroscopic analysis. For a detailed discussion of this stereochemical analysis the interested reader is referred to the further reading at the end of the chapter.

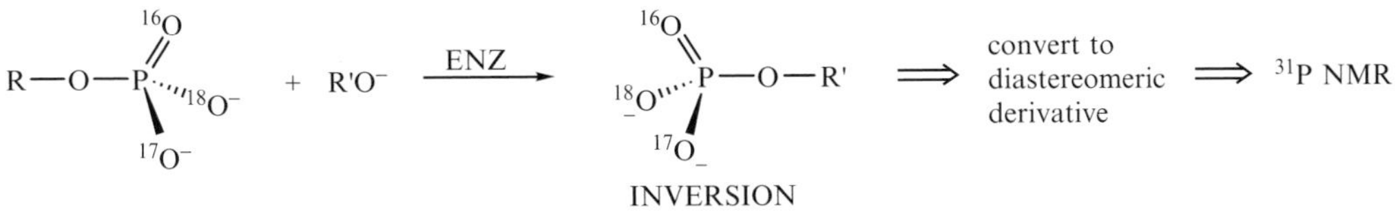

Figure 4.16 Stereochemistry of phosphoryl transfer reaction.

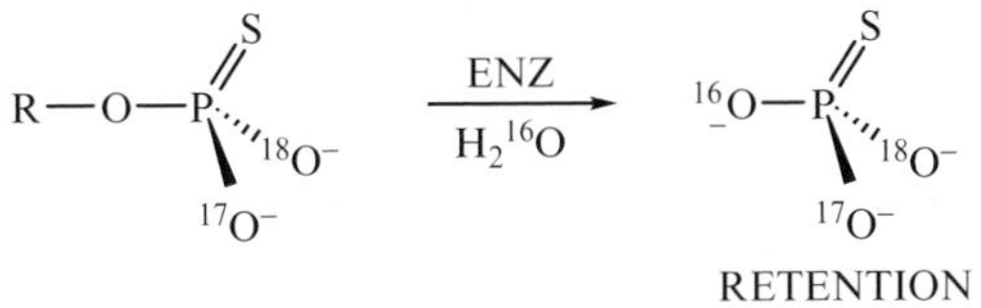

Figure 4.17 Stereochemistry of phosphate release.

4.5 The existence of intermediates in enzymatic reactions

Enzymatic reactions are often multi-step reactions involving a number of transient enzyme-bound intermediates (Figure 4.18). If the enzymatic reaction is very rapid and none of the intermediates are released from the active site, how can we prove the existence of these transient intermediates?

Direct observation

Since the turnover numbers for most enzymes are $> 1 \, s^{-1}$ then it is impractical to observe the formation of intermediates directly. However, in some cases the turnover number can be reduced by changing the temperature or pH, or by using an unnatural substrate, to such an extent that intermediates can be detected directly by NMR spectroscopy.

If no intermediate is detectable by these methods then a faster analytic method can be used in the form of stopped flow methods. Just as stopped flow methods can be used to study rapid enzyme kinetics, in the same way rapid quench methods can be used to isolate intermediates. This method involves mixing enzyme with substrate in a rapid mixing device similar to that shown in Figure 4.8, then, after a fixed time interval of say 100 ms, mixing with a quench reagent such as an organic solvent or a different pH solution.

This type of approach was recently used to identify a tetrahedral intermediate in the reaction of 5-enolpyruvyl-shikimate-3-phosphate (EPSP) synthase, as shown in Figure 4.19 (also see Section 8.5). In this case the intermediate was isolated by quenching 50 mg quantities of enzyme and substrate with neat triethylamine, which was found to stabilise the intermediate.

Trapping

Intermediates in enzymatic reactions that possess enhanced chemical reactivity can sometimes be trapped using a selective chemical reagent. One example

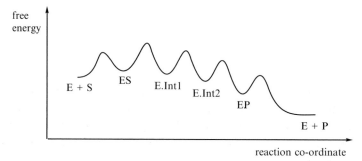

Figure 4.18 Energy profile of a multi-step enzymatic reaction.

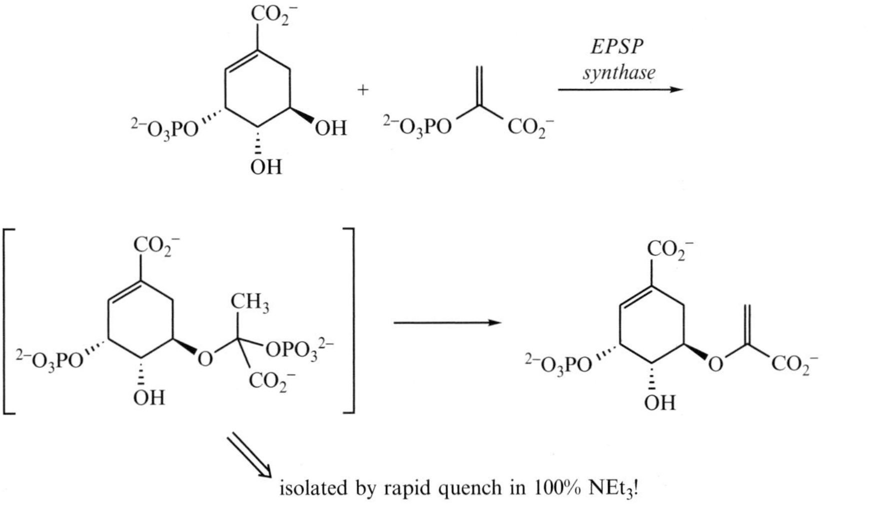

Figure 4.19 Identification of 5-endpyruvyl-shikimate-3-phosphate (EPSP) synthase reaction intermediate. NEt$_3$, triethylamine.

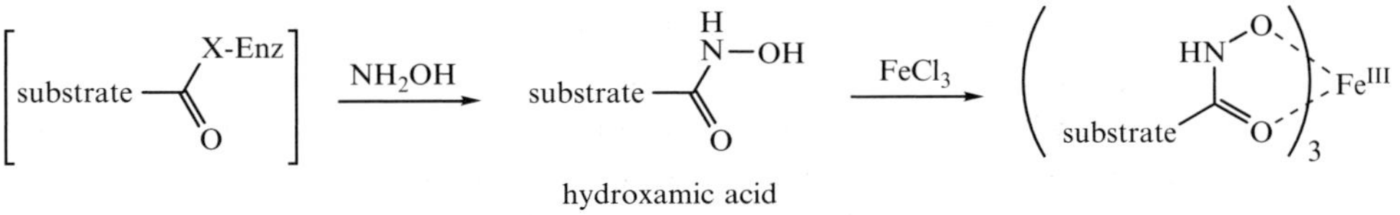

Figure 4.20 Hydroxylamine trapping of an activated intermediate.

already mentioned is the trapping of imine intermediates formed upon reaction of active site lysine residues with carbonyl substrates with sodium borohydride. Another common example is the trapping of activated carbonyl intermediates with hydroxylamine, a potent nitrogen nucleophile. This reaction forms hydroxamic acid products which can be detected spectrophotometrically by treatment with iron(III) chloride solutions, as shown in Figure 4.20.

Chemical inference

The existence of certain intermediates can be inferred by following the fate of individual atoms in the substrate. For example, the existence of an acyl phosphate intermediate in the glutamine synthetase reaction was established by incubating ^{18}O-labelled substrate with enzyme and cofactor adenosine triphosphate (ATP). The inorganic phosphate product was found to contain one atom of ^{18}O, consistent with formation of the acyl phosphate intermediate as shown in Figure 4.21.

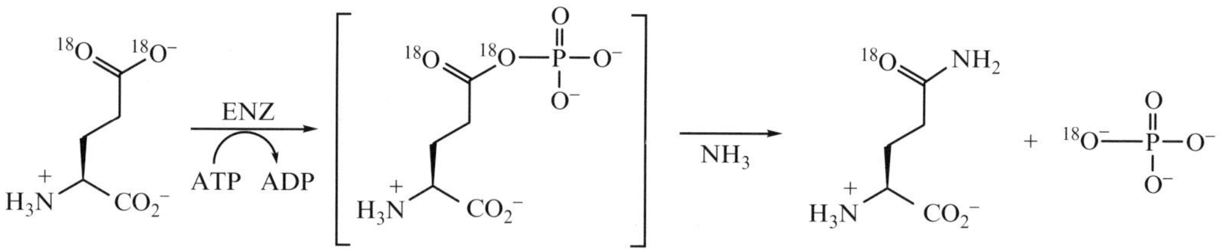

Figure 4.21 Identification of glutamine synthetase intermediate by chemical inference. ADP, adenasine diphosphate; NH3, ammonia.

Isotope exchange

If an enzyme reaction involves the reaction of two species to form an intermediate which is then attacked by a third species, the first step or partial reaction can be analysed in isolation using isotope exchange. Taking the glutamine synthetase reaction as an example, glutamate first reacts with ATP to form adenosine diphosphate (ADP) and an intermediate γ-glutamyl phosphate, which is attacked by ammonia to form glutamine. If the enzyme is incubated with glutamate, ^{14}C-ATP and unlabelled ADP in the absence of ammonia, then the enzyme cannot complete the overall reaction, but it can convert glutamate to enzyme-bound γ-glutamyl phosphate and ^{14}C-ADP. In this case ^{14}C-ADP can be released from the enzyme without releasing the intermediate. The enzyme can then convert unlabelled ADP to ATP via the reverse reaction. Overall this process leads to the 'exchange' of ^{14}C label from ATP to ADP, as illustrated in Figure 4.22.

This method depends on the ability of the enzyme to release ^{14}C-ADP at the intermediate stage. Since many enzymes have well-defined orders of binding of their substrates, this release may be slow or even impossible in the absence of the third substrate. A more subtle method of analysing such exchange is the method of *positional isotope exchange* developed by Rose. This method is illustrated in Figure 4.23 for the glutamine synthetase reaction.

For this method ATP is labelled with ^{18}O at the β,γ-O bridge position. Upon formation of the reaction intermediate the γ-phosphate is transferred to glutamate, allowing the ^{18}O label to scramble amongst the β-phosphate oxygens whilst still bound to the enzyme. Upon reformation of the β,γ-O bridge in the reverse reaction the ^{18}O label will be present not only in the bridge

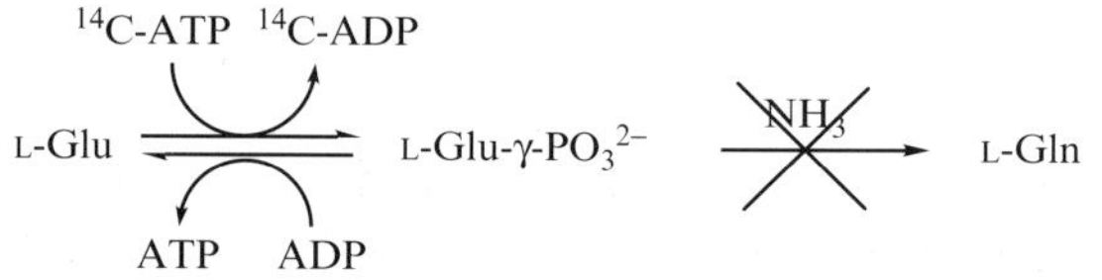

Figure 4.22 Isotope exchange in the glutamine synthetase reaction. L-Gln, glutamine; L-Glu, glutamate; L-Glu-γ-PO$_3^{2-}$, γ-glutamyl phosphate.

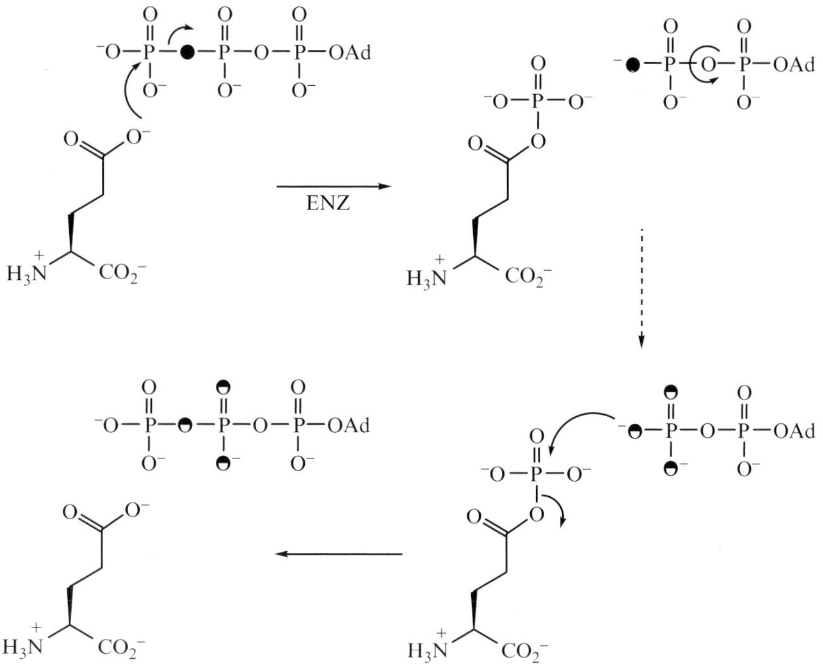

Figure 4.23 Positional isotope exchange.

position but also in the β-phosphate oxygens. Not only can such isotope exchanges be observed, the rate of isotope exchange can be measured and compared with the rate of the overall enzymatic reaction.

The final proof that a certain species is a true intermediate in an enzymatic reaction is by preparing the intermediate by independent chemical synthesis. There are two criteria that must be satisfied for a potential intermediate. The first is *chemical competence*: the intermediate must be converted by the enzyme to the product of the reaction (and also be a substrate for the reverse reaction if the enzymatic reaction is reversible). The second is *kinetic competence*: the rate of conversion to products must be at least as fast as the rate of the overall enzymatic reaction.

4.6   Analysis of transition states in enzymatic reactions

Enzymatic reactions are frequently multi-step reactions. In such reactions the overall rate of the enzymatic reaction is governed by the step having the highest transition state energy – the rate-determining step. If we want to determine a detailed kinetic profile for an enzymatic reaction, it is important to know which is the rate-determining step. Once again we can use isotopic substitution, this time to study the existence of kinetic isotope effects.

If a reaction involves cleavage of a $C-H$ bond in its rate-determining step, then substitution of hydrogen for deuterium leads to a reduction in rate for that step, and hence the overall reaction. This effect arises because the $C-D$ bond is slightly stronger than the $C-H$ bond, because the $C-D$ bond has a lower zero point energy, as shown in Figure 4.24.

There is a larger activation energy for cleavage of a $C-D$ bond than for cleavage of a $C-H$ bond. Substitution of a hydrogen that is removed in an enzymatic reaction with deuterium and measurement of the k_{cat} and K_M values for the deuteriated substrate can, therefore, provide information about the thermodynamic profile of the reaction. Kinetic isotope effects are commonly observed also in organic reactions, for exactly the same reasons; however, the analysis of isotope effects in enzymatic reactions can be complicated in cases where there are a number of transition states.

Four typical scenarios are shown in Figure 4.25:

(a) If the step involving $C-H$ cleavage has a significantly higher transition state energy than the other steps, then substitution for a $C-D$ bond will give a substantial kinetic isotope effect ($k_H/k_D \sim 6-7$).
(b) If the step involving $C-H$ cleavage is of similar transition state energy to the other steps, then this step is only partially rate determining, and a kinetic isotope effect of $k_H/K_D \sim 2-3$ is observed.
(c) If $C-H$ cleavage occurs after the rate-determining step, then no kinetic isotope effect is observed.
(d) Finally, if a partially rate-determining $C-H$ cleavage occurs before the rate-determining step, then a small kinetic isotope effect will be observed on k_{cat}/K_M, but may not be observed on k_{cat}. Remember that k_{cat}/K_M is the bimolecular rate constant for the reaction of free enzyme with free substrate, whereas k_{cat} is the unimolecular rate constant for conversion of

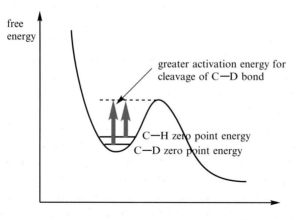

Figure 4.24 Origin of deuterium kinetic isotope effect.

(a) C—H cleavage rate determining $\dfrac{k_H}{k_D} = 6\text{-}7$

$[C\text{----}H]^{\ddagger}$

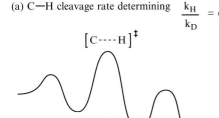

(b) C—H cleavage partially rate determining $\dfrac{k_H}{k_D} = 2\text{-}3$

$[C\text{---}H]^{\ddagger}$

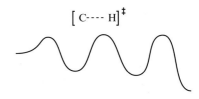

(c) C—H cleavage after rate determining step

no KIE observed

$[C\text{---}H]^{\ddagger}$

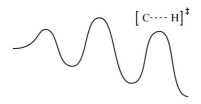

(d) C—H cleavage before rate determining step

KIE observed on k_{cat}/K_M ([S] $<<$ K_M)

$[C\text{----}H]^{\ddagger}$

$[S] >> K_M$

ES

$[S] << K_M$

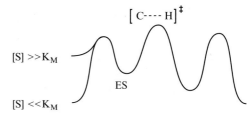

Figure 4.25 Expression of kinetic isotope effects (KIE).

saturated ES complex. In the latter case where the enzyme is fully saturated with substrate an early C−H cleavage step may not be kinetically so significant.

In the early 1980s, measurement of kinetic isotope effects for the yeast and horse alcohol dehydrogenase-catalysed reactions was found to give anomously high secondary kinetic isotope effects ($k_H/k_D = 10.2$). These values are too high to be explained by conventional semi-classical mechanics, thus it was proposed that quantum mechanical tunnelling of the hydrogen transferred onto NAD$^+$ was taking place. Since the de Broglie wavelength for ^{1}H is 0.5 Å, it is quite feasible for hydrogen tunnelling to occur during the movement of hydrogen atoms in enzymatic transition states, and hence to 'tunnel through' a high activation energy barrier, and accelerate the reaction.

Kinetic isotope effects are also observed when the isotopic substitution does not involve any of the bonds broken in the reaction, but does involve a change of orbital hybridisation in a rate-determining step. Thus, if an sp^3-hybridised carbon bearing a deuterium atom changes to an sp^2-hybridised carbon bearing the deuterium atom in the rate-determining step of a reaction, a smaller *secondary deuterium isotope effect* will be observed ($k_H/k_D = 1.1$–1.4). An example is the ketosteroid isomerase reaction we met in Chapter 3 (Figure 3.14), illustrated in Figure 4.26. With 4*R*-[4-^{2}H]-ketosteroid as a substrate the hydrogen being

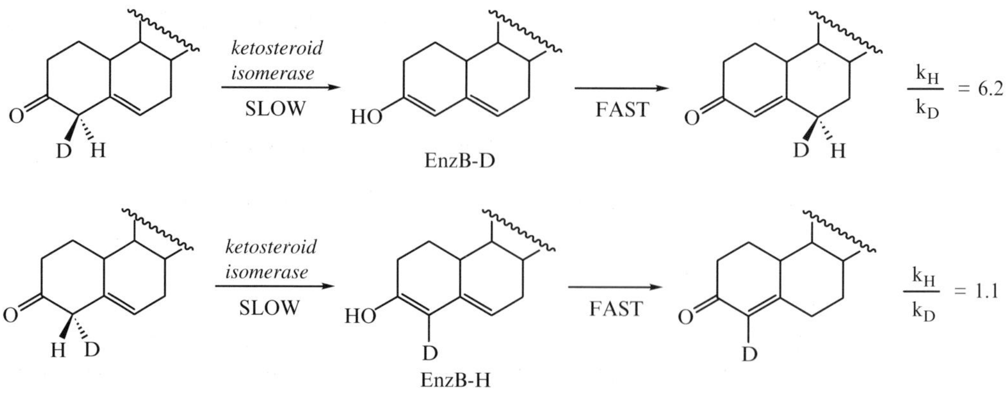

Figure 4.26 Substrate kinetic isotope effects in the ketosteroid isomerase reaction.

abstracted is replaced by deuterium, and a primary kinetic isotope effect of 6.2 is observed on k_{cat}. However, with $4S$-[4-^{2}H]-ketosteroid a secondary kinetic isotope effect of 1.1 is observed, due to re-hybridisation of C-2 in the rate-determining step.

If an enzymatic reaction is carried out in ^{2}H$_2$O rather than ^{1}H$_2$O a solvent kinetic isotope effect is observed if there is a proton transfer from water or a water-exchangeable group in the rate-determining step. For example, in the ketosteroid isomerase reaction there is a D$_2$O solvent isotope effect of 1.6 on k_{cat}, due to replacement of the phenolic proton of Tyr-14 by deuterium, as shown in Figure 4.27.

D$_2$O solvent isotope effects also arise where there is base-catalysed attack of water in a rate-determining step. However, D$_2$O solvent isotope effects can in some cases arise for a number of reasons far removed from active site catalysis, so they need to be interpreted with caution. Finally, kinetic isotope effects are not restricted only to cleavage of C$-$H bonds: rate-determining cleavage of C$-$O, C$-$N and C$-$C bonds can be studied using ^{18}O-, ^{15}N- and ^{13}C-labelled substrates. In these cases the difference in zero point energy between heavy isotopes is very much smaller, so the observed effects are typically

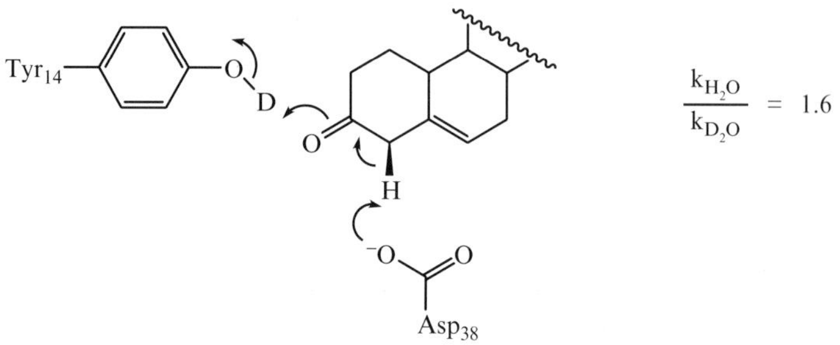

Figure 4.27 D$_2$O solvent isotope effect in the ketosteroid isomerase reaction.

< 1.1. For a discussion of heavy atom isotope effects the reader is referred to specialist references.

4.7   Determination of active site catalytic groups

As well as examining the molecular details of an enzymatic reaction, it is equally important to study the groups present in the enzyme active site which carry out the catalysis.

One convenient method for obtaining clues regarding active site catalytic groups is to analyse the variation of enzyme activity with pH: a pH/rate profile. Thus, if there are acidic and basic groups involved in the enzyme mechanism, they must be in the correct ionisation state in order for the enzyme to operate efficiently. For example, the ketosteroid isomerase reaction illustrated above (Figure 4.27) has the pH/rate profile shown in Figure 4.28, from which the pK_a value of 4.7 for the active site Asp-38 was first deduced.

The second method that can be used to identify active site groups is by covalent modification. There are a series of chemical reagents available which will react in a fairly specific way with different amino acid side chains, shown in Table 4.3. Thus, if an enzyme is inactivated upon treatment with diethyl pyrocarbonate, then this provides a clue that there may be an essential histidine residue at the active site of the enzyme. However, residues identified by such methods are not necessarily catalytic groups: they may simply be residues in the vicinity of the active site which when covalently modified block the entrance to the active site sufficiently to inactivate the enzyme.

A related technique involving substrate analogues is known as affinity labelling. A substrate analogue is synthesised containing a reactive functional group (e.g. halogen substituent, epoxide, etc.) in a part of the molecule. The substrate analogue is recognised by the enzyme and binds to the active site in the same way as the natural substrate, but then alkylates an essential

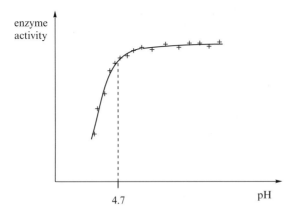

Figure 4.28 pH/rate profile of ketosteroid isomerase.

Table 4.3 Group specific reagents for active site amino acid modification.

Amino acid	Group modified	Reagent	Modification reaction
Cysteine	S	Iodoacetate, iodoacetamide	Alkylation
		DTNB	Disulphide formation
		p-Hydroxymercuribenzoate	Metal complexation
Histidine	N	Diethyl pyrocarbonate	Acylation
Lysine	N	Succinic anhydride	Acylation
Arginine	N	Phenylglyoxal	Heterocycle formation
Asp, Glu	O	EDC (water-soluble carbodi-imide) + amine	Amide formation
Tyrosine	O	Tetranitromethane	Nitration
Tryptophan	Indole	*N*-bromosuccinimide	Oxidation

DTNB, 5,5'-dithiobis-(2-nitrobenzoic acid); EDC, 1-ethyl-3-(3-dimethylaminopropyl)carbodi-imide.

active site residue. The enzyme is then irreversibly inactivated, since the active site is blocked. The advantage of this method over the group specific reagents mentioned above is that affinity labelling is more selective in its site of action, due to the precise positioning of the substrate analogue at the enzyme active site. One example of this method is the inactivation of the serine proteases by chloromethyl ketone substrate analogues, which will be described in Section 5.2.

If a successful irreversible inhibitor is found, the site of action of the affinity label can be determined using a radiolabelled inhibitor as shown in Figure 4.29. The enzyme inactivated with labelled inhibitor contains a modified active site

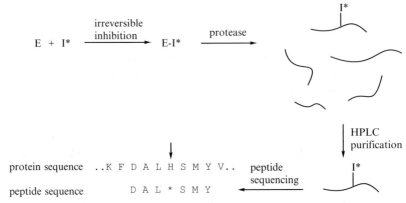

Figure 4.29 Identification of an active site residue by peptide mapping. HPLC, high-performance liquid chromatography.

residue bearing a radioactive label. In order to identify this active site residue the inactivated enzyme is broken down into peptide fragments using a protease enzyme, and the peptide fragment containing the ^{14}C label is sequenced. At the position in the peptide sequence containing the radiolabelled inhibitor a non-standard amino acid is found and the radioactive label is released from the peptide. This method is known as peptide mapping.

The covalent attachment of an inhibitor or substrate to an enzyme can also be analysed by 'weighing' the protein. The technique of electrospray mass spectrometry can be used to determine accurately the molecular weight of pure proteins of up to 50 kDa to an accuracy of ± 1 Da! The molecular weight of a covalently modified enzyme can, therefore, reveal the molecular weight of the attached small molecule.

Finally, with the advent of modern molecular biology techniques it has become possible to specifically replace individual amino acids in an enzyme by site-directed mutagenesis. This method involves specifically altering the sequence of the gene encoding the enzyme in such a way that the triplet codon encoding the amino acid of interest is changed to that of a non-functional amino acid such as alanine. The mutant enzyme can then be purified and tested for enzymatic activity. In this way the precise role of individual amino acids implicated by modification studies or sequence alignments can be explored.

The ultimate solution to identifying active site amino acid groups is to solve the three-dimensional structure of the whole enzyme. This is most commonly done by X-ray crystallography, which depends on obtaining a high-quality crystal of pure enzyme that is suitable for X-ray diffraction. Recent developments in multi-dimensional NMR spectroscopy have allowed the structure determination of small proteins up to 20 kDa in size. Most of the detailed examples that we shall meet in later chapters have been analysed both by X-ray crystallography and by several of the above methods.

Problems

(1) Using the data given in Table 4.2, calculate the turnover number (k_{cat}) for the enzyme being purified. Assume that the final purified enzyme is 100% pure, and that the enzyme contains one active site per monomer. The subunit molecular weight is 28 kDa.

(2) From the data below obtained from the rate of an enzyme-catalysed reaction at a range of substrate concentrations, calculate the K_M and v_{max} of the enzyme for this substrate. If 1.65 μg of enzyme was used for each assay, and if the molecular weight of the enzyme is 36 000, work out the turnover number for the enzyme for this substrate. Hence calculate the catalytic efficiency (k_{cat}/K_M) for this substrate.

[S] (mM)	Rate of product formation (nmol min^{-1}; duplicate assays)
1.8	4.75, 4.44
0.9	3.55, 3.20
0.4	2.15, 2.19
0.2	1.28, 1.32
0.1	0.74, 0.76

(3) Some enzymes are inhibited in the presence of high concentrations of substrate. This behaviour can be rationalised by the model below, involving the formation of a non-productive ES_2 complex. Starting from this model, use the steady state approximation for ES and ES_2 to construct a rate equation of the form given below. By considering the behaviour at low and high substrate concentrations, sketch the expected dependence of v versus [S].

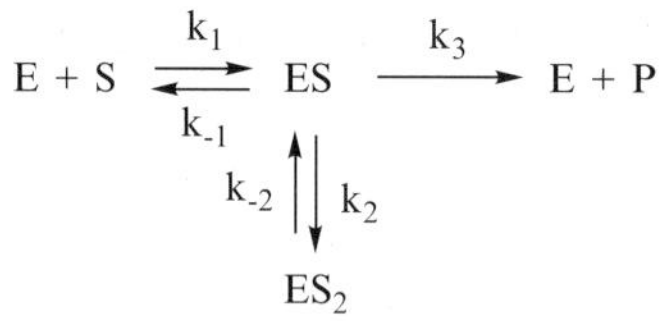

$$E + S \underset{k_{-1}}{\overset{k_1}{\rightleftharpoons}} ES \xrightarrow{k_3} E + P$$

$$ES \underset{k_{-2}}{\overset{k_2}{\rightleftharpoons}} ES_2$$

$$Rate = \frac{k_3[E]_0[S]}{K_M + [S] + K_2[S]^2} \quad \text{where } K_2 = k_2/k_{-2}$$

(4) A carbon–phosphorus lyase activity has been found which catalyses the reductive cleavage of ethyl phosphonate to ethane, as shown below. The stereochemistry of this reaction was elucidated using a chiral methyl group approach. Incubation of [$1R$-^{2}H, ^{3}H]-ethyl phosphonate with enzyme gave a ^{3}H-labelled ethane product, which was converted via halogenation and oxidation into ^{3}H-labelled acetic acid. Analysis using the method described in the text revealed that the acetate derivative had predominantly the S configuration. Incubation of [$1S$-^{2}H, ^{3}H]-ethyl phosphonate with enzyme and analysis by the same method gave $2R$-acetate. Deduce whether the reaction proceeds with retention or inversion of configuration, and comment on this result.

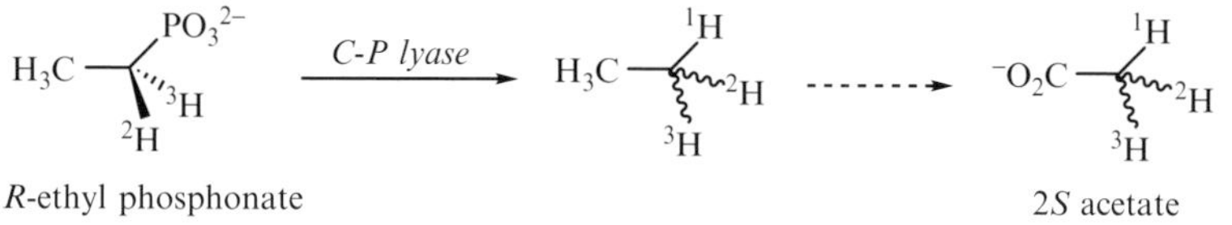

1R-ethyl phosphonate   2S acetate

(5) The enzyme phosphonoacetaldehyde hydrolase catalyses the conversion of phosphonoacetaldehyde to phosphate and acetaldehyde, as shown below. The enzyme requires no cofactors, but is inactivated by treatment with phosphonoacetaldehyde and sodium borohydride. Deduce which amino acid side chain is involved in the catalysis and suggest a possible mechanism.

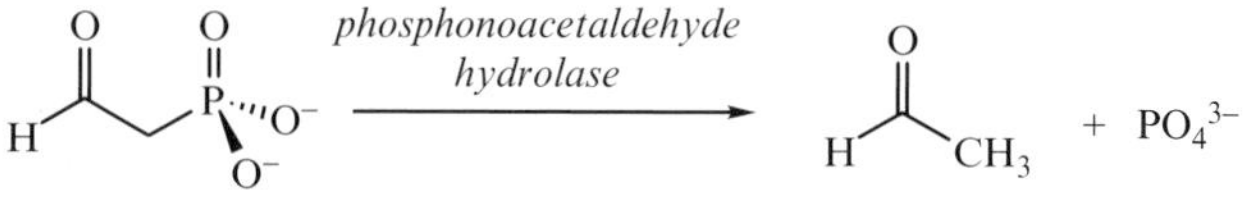

$[^{17}O,\ ^{18}O]$-Thiophosphonoacetaldehyde was prepared with the stereochemistry shown below, and incubated with the enzyme in $H_2^{16}O$. The resulting thiophosphate was analysed and found to have the S configuration. Deduce whether the reaction proceeds with retention of inversion of configuration at the phosphorus centre. Comment on the implications for the enzyme mechanism.

The same reaction is catalysed by aniline ($PhNH_2$), but at a much slower rate. Using the labelled substrate for the aniline-catalysed process, the thiophosphate product was found to have the R configuration. Explain these observations.

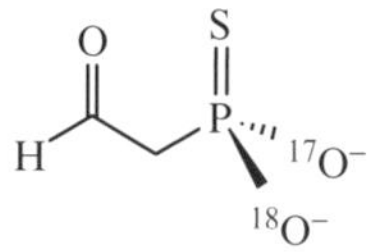

(6) How would you attempt to obtain further evidence for the intermediate implied in Problem 5?

Further reading

Enzyme purification

R.K. Scopes (1987) *Protein purification: Principles and Practice.* Springer-Verlag, New York.

Enzyme kinetics

I.H. Segel (1993) *Enzyme Kinetics*. Wiley-Interscience, New York.
A. Fersht (1985) *Enzyme Structure and Mechanism*, 2nd edn. Freeman, New York.

Stereochemistry of enzymatic reactions

H.G. Floss & S. Lee (1993) Chiral methyl groups: small is beautiful. *Acc. Chem. Res.*, **26**, 116–22.
H.G. Floss & M.D. Tsai (1979) Chiral methyl groups. *Adv. Enzymol.*, **50**, 243–302.
J.A. Gerlt, J.A. Coderre & S. Mehdi (1984) Oxygen chiral phosphate esters. *Adv. Enzymol.*, **55**, 291–380.
J.R. Knowles (1980) Enzyme-catalysed phosphoryl transfer reactions. *Annu. Rev. Biochem.*, **49**, 877–920.
K.H. Overton (1979) Concerning stereochemical choice in enzymic reactions. *Chem. Soc. Rev.*, **8**, 447–73.
C.T. Walsh (1979) *Enzymatic Reaction Mechanisms*. Freeman, San Francisco.

Intermediates in enzymatic reactions

K.S. Anderson & K.A. Johnson (1990) Kinetic and structural analysis of enzyme intermediates: lessons from EPSP synthase. *Chem. Rev.* **90**, 1131–49.
P.D. Boyer (1978) Isotope exchange probes and enzyme mechanisms. *Acc. Chem. Res.*, **11**, 218–24.
I.A. Rose (1979) Positional isotope exchange studies on enzyme mechanisms. *Adv. Enzymol.*, **50**, 361–96.

Isotope effects in enzymatic reactions

N.P. Botting (1994) Isotope effects in the elucidation of enzyme mechanisms. *Nat. Prod. Reports*, **11**, 337–53.
W.W. Cleland (1987) The use of isotope effects in the detailed analysis of catalytic mechanisms of enzymes. *Bioorg. Chem.*, **15**, 283–302.
D.B. Northrop (1981) The expression of isotope effects on enzyme-catalysed reactions. *Annu. Rev. Biochem.*, **50**, 103–32.
L. Xue, P. Talalay, & A.S. Mildvan (1990) Studies of the mechanism of the Δ^5-3-ketosteroid isomerase reaction by substrate, solvent, and combined kinetic deuterium isotope effects on wild-type and mutant enzymes. *Biochemistry* **29**, 7491–500.

Proton tunnelling in enzymatic reactions

Y. Cha, C.J. Murray, and J.P. Klinman (1989) *Science*, **243**, 1325.
A. Kohen & J.P. Klinman (1998) *Acc. Chem. Res.*, **31**, 397.

Covalent modification of enzymes

T.E. Creighton (ed.) (1989) *Protein Function – a Practical Approach*. IRL Press, Oxford.

5 Enzymatic Hydrolysis and Group Transfer Reactions

5.1 Introduction

Hydrolysis reactions are fundamental to cellular metabolism. In order to break down biological foodstuffs into manageable pieces that animals can utilise for energy, they must have enzymes capable of hydrolysing biological macromolecules. Thus, many of the hydrolytic enzymes that we shall meet in this chapter are involved in digestive processes. However, that is by no means the only area in which we shall encounter hydrolase enzymes.

We shall look in turn at enzymes that hydrolyse each of the three major classes of biological macromolecules: polypeptides, polysaccharides and nucleic acids. More detailed discussions of the structures and chemistry of these macromolecules can be found in most advanced chemistry or biochemistry texts. The classes of enzymes that hydrolyse these molecules are shown in Table 5.1:

Table 5.1 Classes of hydrolase and group transfer enzymes.

Group transferred			Hydrolase	Transferase
Acyl	$\overset{O}{\underset{}{\overset{\|\|}{-C}}}-NHR$ $\longrightarrow$	$\overset{O}{\underset{}{\overset{\|\|}{-C}}}-OH$ + RNH_2	Protease Peptidase Amidase	Transpeptidase
	$\overset{O}{\underset{}{\overset{\|\|}{-C}}}-OR$ $\longrightarrow$	$\overset{O}{\underset{}{\overset{\|\|}{-C}}}-OH$ + ROH	Esterase Lipase	Acyl transferase
Glycosyl	(sugar ring) OR $\longrightarrow$	(sugar ring) OH + ROH	Glycosidase	Glycosyl transferase
Phosphoryl	$\overset{O}{\underset{O^-}{\overset{\|\|}{-P}}}-OR$ $\longrightarrow$	$\overset{O}{\underset{O^-}{\overset{\|\|}{-P}}}-OH$ + ROH	Phosphatase Nuclease Phosphodiesterase	Kinase Phosphotransferase

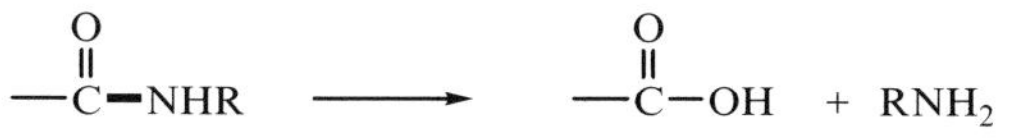

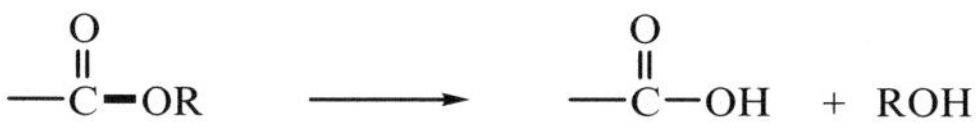

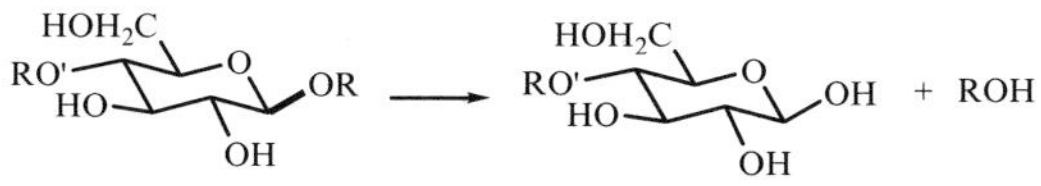

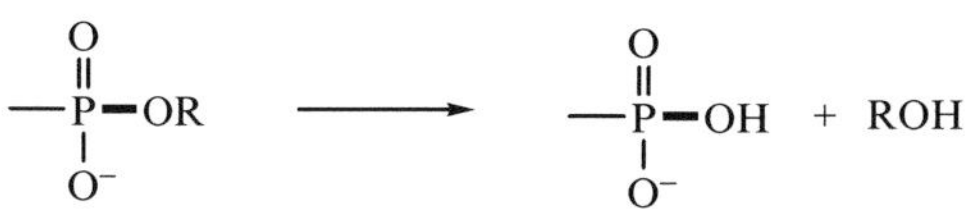

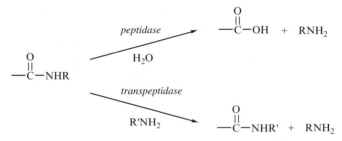

Figure 5.1 Peptidase versus transpeptidase.

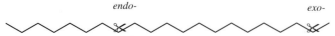

Figure 5.2 *Exo-* versus *endo*-cleavage

the polypeptides are hydrolysed by peptidase (or protease) enzymes; polysaccharides by glycosidases; and nucleic acids by nucleases.

The group transferases are related in function to the hydrolases, but carry out quite distinct reactions. In each of the above classes of hydrolases a group is being cleaved and transferred to the hydroxyl group of water: for polypeptides an acyl group; for polysaccharides a glycosyl group; and for nucleic acids a phosphoryl group. Transferases simply transfer this group to an acceptor other than water. For example, there are a few transpeptidases which cleave an amide bond and transfer the acyl group to another amino group, forming a new amide bond (see Figure 5.1). In the same way glycosyl transferases and phosphoryl transferases transfer glycosyl and phosphoryl groups to acceptor substrates.

One final piece of terminology regards the position of cleavage of a very long polymeric biological macromolecule. Enzymes which cleave such biological polymers either cleave progressively from the end of the chain, which is known as *exo*-cleavage, or they cleave at specific points in the middle of the chain, which is known as *endo*-cleavage (see Figure 5.2). In the case of *exo*-cleavage the end which is cleaved is specified, for example $5' \rightarrow 3'$-exonuclease.

This chapter will deal with each of the major classes of peptidases, glycosidases, and nucleases, and we will focus on the human immunodeficiency virus 1 (HIV-1) protease as a topical example to examine in more detail. We will also examine other examples of acyl group transfer and methyl group transfer which are of considerable biological significance.

5.2   The peptidases

Peptidases are responsible for hydrolysing the amide bonds found in the polypeptide structures of proteins, hence they are often known as proteases or proteinases. They have a very important role in the digestive systems of all animals for the breakdown of the protein content of food and are produced in large quantities in the stomach and pancreas. Different types of peptidases are produced elsewhere in the body for a large assortment of hydrolytic purposes, notably their key involvement in the blood coagulation cascade. Other proteases are produced in a wide range of species, including bacteria, yeasts, and plants, of which a selection are listed in Table 5.2. Note that there is an active

Table 5.2 Some commercially available proteases.

Name	Class	Exo/endo	Specificity	Source	pH$_{opt}$
Bromelain	Cys	Endo	X–X	Pineapple	6.0
Carboxypeptidase A	Metallo	Exo (C)	X–C (not Arg, Lys)	Bovine pancreas	7–8
Carboxypeptidase B	Metallo	Exo (C)	X–C (Arg/Lys)	Pig pancreas	7–9
Carboxypeptidase P	Ser	Exo (C)	X–C	*Penicillium*	4–5
Carboxypeptidase Y	Ser	Exo (C)	X–C	Yeast	5.5–6.5
Cathepsin C (dipeptidase)	Cys	Exo (N)	N–Gly/Pro–X	Bovine spleen	4–6
Cathepsin D	Asp	Endo	Phe/Leu–X	Bovine spleen	3–5
Chymotrypsin A	Ser	Endo	Aro–X	Bovine pancreas	7.5–8.5
Clostripain	Cys	Endo	Arg–X	*Clostridium*	7.1–7.6
Elastase	Ser	Endo	Neutral aa–X	Porcine pancreas	7.8–8.5
Factor X	Ser	Endo	Arg–X	Bovine plasma	8.3
Leucine aminopeptidase	Metallo	Exo (N)	N(not Arg/Lys)–X	Porcine kidney	7.5–9
Papain	Cys	Endo	Arg/Lys–X	Papaya plant	6–7
Pepsin	Asp	Endo	Hyd–X	Porcine stomach	2–4
Proteinase K	Ser	Endo	X–Aro/Hyd	*Tritirachium album*	7.5–12
Renin	Asp	Endo	His–Leu–X	Porcine kidney	6.0
Subtilisin Carlsberg	Ser	Endo	Neutral/acidic aa–X	*Bacillus subtilis*	7–8
Thermolysin	Metallo	Endo	X–Hyd	*B. thermoproteolyticus*	7–9
Thrombin	Ser	Endo	Arg–Gly	Bovine plasma	7–8
Trypsin	Ser	Endo	Lys/Arg–X	Bovine pancreas	8.5–8.8
V8 Protease	Ser	Endo	Asp/Glu–X	*Staphylococcus aureus*	7.8

aa, amino acid; Aro, aromatic amino acid (Phe/Tyr/Trp); C, C-terminal amino acid; Hyd, hydrophobic amino acid (Leu/Ile/Val/Met); N, N-terminal amino acid; X, any amino acid.

non-specific protease called bromelain present in fresh pineapple, which is why fresh pineapple will attack your gums if you do not brush your teeth!

There are four main classes of peptidase enzyme, classified according to the groups found at their active site which carry out catalysis. They are as follows: the serine proteases; the cysteine proteases; the metalloproteases and the aspartyl (or acid) proteases. Table 5.2 shows the more common commercially available proteases. The table shows to which mechanistic group they belong, whether they are *exo-* or *endo-*proteases, and what is their preferred site of cleavage. We shall now consider each class of protease in turn.

The serine proteases

These enzymes are characterised by an active site serine residue which participates covalently in catalysis. The active site serine is assisted in catalysis by a histidine and an aspartate residue, the three residues acting in concert as a 'catalytic triad'. The best characterised of the serine proteases is α-chymotrypsin, a 241-amino acid endoprotease which shows specificity for cleavage after aromatic amino acids (phenylalanine, tyrosine or tryptophan). This selectivity arises from a favourable hydrophobic interaction between the aromatic side chain of the substrate and a hydrophobic binding pocket situated close to the catalytic site, as illustrated in Figure 5.3.

The catalytic mechanism of chymotrypsin is shown in Figure 5.4. It is a classic example of covalent catalysis, in which the active site Ser-195 attacks the amide carbonyl to form a tetrahedral oxyanion intermediate. Attack of Ser-195 is made possible by base catalysis from His-57, generating an imidazolium cation which is stabilised by the carboxylate of Asp-102. Selective stabilisation of the high energy oxyanion intermediate takes place via formation of two

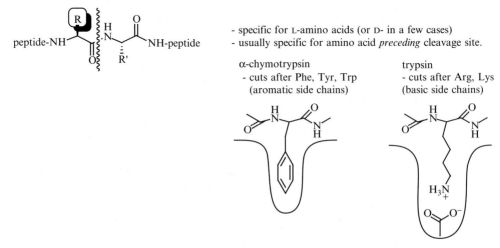

Figure 5.3 Specificity of endopeptidases.

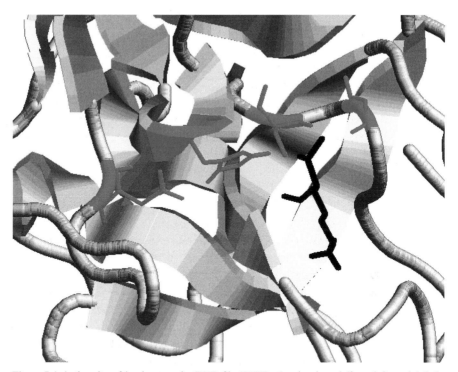

Figure 5.4 Active site of bovine trypsin (PDB file 1BTX), showing in red (from left to right) the catalytic triad Asp-102, His-57 and Ser-195; the oxyanion hole formed by Gly-193; and a bound boronic acid inhibitor in black.

hydrogen bonds between the tetrahedral oxyanion and the backbone amide N−H bonds of Ser-195 and Gly-193, as shown in Figure 5.5. Note that these interactions are specific for the oxyanion intermediate and are not formed with the bound substrate, leading to reduction of the activation energy for the overall step (since this is a non-isolable high-energy intermediate that is similar to the transition state for the first step, the transition state stabilisation logic of Section 3.4 applies in this case).

Protonation of the departing nitrogen by His-57 gives the amine product and leads to the formation of an acyl enzyme intermediate. There is considerable experimental evidence for the existence of this covalent intermediate from kinetic studies, trapping experiments and X-ray crystallography. Histidine-57 then acts as a base to deprotonate water attacking the acyl enzyme intermediate, leading to a second tetrahedral oxyanion intermediate which is once again stabilised by hydrogen bond formation. Protonation of the departing serine oxygen by His-57 leads to release of the carboxylic acid product and completion of the catalytic cycle.

The use of the hydroxyl group as a nucleophile in this reaction is quite remarkable since an alcohol would normally have a pK_a of about 16, whereas the pK_a of histidine is normally in the range 6–8. Effectively this enzyme is using

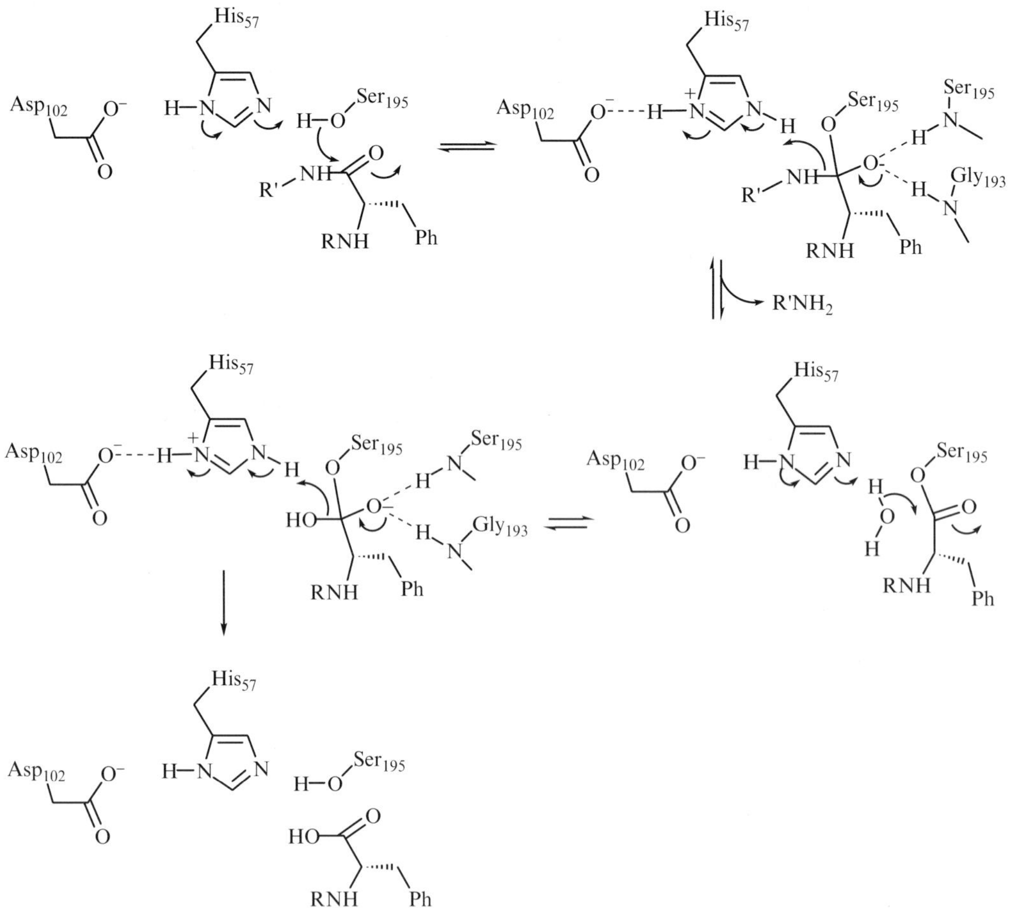

Figure 5.5 Mechanism for α-chymotrypsin.

a relatively weak imidazole base to generate a potent alkoxide nucleophile. This process is made thermodynamically favourable by participation of Asp-102 in stabilising the imidazolium cation of His-57. Mutant enzymes have been generated using site-directed mutagenesis in which Asp-102 has been replaced by asparagine – these enzymes are 10^4-fold less active, showing the catalytic importance of this interaction.

Serine proteases are specifically inactivated by two classes of inhibitors: organophosphorus and chloromethyl ketone substrate analogues. Mechanisms of inactivation are shown in Figure 5.6. Organophosphorus inhibitors are attacked by the active site serine, generating stable tetrahedral phosphate esters which closely resemble the tetrahedral transition state of the normal enzymatic reaction, and hence are bound tightly by the enzyme. Chloromethyl ketone analogues were found to modify the active site histidine base, in this case

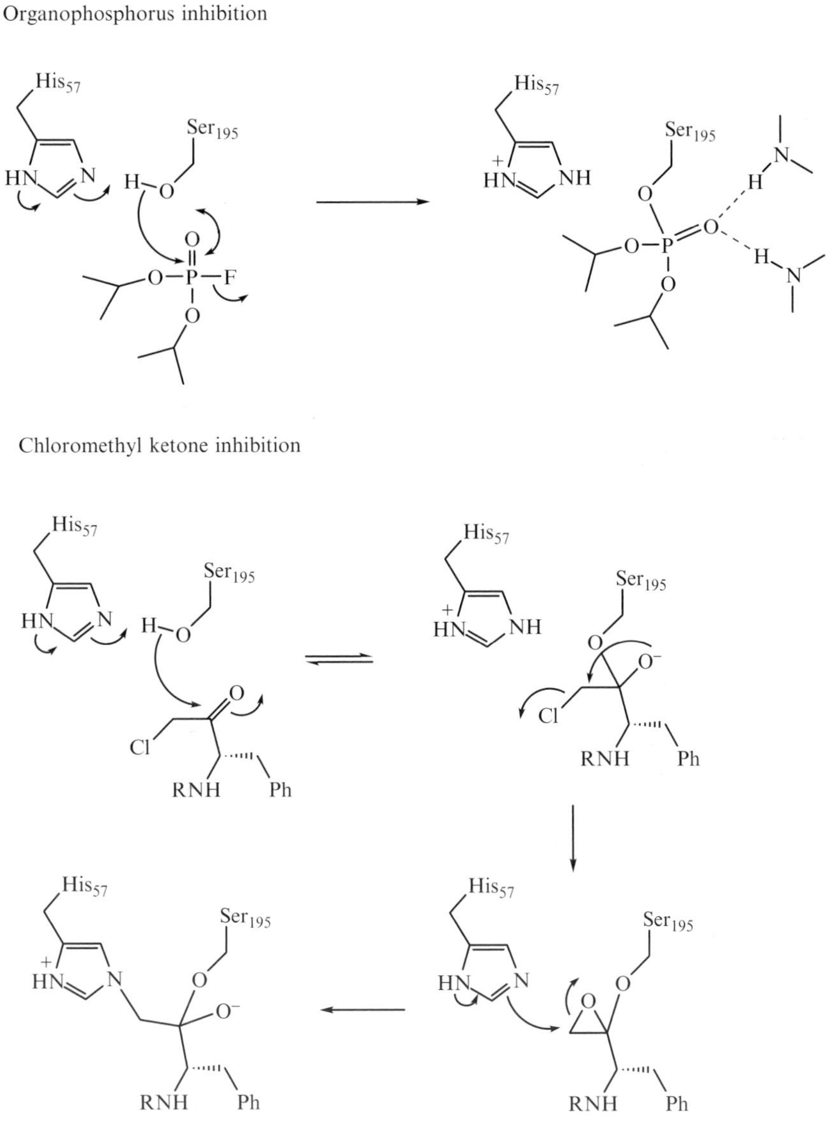

Organophosphorus inhibition

Chloromethyl ketone inhibition

Figure 5.6 Inhibitors of serine proteases.

His-57, rather than the apparently more reactive serine residue. A stereochemical analysis of this inactivation reaction has revealed that it proceeds with retention of configuration at the chlorine-bearing carbon, suggesting that a double inversion is taking place. This can be rationalised by attack of the active site serine on the carbonyl group and displacement of chloride by the resulting oxyanion, generating an enzyme-bound epoxide. Opening of the epoxide by the neighbouring histidine leads to covalent modification of the histidine residue.

Other serine proteases such as trypsin, elastase and subtilisin have very similar active site structures (Figure 5.7) and employ the same type of mechanism, even though in the case of subtilisin the primary sequence of the enzyme bears no resemblance to that of chymotrypsin. This may be a rare example of *convergent evolution*, where the same active site structure has arisen from two evolutionary origins, suggesting that this particular three-dimensional align-

(a)

(b)

(c)

Figure 5.7 Comparison of the structures of: (a) α-chymotrypsin (PDB file 5CHA, catalytic triad Asp-102, His-57, Ser-195); (b) trypsin (PDB file 1BTX, catalytic triad Asp-102, His-57, Ser-195); and (c) subtilisin Carlsberg (PDB file 1AV4, catalytic triad Asp-32, His-64, Ser-125). In (a) and (b) the specificity loops are shown in red. Selectivity for a basic side chain in trypsin is provided by Asp-189, which is found in α-chymotrypsin as Ser-189.

ment of functional groups is especially adept for this type of catalysis. The specificity of trypsin is for cleavage after basic amino acids such as lysine and arginine. This specificity is provided by a similar specificity pocket to that of chymotrypsin in which there is the carboxylate side chain of Asp-189 at the bottom of the pocket which forms a favourable electrostatic interaction with the basic side chains of lysine and arginine-containing substrates (see Figure 5.3).

There are several other classes of enzyme which also contain serine catalytic triads. The $\alpha\beta$-hydrolase family of esterase and lipase enzymes, discussed in Section 5.3, also contain active site serine groups, and use the same type of mechanism as chymotrypsin (which itself is capable of hydrolysing esters as well as amides).

The cysteine proteases

This family of proteins is characterised by an active site cysteine residue whose thiol side chain is also involved in covalent catalysis. The cysteine proteases are less commonly used for digestive purposes and are more often found in intracellular proteases used for post-translational processing of cellular proteins. The active site thiol is prone to oxidation, which means that these enzymes must be purified and handled in the presence of mild reducing agents. The active site cysteine is easily modified by cysteine-directed reagents such as p-chloromercuribenzoate, an organomercury compound which functions by forming a strong mercury–sulphur bond.

The best characterised member of this family is papain, a 212-amino acid endoprotease found in papaya plants. The preferred cleavage site is following basic amino acids such as arginine and lysine. The active site of papain contains the nucleophilic Cys-25 and an active site base, His-159, as shown in Figure 5.8. There is good evidence that Cys-25 acts as a nucleophile to attack the amide bond, generating a covalent thioester intermediate. There is evidence from X-ray crystallography and modification studies that Cys-25 is deprotonated by His-159 as it attacks the amide substrate. As in the case of the serine proteases, a high-energy oxyanion intermediate is formed which is specifically stabilised by hydrogen bonding to the backbone amide N−H bonds of Cys-25 and Gln-17. Breakdown of the thioester intermediate by base-catalysed attack of water leads to formation of the carboxylic acid product, as shown in Figure 5.9.

Analysis of the active site histidine residue by ^{1}H nuclear magnetic resonance (NMR) spectroscopy has revealed that in the active form of the enzyme the imidazole ring is in fact protonated, suggesting that in this case the resting state of the enzyme contains an imidazolium–thiolate ion pair. This is possible in the case of the cysteine proteases since the pK_a of the thiol side chain of cysteine is only 8–9, and stabilisation of the ion pair by active site electrostatic interactions seems likely.

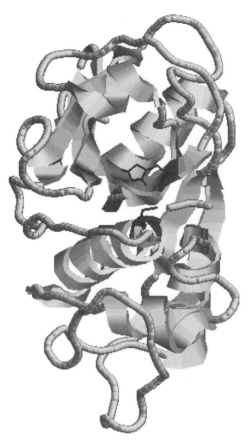

Figure 5.8 Structure of papain (PDB file 1POP), showing in red the catalytic residues Cys-25 and His-159.

Given the similarities in mechanism between the serine proteases and the cysteine proteases, it is interesting to note that the serine proteases may have evolved from a forerunner of the cysteine proteases. This hypothesis was put forward by Brenner, who observed that in some cases the triplet codons encoding the active site serine have the UCX codon, whilst others have the AGC/AGU codon for serine. The key point is that these two sub-families of enzymes cannot be directly related in evolutionary terms, since two simultaneous nucleotide base changes would be required to change from one to the other, which is statistically improbable. However, they could *both* have evolved from the UGC/UGU cysteine codon via one-base changes as shown in Figure 5.10. So just on the basis of this one observation it seems probable that a cysteine-containing enzyme was the forerunner of the serine-containing enzymes. This makes a lot of sense in terms of the catalytic chemistry, since the thiol side chain of cysteine has a lower pK_a, is more nucleophilic and is a better leaving group than the hydroxyl group of serine. So why do the serine proteases

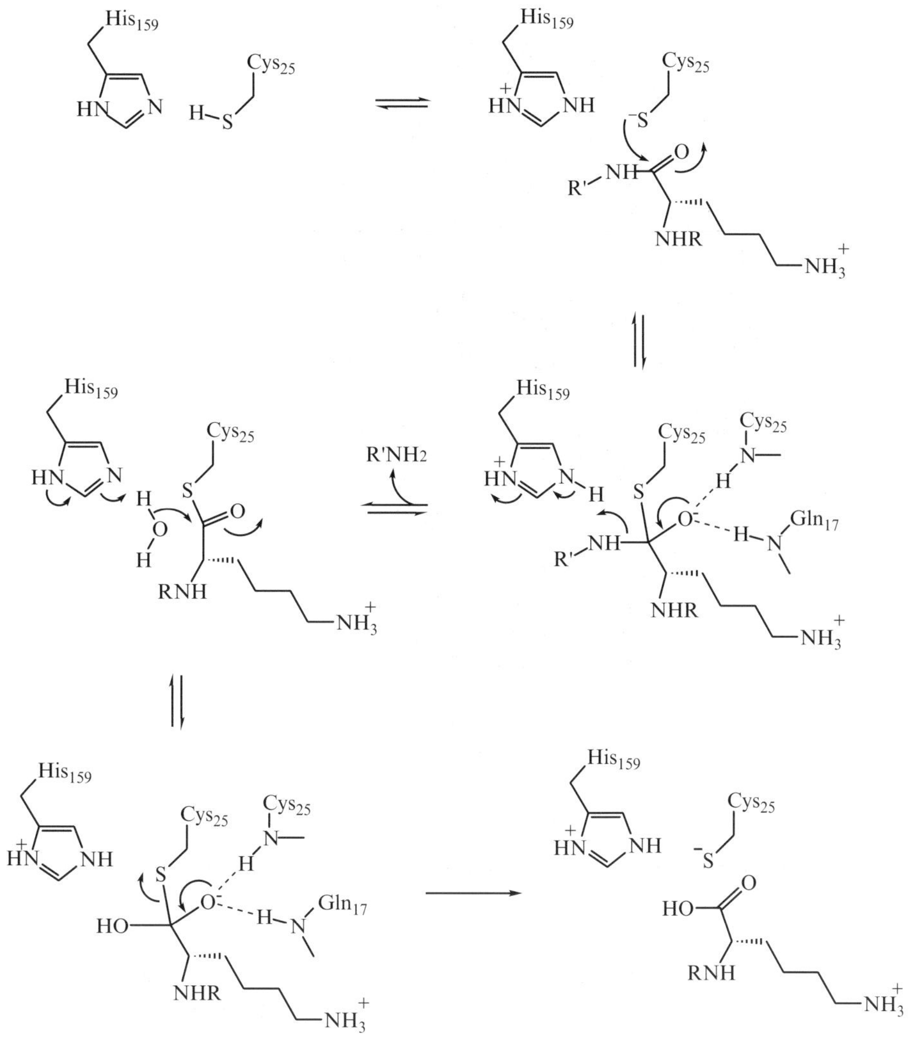

Figure 5.9 Mechanism for papain.

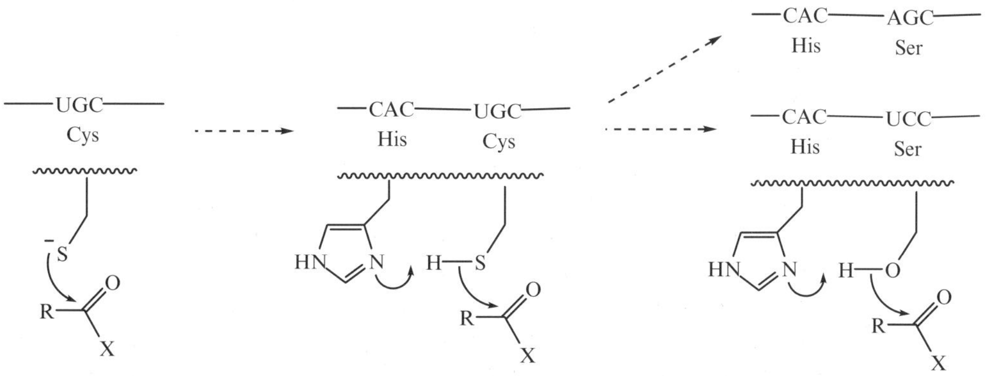

Figure 5.10 Possible evolution of serine proteases.

now predominate? Perhaps, as mentioned above, because they are more stable in an oxygen atmosphere.

The metalloproteases

The metalloproteases are characterised by a requirement for an active site metal ion, usually Zn^{2+}, which is involved in the catalytic cycle. These enzymes can be readily distinguished from the other classes by treatment with metal chelating agents such as ethylene diamine tetra-acetic acid (EDTA) or 1,10-phenanthroline (see Figure 5.11), leading to removal of the metal ion cofactor and inactivation.

The best characterised members of this family are carboxypeptidase A, a 307-amino acid exopeptidase from bovine pancreas which cleaves the C-terminal residue of a peptide chain (not arginine, lysine or proline); and thermolysin, a 35-kDa endopeptidase from *Bacillus thermoproteolyticus* which cleaves before hydrophobic amino acids such as leucine, isoleucine, valine or phenylalanine. Both enzymes contain a single Zn^{2+} ion at their active sites, which in the resting state of the enzyme is co-ordinated by three protein ligands and one solvent water molecule. There is evidence to suggest that in both enzymes when the substrate is bound the water molecule is displaced by the carbonyl oxygen of the amide bond to be hydrolysed, which is thus activated towards nucleophilic attack by Zn^{2+} Lewis acid catalysis.

The mechanism of amide bond hydrolysis has been well studied in both enzymes, with slightly different results emerging. Both enzymes contain an active site glutamate, which in theory could either act as a nucleophile to attack the amide carbonyl, or act as a base to deprotonate an attacking water molecule. In thermolysin the active site glutamate (Glu-143) is positioned 3.9 Å away from the amide carbonyl (see Figure 5.12a), too far away to act as a nucleophile, but far enough to accommodate an intervening water molecule, which is thought to be activated by co-ordination to the Zn^{2+} cofactor. There is evidence from X-ray crystallography to suggest that Glu-143 acts as a base to deprotonate an attacking water molecule, forming an oxyanion intermediate which is stabilised by specific hydrogen bonds. Breakdown of this intermedi-

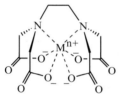

Ethylene diamine tetra-acetic acid (EDTA)   1,10-Phenanthroline

Figure 5.11 Metal chelating agents.

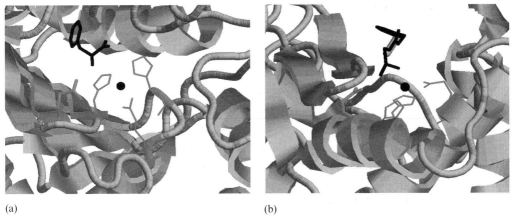

(a)   (b)

Figure 5.12 Comparison of the active sites of: (a) thermolysin (PDB file 5TMN, active site Glu-143); (b) carboxypeptidase A (PDB file 2CTC, active site Glu-270). In each structure the active site glutamic acid residue is (on left hand side of each figure) highlighted in red. The Zn^{2+} cofactor and its ligands are illustrated. A bound substrate analogue is drawn in black.

ate using Glu-143 to transfer a proton to the departing nitrogen gives the hydrolysis products.

In the active site of carboxypeptidase A, the active site Glu-270 is positioned somewhat closer to the substrate, 2.5 Å away from the amide carbonyl (see Figure 5.12b), close enough to act either as a nucleophile or as a base. Evidence for the existence of an anhydride intermediate, formed by nucleophilic attack of Glu-270 on the amide substrate, has come from the treatment of carboxypeptidase A with $NaB(CN)^3H_3$ in the presence of substrate leading to the incorporation of ^3H label into the protein and the isolation of ^3H-hydroxynorvaline in the modified enzyme (see Figure 5.13). Further support for an anhydride intermediate in the carboxypeptidase A reaction has also emerged from resonance Raman spectroscopic studies of the enzyme-catalysed reaction, showing bands at $1700-1800\,cm^{-1}$ characteristic of anhydrides. Nucleophilic and base-catalysed mechanisms for carboxypeptidase A and thermolysin, respectively, are shown in Figure 5.14.

Thermolysin is strongly inhibited (K_i 28 nM) by a natural product phosphoramidon, a monosaccharide derivative containing a phosphonamidate functional group (Figure 5.15). The phosphonamidate group binds to the

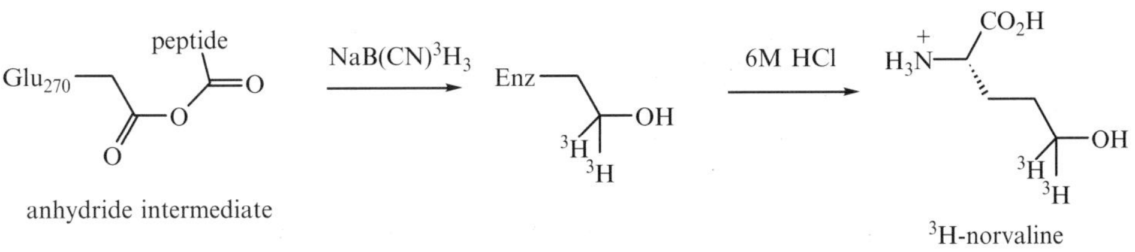

Figure 5.13 Sodium cyanoborohydride trapping of putative anhydride intermediate.

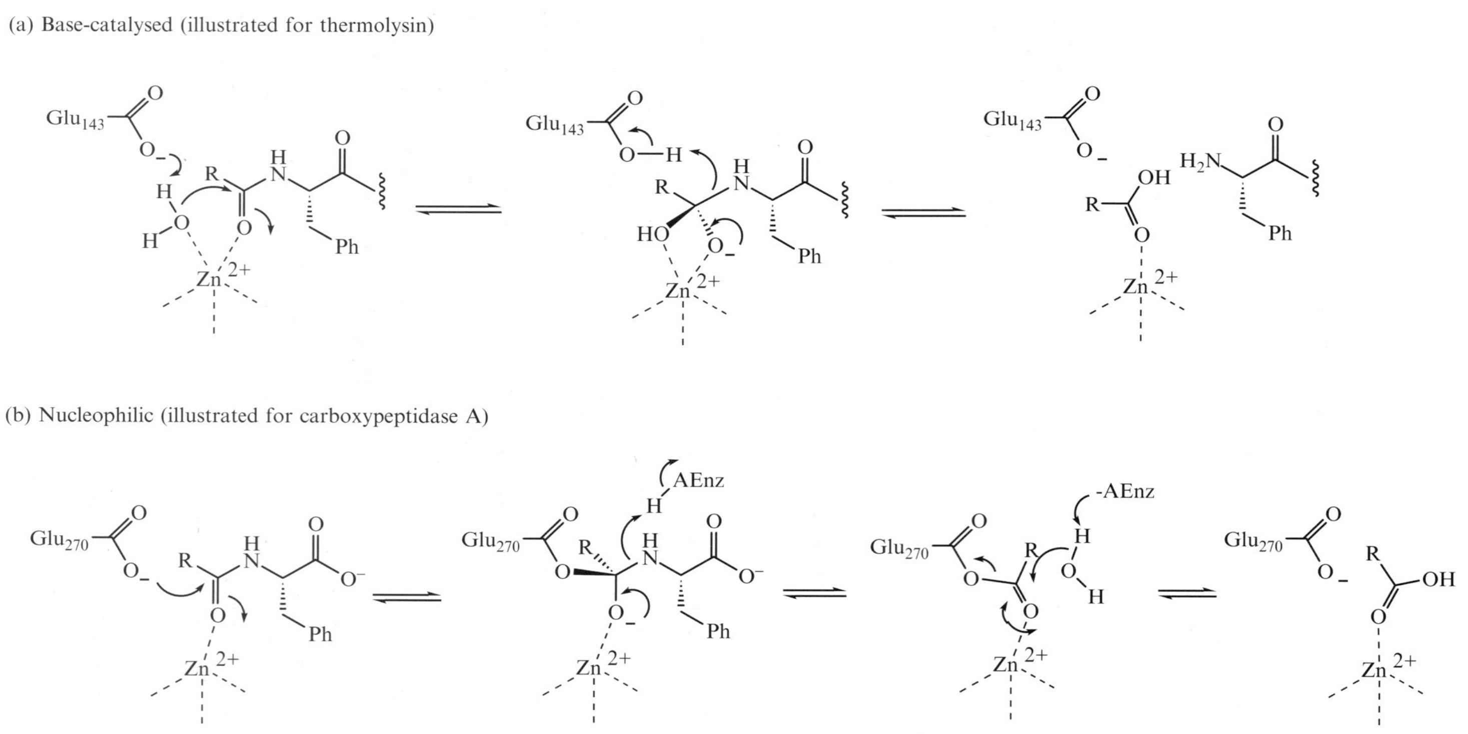

(a) Base-catalysed (illustrated for thermolysin)

(b) Nucleophilic (illustrated for carboxypeptidase A)

Figure 5.14 Possible base-catalysed (a) and nucleophilic (b) mechanisms for thermolysin and carboxypeptidase A. R, peptide chain.

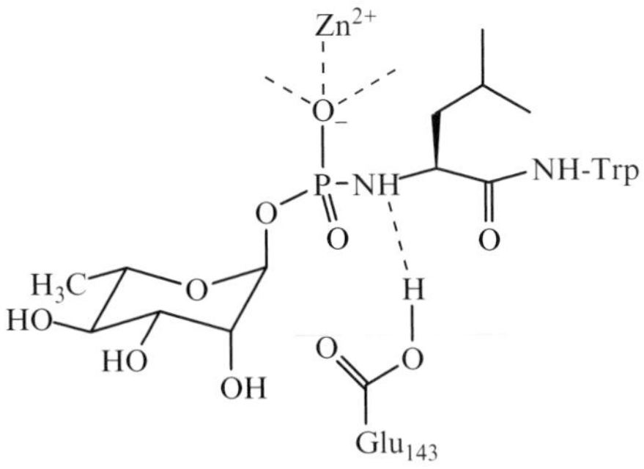

Figure 5.15 Inhibition of thermolysin by phosphoramidon.

active site zinc, acting as an analogue of the oxyanion tetrahedral intermediate, and is therefore bound extremely tightly by the enzyme. Further phosphona-midate inhibitors have now been devised and synthesised which bind even more tightly.

A further class of metalloproteases contain two divalent metal ions, arranged in a binuclear metal cluster. One example of this family of enzymes is leucine aminopeptidase. In these enzymes it is thought that one metal centre activates the nucleophilic water molecule, while the second metal centre activates the amide carbonyl group via Lewis acid activation.

The aspartyl proteases

The aspartyl (or acid) proteases are characterised by the presence of two active site aspartate residues whose carboxylate side chains are involved in catalysis. They are probably the smallest class of protease enzymes, but have recently come to prominence through the discovery that the HIV-1 virus contains an essential aspartyl protease enzyme, which we shall examine below. The aspartyl proteases are characteristic in their ability to operate at low pH: the enzyme pepsin operates in the range 2–4, which makes it well suited for operating in the acidic environment of the stomach. This pH requirement will become apparent when we examine the mechanism of these enzymes, since one of the two aspartate residues must be protonated for activity.

This family of enzymes are also characterised by their inhibition by low levels (1 μg ml$^-$) of pepstatin (Figure 5.16), a modification of a naturally occurring peptide containing the unusual amino acid statine (see Figure 5.20 for an illustration of the mechanism of inhibition by this class of inhibitor).

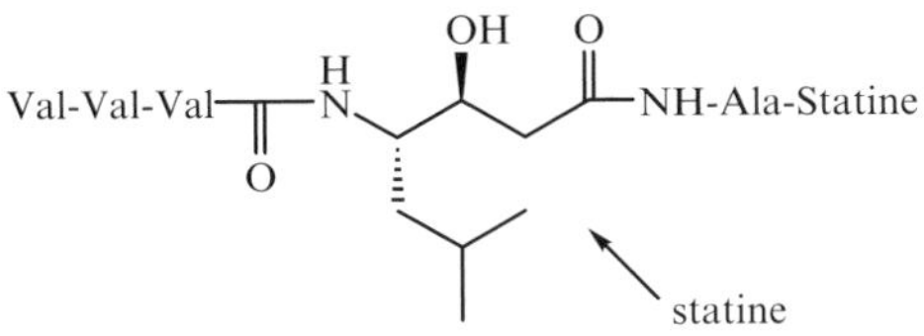

statine   **Figure 5.16** Pepstatin.

CASE STUDY: HIV-1 protease

The discovery in 1983 that human immunodeficiency virus (HIV) is the causative agent of acquired immunodeficiency syndrome (AIDS) prompted a huge research effort into this virus. The virus was found to contain an essential aspartyl protease (known as 'HIV protease') which is required for the cleavage of two 55-kDa and 160-kDa precursor polypeptides produced from the *gag* and *pol* genes of HIV-1. The cleavage products of these precursor polypeptides are the structural proteins and retroviral replication enzymes required for the assembly of new HIV-1 virions and completion of the viral life cycle. This enzyme, therefore, represented an immediate target for anti-HIV therapy: if inhibitors could be devised for the HIV-1 protease, the virus would be unable to synthesise its essential proteins and the life cycle would be blocked.

The HIV-1 protease has been overexpressed, purified and crystallised, and several research groups have solved its X-ray crystal structure to high resolution. The enzyme is a 99-kDa homodimer that is similar in structure to other members of the family such as pepsin. Its active site lies at the interface of the two sub-units, and the two active site carboxylate residues are Asp-25 from one sub-unit and the complementary Asp-25′ from the other sub-unit, as shown in Figure 5.17. Preferred cleavage sites for the HIV-1 protease are at Aro–proline sites, where Aro is an aromatic amino acid (tyrosine, phenylalanine, tryptophan). Using synthetic substrates containing a tyrosine–proline cleavage site, enzyme-catalysed exchange of ^{18}O from $H_2^{18}O$ into the amide carbonyl has been

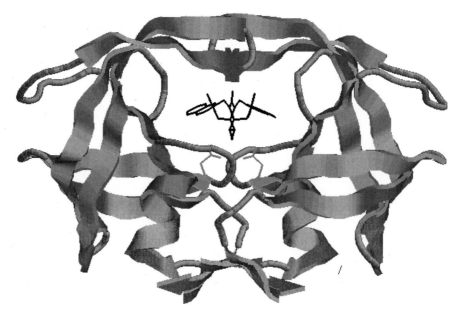

Figure 5.17 Structure of HIV-1 protease (PDB file 1HVR), showing the catalytic aspartic acid residues Asp-25 and Asp-25′ in red, and a bound inhibitor in black.

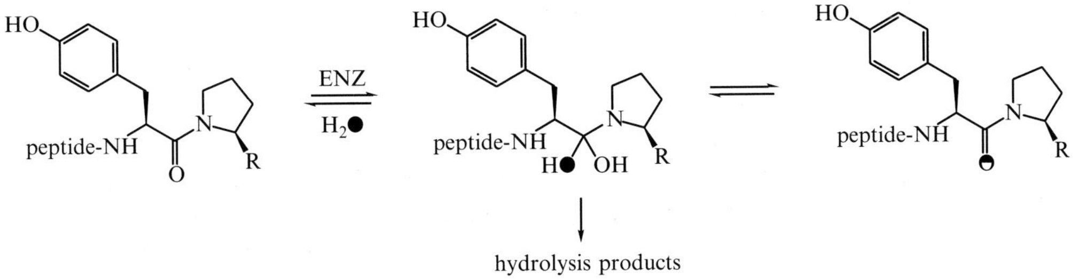

Figure 5.18 ^{18}O exchange via a hydrate intermediate.

observed at up to 10% of the rate of the forward reaction. Since the reaction is in practice irreversible, this exchange is consistent with the reversible formation of a hydrated intermediate, as shown in Figure 5.18.

D$_2$O solvent isotope effects of 1.5–3.2 have been measured under a range of conditions, consistent with base-catalysed attack of water being the first step of the mechanism. Finally, small inverse ^{15}N isotope effects have been measured for the departing nitrogen atom, suggesting protonation of nitrogen in the rate-determining step of the mechanism. The proposed mechanism is illustrated in Figure 5.19.

According to this mechanism Asp-25 acts as a base to deprotonate an attacking water molecule, with Asp-25′ acting as a general acid, forming the hydrated intermediate. Breakdown of this intermediate with protonation of the departing nitrogen atom by the protonated Asp-25 completes the catalytic cycle.

Most of the research effort on the HIV-1 protease has been directed towards synthesising potent inhibitors. Given the precedented inhibition of pepsin and other aspartyl proteases by pepstatin, a range of substrate analogues containing statine-like groups have been synthesised and found to act as potent inhibitors for the HIV-1 protease. It is thought that the statine unit mimics the tetrahedral intermediate formed in the reaction, as shown in Figure 5.20.

Administration of human T-lymphocyte cells that are infected with the HIV-1 virus with such inhibitors has demonstrated that these compounds

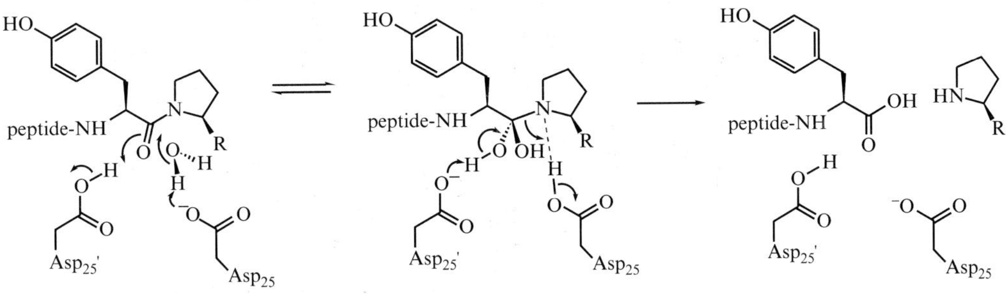

Figure 5.19 Mechanism for HIV-1 protease.

Transition state

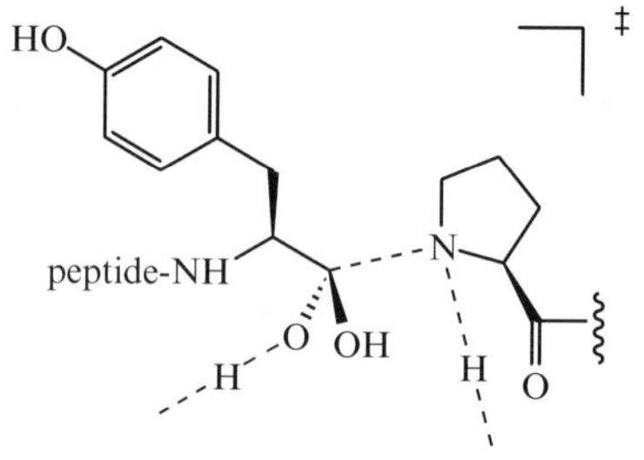

Inhibitor JG-365
$K_i = 0.66$ nM

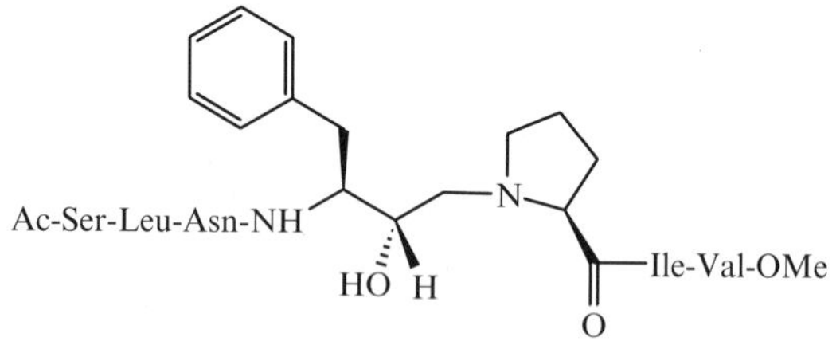

Figure 5.20 Transition state inhibitor for HIV-1 protease.

have potent anti-viral properties *in vivo*, so this stategy represents a realistic hope for development of anti-HIV therapy. Figure 5.17 shows the X-ray crystal structure of the above inhibitor bound to the active site of HIV-1 protease. This type of high-resolution data provides a good model for the development of further inhibitors of this enzyme.

5.3 Esterases and lipases

An important part of food digestion is the breakdown of fats, oils and lipid content in food. Lipids are largely made up of glycerol esters of long-chain fatty acids. The digestive system of animals contains high levels of esterase and lipase enzymes which hydrolyse the ester functional groups of fats and oils. Lipases are often fat-soluble enzymes which are able to operate at the lipid–water surface which would otherwise present a physical barrier for a soluble esterase enzyme.

Several members of the family of esterase and lipase enzymes contain active site serine groups, and proceed via the same type of mechanism as chymotrypsin (which is capable of hydrolysing esters as well as amides). Two notable examples are the enzymes pig liver esterase and porcine pancreatic lipase which are commercially available in large quantities. These enzymes have found a number of applications in organic synthesis due to their ability to hydrolyse a wide range of ester substrates with high stereospecificity. Examples of resolution reactions catalysed by these enzymes were illustrated in Figures 3.2 and 3.3.

Since these enzymes proceed through a covalent acyl enzyme intermediate, they are able to catalyse transesterification reactions in which an acyl group is

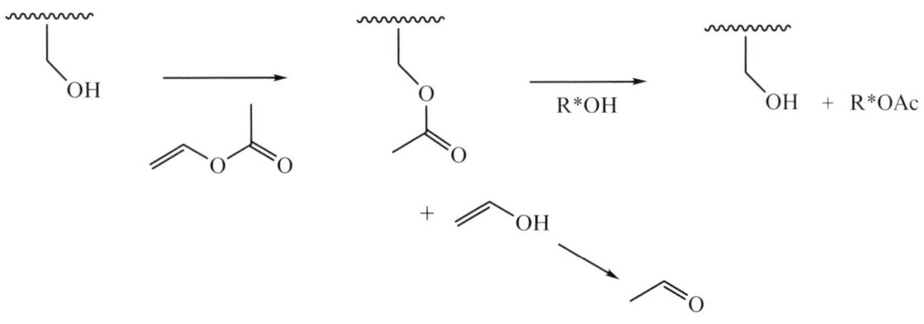

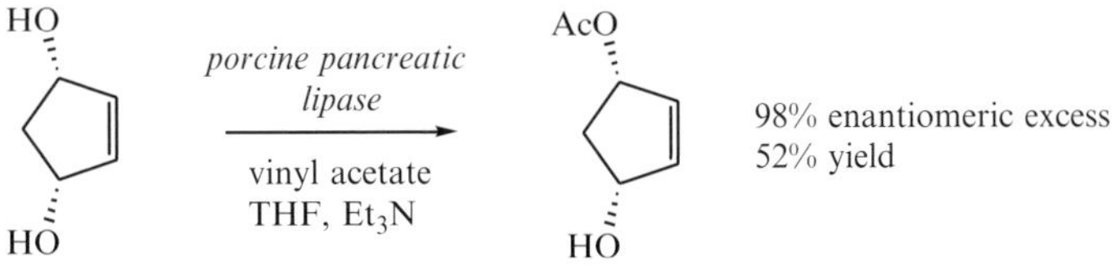

Figure 5.21 Enantioselective acylation reaction using a serine esterase. Et$_3$N, triethylamine; THF, tetrahydrofuran.

transferred to an alcohol acceptor. High-yielding regio- and enantio-specific acylation reactions have been developed using vinyl acetate as a solvent. For-mation of the acetyl–enzyme intermediate is accompanied by formation of the enol form of acetaldehyde, which rapidly tautomerises to give acetaldehyde, making the reaction irreversible. Attack of the alcohol functional group then gives the acetylated product. An example is illustrated in Figure 5.21 in which an achiral *meso*-substrate is regioselectively acylated to give a single enantio-meric product.

5.4   Acyl transfer reactions in biosynthesis: use of coenzyme A (CoA)

How do living systems synthesise the amide bonds found in proteins, or the ester functional groups found in lipids, oils and other natural products? The general strategy, shown in Figure 5.22, is to make an activated acyl derivative containing a good leaving group, and then to carry out an acyl transfer reaction.

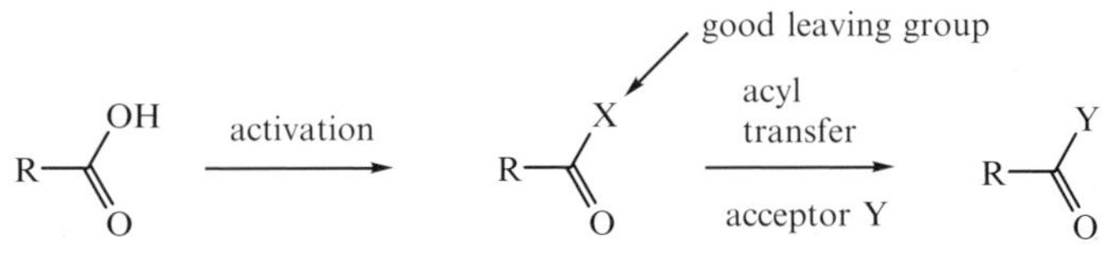

Figure 5.22 Use of an activated acyl group for acyl transfer.

The aminoacyl group of amino acids is activated and transferred during the assembly of the polypeptide chains of proteins by ribosomes. Amino acid activation is carried out by adenosine triphosphate (ATP)-dependent amino acyl transfer RNA (tRNA) synthetase enzymes. Each individual amino acid is converted into an acyl adenylate mixed anhydride derivative, followed by transfer of the aminoacyl group onto a specific tRNA molecule. The aminoacyl-tRNA ester is then bound to the ribosome and the free amine used to form the next amide bond in the sequence of the protein (see Figure 5.23).

Activation and transfer of acyl groups is a common process found in fatty acid biosynthesis, polyketide natural product biosynthesis, and the assembly of a variety of amide and ester functional groups in biological molecules. For the majority of these processes a special cofactor is used – coenzyme A (CoA). The structure of CoA contains a primary thiol group which is the point of attachment to the acyl group being transferred, forming a thioester linkage shown in Figure 5.24.

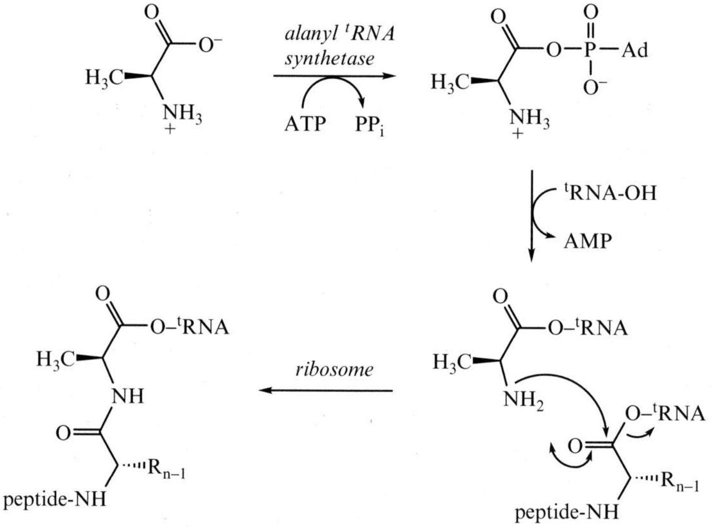

Figure 5.23 Acyl transfer reactions in protein biosynthesis. PP_i, inorganic pyrophosphate.

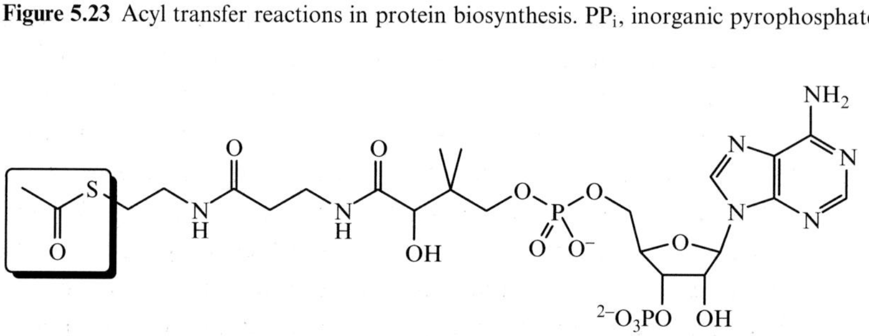

Figure 5.24 Structure of acetyl CoA.

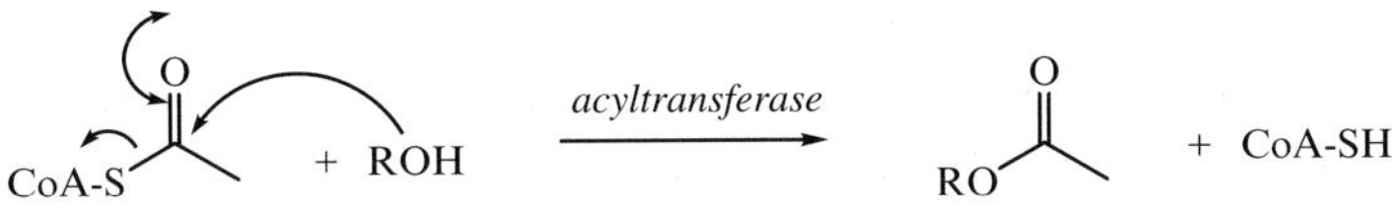

Figure 5.25 Acyl transfer using acetyl CoA.

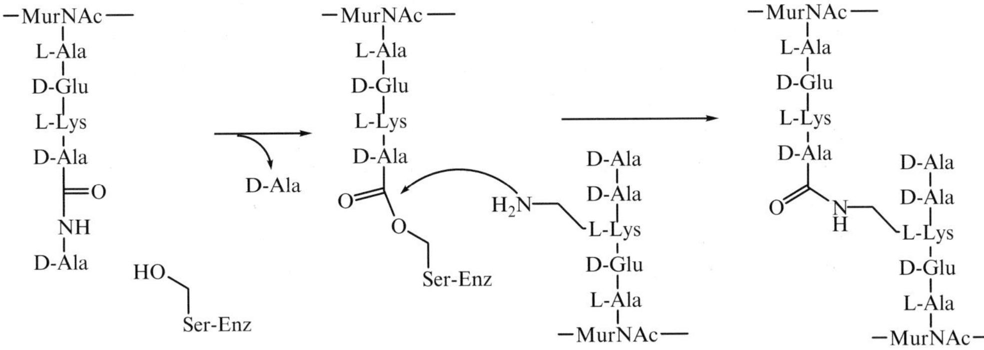

Figure 5.26 Cross-linking of peptidoglycan catalysed by D,D-transpeptidase.

Coenzyme A is well suited to carry out acyl transfer reactions, since thiols are inherently more nucleophilic than alcohols or amines. Thiols are also better leaving groups (pK$_a$ 8–9), which explains why the hydrolysis of thioesters under basic conditions is more rapid than ester hydrolysis. Acetyl CoA is used by acyltransferase enzymes to transfer its acetyl group to a variety of acceptors, which can be alcohols, amines, carbon nucleophiles or other thiol groups (see Figure 5.25). We shall encounter specific examples of the use of acetyl CoA in later chapters.

Finally, there are a small number of transpeptidase enzymes that transfer the acyl group of a peptide chain onto another amine acceptor. One important example is the transpeptidase enzyme involved in the final step of the assembly of peptidoglycan – a major structural component of bacterial cell walls. Peptidoglycan consists of a polysaccharide backbone of alternating N-acetyl-glucosamine (GlcNAc) and N-acetyl-muramic acid (MurNAc) residues, from which extend pentapeptide chains which contain the unusual D-amino acids D-alanine and D-glutamate. The final step in peptidoglycan assembly involves the cross-linking of these pentapeptide side chains, catalysed by a transpeptidase enzyme which contains an active site serine residue analogous to the serine proteases (see Figure 5.26). A covalent acyl enzyme intermediate is formed with release of D-alanine, which can either be hydrolysed to generate a tetrapeptide side chain, or be attacked by the ε-amino side chain of a lysine residue from another chain, generating an amide cross-link. These cross-links add considerable rigidity to the peptidoglycan layer, enabling it to withstand the high osmotic stress from inside the bacterial cell.

5.5   Enzymatic phosphoryl transfer reactions

Phosphate esters are widespread in biological systems: phosphate mono-esters occur as alkyl phosphates, sugar phosphates, and even phosphoproteins; whilst phosphodiester linkages comprise the backbone of the nucleic acids RNA and DNA, and are found in the phospholipid components of biological membranes. Examples of phosphotriesters in biological systems are known, but they are relatively scarce in comparison. Some examples of biologically important phosphate esters are shown in Figure 5.27.

Phosphoryl transfer is therefore of fundamental importance to biological systems for the biosynthesis and replication of nucleic acids, and in the transfer of phosphate groups between small molecules and between proteins. Three mechanisms have been observed for phosphoryl transfer reactions, as shown in Figure 5.28. Mechanism A is a dissociative mechanism (similar to an S_N1 mechanism at carbon), in which a metaphosphate intermediate is formed, prior to attack by a nucleophile. Mechanism B is a concerted, S_N2-like mechanism, with no intermediate, but proceeding via a penta-co-ordinate transition state. Mechanism C is an associative mechanism, in which attack of the nucleophile occurs first, to give a penta-co-ordinate phosphorane intermediate.

Studies of non-enzymatic phosphoryl transfer have shown that phosphate mono-ester dianions usually react via a concerted mechanism (mechanism B), and with hindered nucleophiles can access a more dissociative mechanism (mechanism A). Phosphate mono-ester anions proceed via a mechanism involv-

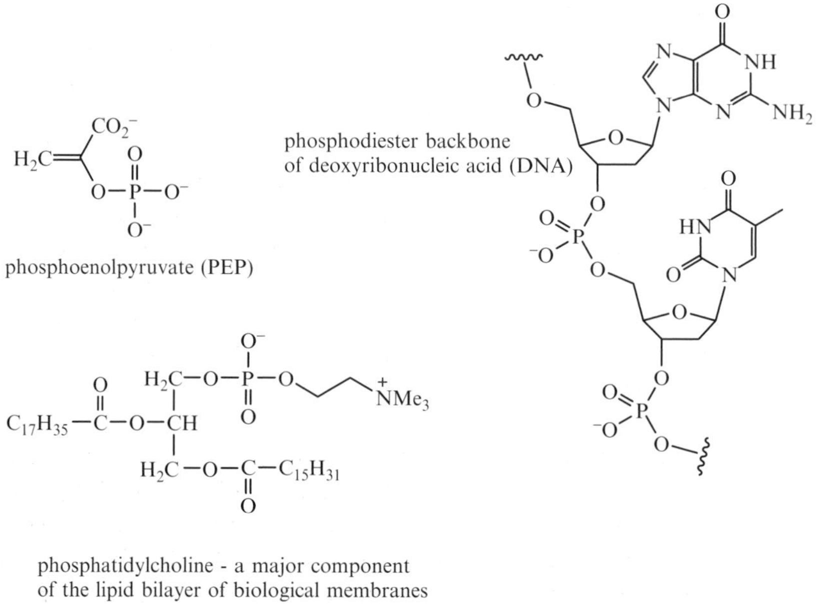

phosphoenolpyruvate (PEP)

phosphodiester backbone
of deoxyribonucleic acid (DNA)

phosphatidylcholine - a major component
of the lipid bilayer of biological membranes

Figure 5.27 Examples of biologically important phosphates.

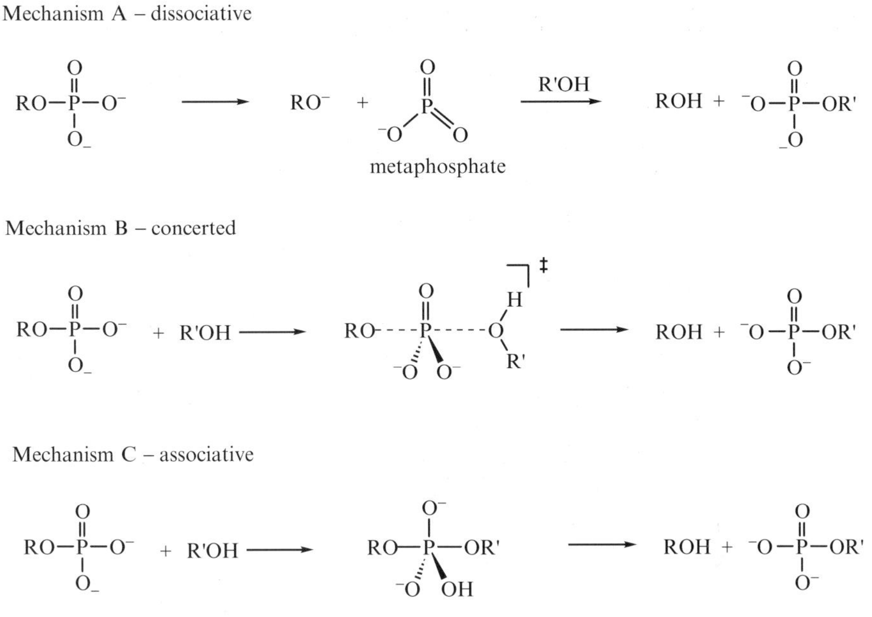

Mechanism A – dissociative

metaphosphate

Mechanism B – concerted

Mechanism C – associative

Figure 5.28 Mechanisms for phosphoryl transfer.

ing transfer of a proton onto the departing oxygen, and concerted phosphoryl transfer. Phosphate diesters and triesters proceed via progressively more associative mechanisms, either concerted (mechanism B) in the presence of a good leaving group, or fully associative (mechanism C) in the absence of a good leaving group.

Enzymatic hydrolysis of phosphate mono-esters is carried out by a family of phosphatase enzymes. Of these enzymes the bovine alkaline phosphatase has been best studied, due to its availability and its broad substrate specificity: it will hydrolyse a very wide range of phosphate mono-esters. Analysis of the stereochemistry of the alkaline phosphatase reaction using the chiral phosphate methodology described in Section 4.4 revealed that this reaction proceeds with retention of stereochemistry at phosphorus. Rapid kinetic analysis has revealed that a stoichiometric amount of the alcohol product ROH is released prior to phosphate release, and that the k_{cat} for a range of substrates is independent of the nature of ROH. These data suggest the existence of a phospho–enzyme intermediate, whose breakdown is rate-determining. Incubation of enzyme with $RO^{32}PO_3^{2-}$ gave enzyme labelled with ^{32}P, which upon tryptic digestion and sequencing was found to be localised on a unique serine residue, Ser-102. The active site of alkaline phosphatase, shown in Figure 5.29, contains two Zn^{2+} ions, with a separation of 3.9 Å. One zinc centre is used to bind the phosphate mono-ester substrate, the other to activate Ser-102 for nucleophilic attack, as shown in Figure 5.30. The phosphate oxygens are co-ordinated by Arg-166, which provides transition state stabilisation for the phosphotransfer reaction.

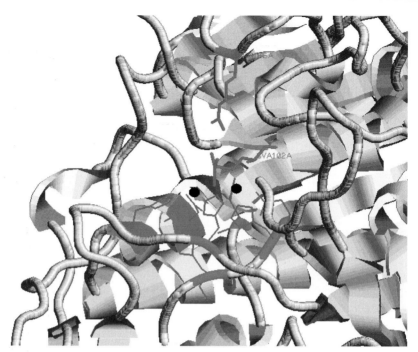

Figure 5.29 Structure of alkaline phosphatase (PDB file 1B8J), in which the catalytic Ser-102 has been derivatised with vanadate, representing the phospho–serine intermediate. Modified Ser-102 and Arg-166 are highlighted in red. The two active site zinc ions are shown in black.

Protein tyrosine phosphatases, which catalyse the dephosphorylation of phosphotyrosine peptides, also contain an active site arginine residue which provides transition state stabilisation for phosphoryl transfer. [15]N kinetic isotope effects measured for these enzymes are consistent with extensive bond cleavage to the leaving group in the transition state, consistent with a dissociative transition state (mechanism A, Figure 5.28). Therefore, the active site arginine residue appears to stabilise the metaphosphate intermediate 'in flight'.

Enzymes that cleave the phosphodiester backbone of RNA and DNA are known as nucleases. Some nucleases are relatively non-specific, for example an $3' \rightarrow 5'$ exonuclease enzyme found in snake venom which will digest single-stranded DNA from the 3' end successively. However, the endonuclease enzymes usually have highly specific cleavage sites, for example the restriction endonuclease *Eco*R1 (produced by *Escherichia coli*) cleaves at a specific six-base sequence 5'-GAATTC-3', making cuts on both strands of the DNA as shown in Figure 5.31.

The best characterised nuclease enzyme is ribonuclease A, a 14-kDa protein which was the first protein to be reversibly unfolded and re-folded in the absence of other proteins. This enzyme cleaves RNA at sites immediately following pyrimidine bases (cytidine or uridine), leaving a

Arg166
HN
H2N
NH2
17O
18O
R
O
P
S
Ser102
O
H
Zn2+
‡
OH
HO

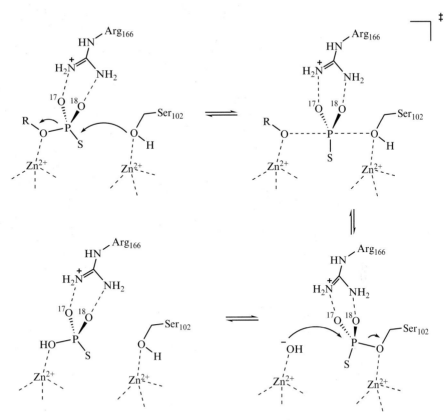

Figure 5.30 Mechanism and stereochemistry of alkaline phosphatase.

(1) Ribonuclease A - cleavage of RNA after pyrimidine (cytidine or uridine)

$$5'\text{-....X-p-Y-p-C-p-Z-p....-3'} \xrightarrow{\text{RNase A}} 5'\text{....X-p-Y-p-C-3'-}OPO_3^{2-} + 5'\text{-HO-Z-p....3'}$$

(2) Restriction endonucleases - cleavage of dsDNA at specific 6-base recognition site

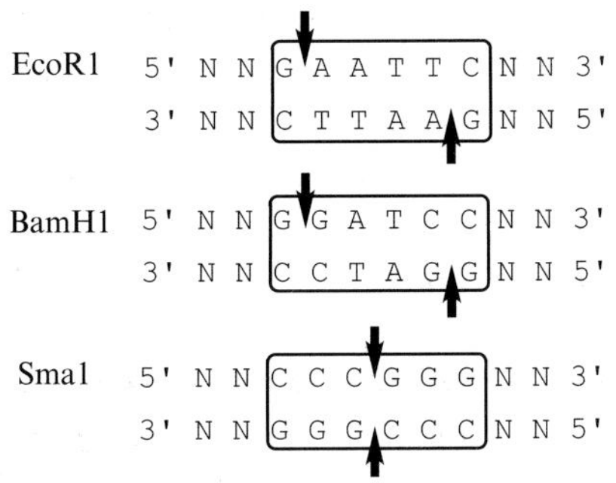

EcoR1
5' N N G A A T T C N N 3'
3' N N C T T A A G N N 5'
BamH1
5' N N G G A T C C N N 3'
3' N N C C T A G G N N 5'
Sma1
5' N N C C C G G G N N 3'
3' N N G G G C C C N N 5'

Figure 5.31 Specificity of phosphodiesterases.

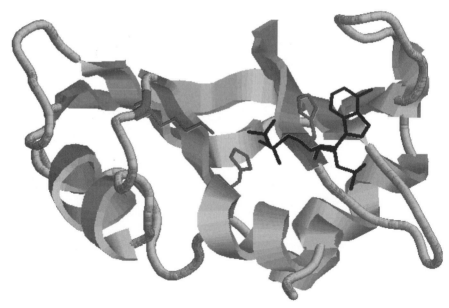

Figure 5.32 Structure of ribonuclease A (PDB file 1AFL), showing the catalytic residues His-12, His-119 and Lys-41 in red, and a bound inhibitor in black.

pyrimidine 3′-phosphate product. Covalent modification studies using iodoacetate and iodoacetamide implicated two histidine residues, His-12 and His-119, as being involved in catalysis. When the X-ray crystal structure of the enzyme was solved, these residues were found on opposite sides of the active site, with the side chain of Lys-41 also occupying a prominent position (see Figure 5.32). The currently accepted mechanism for ribonuclease A is shown in Figure 5.33. The mechanism involves participation of the 2′-hydroxyl of the pyrimidine residue, which is deprotonated by His-112 and attacks the phosphodiester via an associative mechanism to form a divalent transition state stabilised by Lys-41. Breakdown of this transition state using His-119 as an acid leads to a cyclic phosphodiester intermediate which, due to internal strain, is much more reactive than the substrate. Acid–base catalysis by the two histidine groups completes the mechanism via a second divalent transition state.

5.6   Adenosine 5′-triphosphate

Enzymatic phosphoryl transfer reactions usually involve the transfer of phosphoryl groups from a 'high-energy' phosphoric anhydride species to an acceptor which can be an alcohol, a carboxylic acid or another phosphate. The most common source of phosphoryl groups for such transfer reactions is the coenzyme adenosine 5′-triphosphate or ATP. This is the nucleoside triphosphate derivative of adenosine, which is one of the components of RNA. However, in addition to its role in RNA it is used as a coenzyme by a wide range of enzymes.

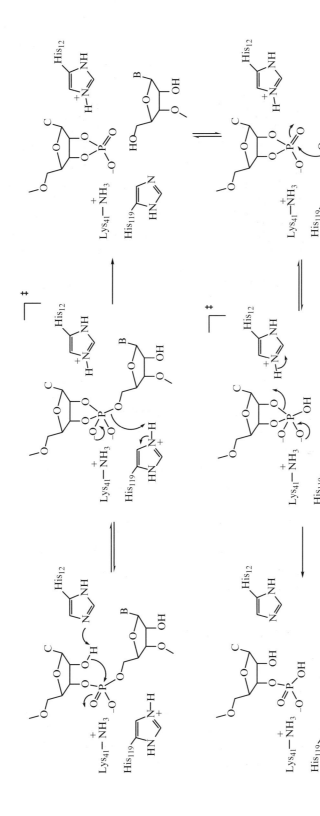

Figure 5.33 Mechanism for ribonuclease A.

Adenosine 5′-triphosphate is a thermodynamically unstable molecule, since the hydrolysis of its phosphoric anhydride linkages is thermodynamically highly favourable, hence its designation as a 'high-energy' source of phosphate. Yet ATP is reasonably stable in aqueous solution – why? The explanation is that although the hydrolysis of ATP is thermodynamically favourable it is kinetically unfavourable, particularly the hydrolysis of the phosphodiester groups.

The triphosphate group of ATP can be cleaved at a number of different points, shown in Figure 5.34, leading to the transfer of either a phosphoryl group, a pyrophosphoryl group or an adenosine phosphoryl group. The most common transfer is of a single phosphoryl unit, leaving behind adenosine 5′-diphosphate (ADP). This process is used by kinase enzymes in the phosphorylation of alcohols, and is used by a number of ligase enzymes to activate carboxyl groups as acyl phosphate intermediates (see Figure 5.35).

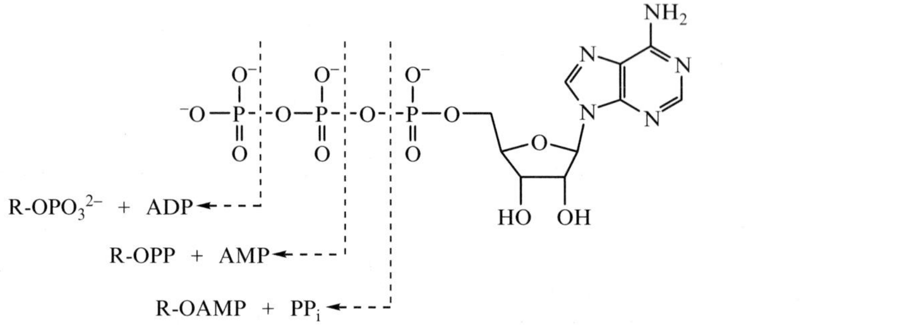

Figure 5.34 Structure of ATP and three common modes of phosphoryl transfer.

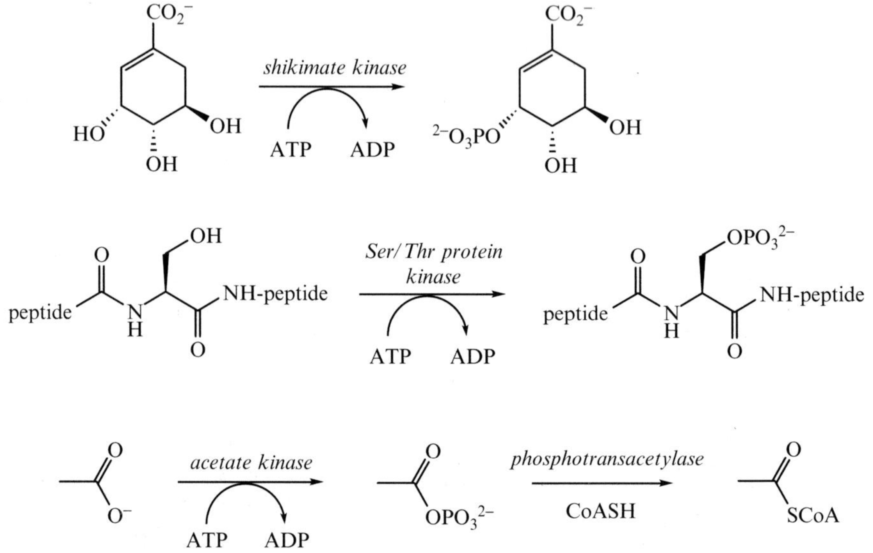

Figure 5.35 Phosphorylation of alcohols and carboxylates by ATP.

One example of the use of ATP to activate a carboxylic acid derivative is in the biosynthesis of acetyl CoA. This is carried out in bacteria by the action of two enzymes: acetate kinase, which activates acetate as acetyl phosphate using ATP as a cofactor; and phosphotransacetylase, which transfers the acetyl group onto coenzyme A (see Figure 5.35). Phosphorylation of serine and tyrosine residues on the surface of proteins is a very important reaction in cell signalling pathways. These reactions are catalysed by the serine–threonine and tyrosine–protein kinases, which also use ATP as coenzyme.

5.7   Enzymatic glycosyl transfer reactions

Carbohydrates fulfil many important roles in biological systems: polysaccharides are important structural components of plant and bacterial cell walls, and constitute an important part of the human diet; mammalian polysaccharides, such as glycogen, are used as short-term cellular energy stores; whilst the attachment of carbohydrates to mammalian glycoproteins has an important role to play in cell–cell recognition processes.

Hydrolysis of polysaccharides can be achieved in the laboratory using acid hydrolysis, as shown in Figure 5.36, since the glycosidic linkage is an acetal

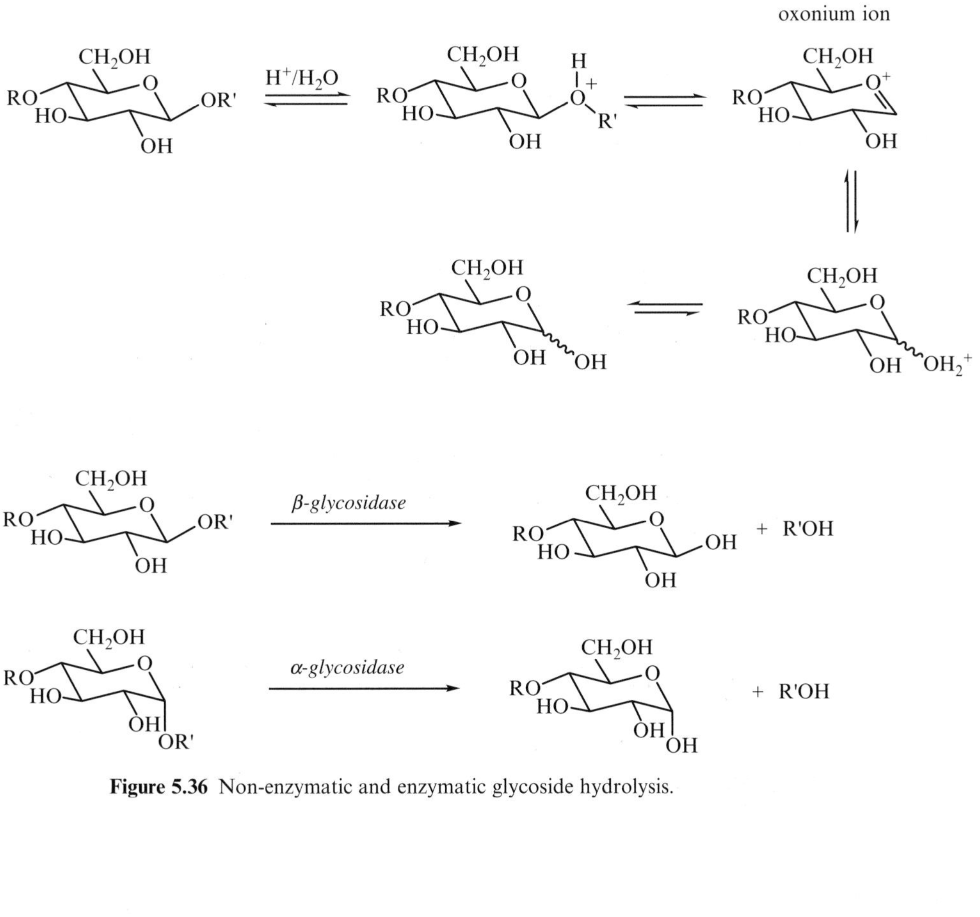

Figure 5.36 Non-enzymatic and enzymatic glycoside hydrolysis.

functional group. Glycosyl transfer enzymes also employ acid catalysis for just the same reason. It will not surprise you to learn that glycosidase enzymes are highly specific: they are specific for cleavage at the glycosidic bond of a particular monosaccharide, and they are also specific for cleavage of either an α- or a β-glycosidic linkage (see Figure 5.36).

The best studied glycosidase enzyme is lysozyme, a mammalian protein found in such diverse sources as egg whites and human tears. It functions as a mild anti-bacterial agent by hydrolysing the glycosidic linkage between MurNAc and GlcNAc residues in the peptidoglycan layer of bacterial cell walls. The active site contains two carboxylic acid groups which are involved in catalysis: Glu-35 and Asp-52, shown in Figure 5.37. The mechanism is illustrated in Figure 5.38.

Upon binding of substrate, Glu-35 donates a proton to the departing GlcNAc C-4 oxygen, promoting cleavage of the glycosidic bond and formation of an oxonium ion intermediate. There has been much debate in the literature over the precise role of Asp-52: whether it stabilises the oxonium ion via an ion pair electrostatic interaction, or whether it becomes covalently attached to the oxonium ion. There is evidence to suggest that covalent attachment does take place reversibly, and upon attack of water on the subsequent oxonium ion intermediate the product is formed with retention of stereochemistry at the glycosidic centre.

When the X-ray crystal structure of lysozyme was solved, it was found that the MurNAc residue whose glycosidic linkage was attacked could not be modelled into the active site of the enzyme – the lowest energy conformation

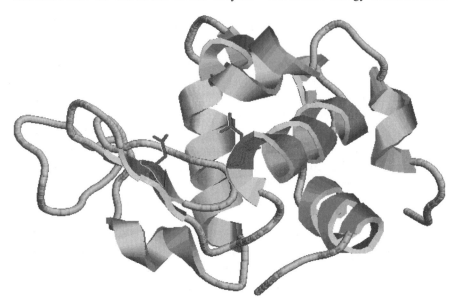

Figure 5.37 Structure of chicken egg lysozyme (PDB file 193L), showing the catalytic residues Glu-35 and Asp-52 in red.

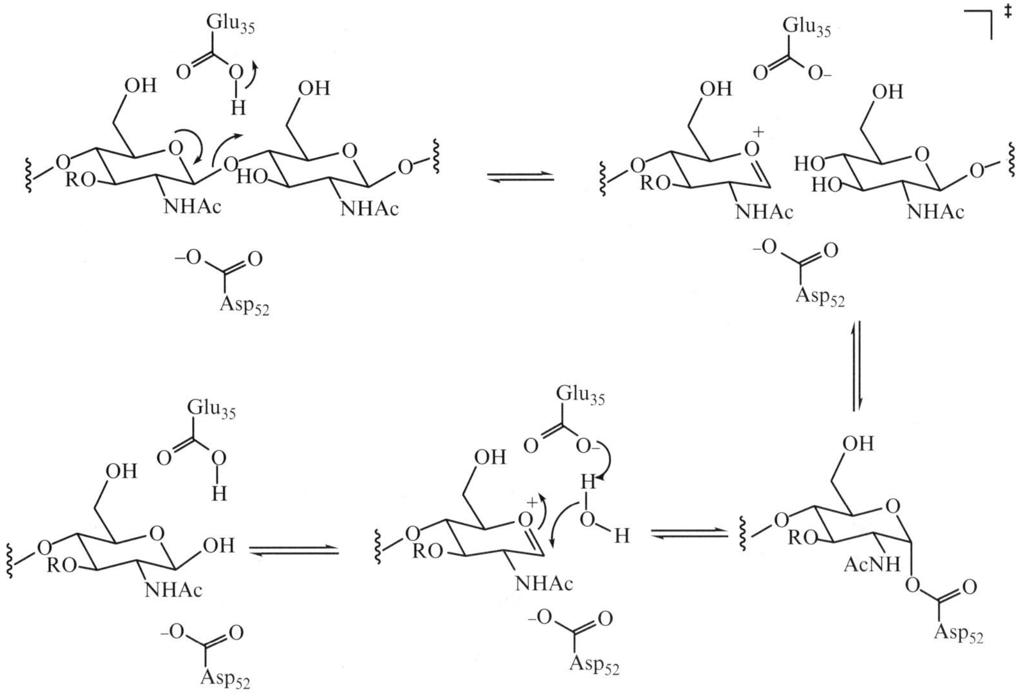

Figure 5.38 Mechanism for lysozyme.

of the substrate simply did not fit! However, if the pyranose ring of this residue adopts a somewhat flattened structure, it is able to bind to the active site, suggesting that the enzyme binds this residue in a strained but more reactive conformation. This is thought to be another example of the use of strain in enzyme catalysis mentioned in Section 3.7. Thus the enzyme uses the binding energy of other residues in the polysaccharide chain to offset the binding of the strained ring, which is then closer in structure and free energy to the flattened oxonium ion intermediate.

There are also glycosidase enzymes which proceed via inversion of stereochemistry at the glycosidic centre. It is thought in these cases that there is no covalent intermediate, and that there is attack directly on the oxonium intermediate on the opposite face of the molecule. This is not strictly an S_N2 displacement, since acetals are not susceptible to nucleophilic attack. However, it may be viewed as a displacement occurring via a dissociative transition state, as shown in Figure 5.39. These enzymes also contain two active site carboxyl groups, but they are situated 10–11 Å apart, far enough to accommodate the nuclophilic water molecule (whereas the two carboxyl groups in lysozyme are 5.5 Å apart).

Glycosyl transferases are involved in the assembly of polysaccharides and oligosaccharides. Glycosidic bonds are assembled through use of an activated glycosyl unit, namely a glycosyl phosphate or glycosyl diphospho-uridine

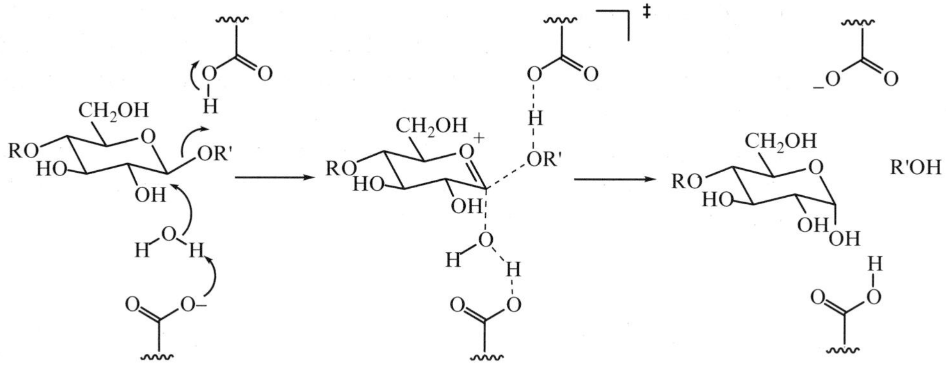

Figure 5.39 Mechanism of an inverting glycosidase.

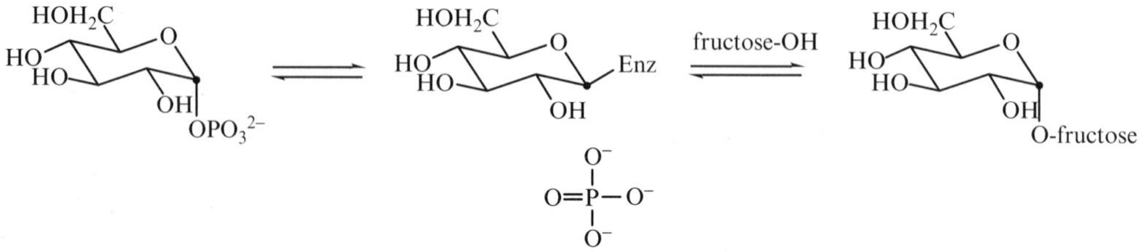

Figure 5.40 Sucrose phosphorylase. Black dot indicates the position of ^{14}C label.

derivative. For example, the disaccharide sucrose that we know as table sugar is synthesised by the enzyme sucrose phosphorylase from glucose-1-phosphate and fructose, as shown in Figure 5.40.

The overall reaction involves displacement of phosphate by fructose with retention of stereochemistry at the glycosyl centre, which by analogy with the above glycosidase enzymes would suggest that this is a double displacement reaction involving a covalent intermediate. This has been confirmed by incubation of the enzyme with sucrose labelled with ^{14}C at C-1 of the glucose residue, whereupon in the absence of inorganic phosphate the ^{14}C label becomes covalently attached to the enzyme. Addition of phosphate converts the ^{14}C–glucosyl–enzyme intermediate into ^{14}C-glucose-1-phosphate. Other glycosyl transferases utilise glycosyl diphospho-uridine activated derivatives, which upon glycosylation release uridine diphosphate (UDP). These glycosyl transfer reactions usually occur with inversion of stereochemistry, via a one-step displacement reaction.

5.8   Methyl group transfer: use of S-adenosyl methionine and tetrahydrofolate coenzymes for one-carbon transfers

The final example of group transfer reaction that we shall meet is that of methyl group transfer, and more generally the transfer of one-carbon units. Many biologically important molecules contain methyl ($-CH_3$) groups attached to oxygen, nitrogen and carbon substituents which have arisen by transfer of a methyl group from Nature's methyl group donor – S-adenosyl methionine (SAM). The methyl group to be transferred is attached to a positively charged sulphur atom, which is a very good leaving group for such a methylation reaction. Analysis of such methyltransferase reactions using the chiral methyl group approach detailed in Section 4.4 has revealed that they proceed with inversion of stereochemistry, implying that the reaction is a straightforward S_N2 displacement reaction (see Figure 5.41).

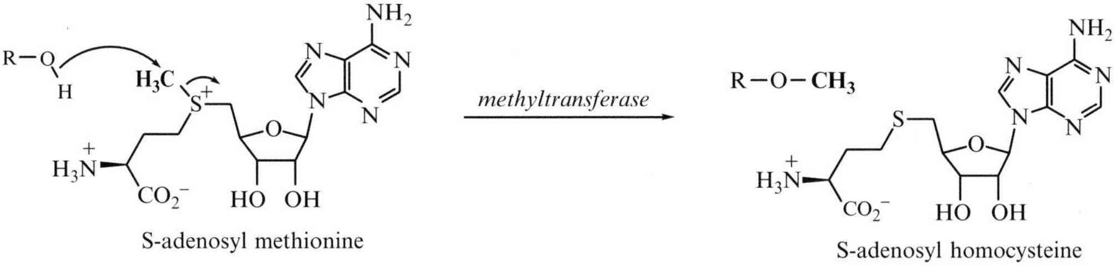

Figure 5.41 Methyl group transfer from S-adenosyl methionine.

As well as oxygen and nitrogen nucleophiles, SAM-dependent methyltransferases also operate on stabilised carbon nucleophiles, providing many of the methyl groups found in polyketide natural products, and in the structure of vitamin B_{12} (see Section 11.2). The by-product of methyltransferase enzymes is S-adenosyl homocysteine, which is recycled to SAM via hydrolysis to adenosine and homocysteine. How is the structure of S-adenosyl methionine assembled? It is synthesised from methionine and ATP by a very unusual displacement of triphosphate, which is subsequently hydrolysed to phosphate and pyrophosphate as shown in Figure 5.42.

As well as transferring methyl groups, Nature is able to transfer methylene ($-CH_2-$) groups and even methyne ($-CH=$) groups using another cofactor – tetrahydrofolate. Tetrahydrofolate is biosynthesised from folic acid, which is an essential element of the human diet. The active part of the molecule as far as one-carbon transfer is concerned are the two nitrogen atoms N_5 and N_{10}. It is

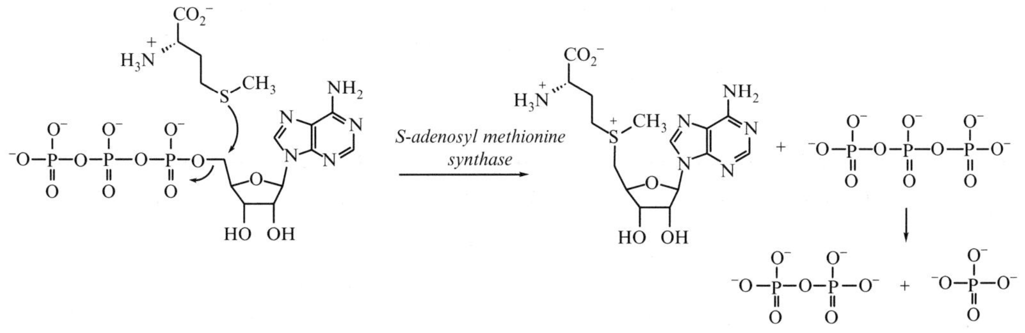

Figure 5.42 Biosynthesis of S-adenosyl methionine.

to these atoms that the one-carbon unit is attached, and there are several different forms of the cofactor, illustrated in Figure 5.43.

The mechanism for transfer of a methylene or methyne involves nucleophilic attack on the imine formed between the one-carbon unit and either N_5 or N_{10}, followed by breakdown of the substrate–coenzyme intermediate. An example is the formation of 5-hydroxymethylcytidine from cytidine, which is illustrated in Figure 5.44.

How is methylene-tetrahydrofolate synthesised in the cell? Not from free formaldehyde, which is toxic to biological systems. Conversion of tetrahydrofolate to methylene-tetrahydrofolate is carried out by serine hydroxymethyltransferase, a pyridoxal phosphate-dependent enzyme which we shall meet in Section 9.5. Suffice it to say that it is the hydroxymethyl group of serine which is transferred, generating glycine as a by-product. Methyne-tetrahydrofolate can be synthesised either by the NADP-dependent oxidation of methylene-tetrahydrofolate, or it can be synthesised by a synthetase enzyme which uses formyl phosphate as an activated one-carbon equivalent, generated from formate by ATP (see Figure 5.45).

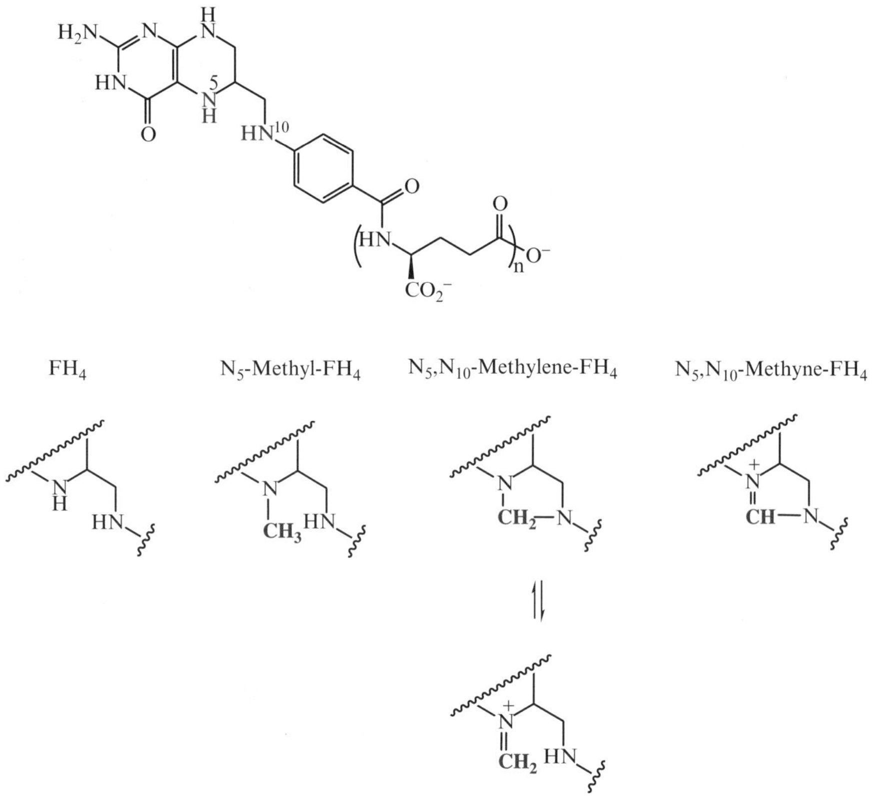

Figure 5.43 Tetrahydrofolate and its one-carbon adducts.

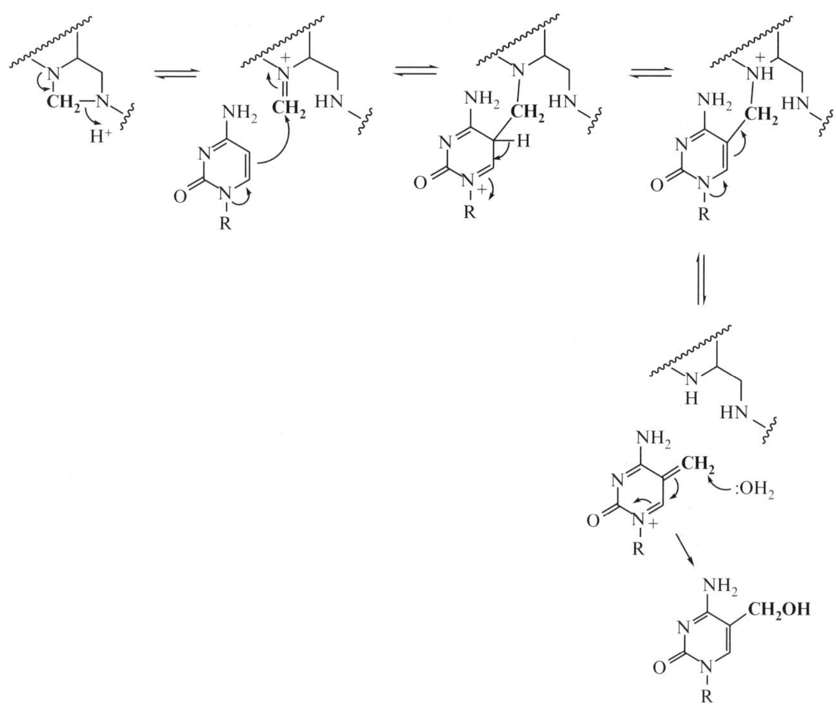

Figure 5.44 Methylene transfer to cytidine.

Thus, we have seen how enzymes are able to transfer a wide variety of carbon- and phosphorus-base groups in biological systems. These reactions are equally important for the breakdown and assembly of biological materials, and form the cornerstone of the family of enzymatic reactions found in living systems.

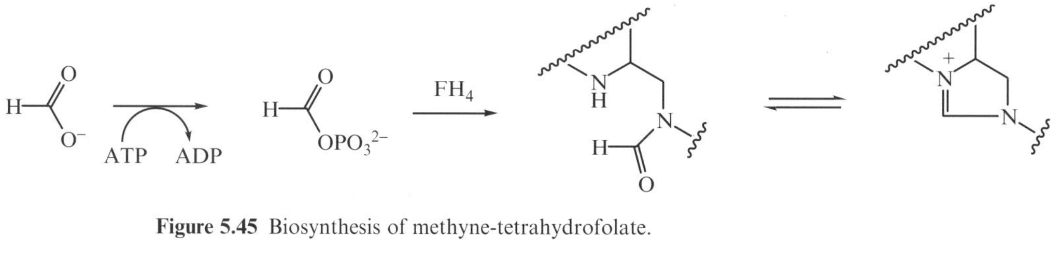

Figure 5.45 Biosynthesis of methyne-tetrahydrofolate.

Problems

(1) Using *para*-nitrophenyl acetate as a substrate for chymotrypsin, a rapid 'burst' of p-nitrophenol is observed, followed by a slower steady state release of p-nitrophenol. The amount of product released in the initial burst is 1 μmol of p-nitrophenol per μmol of enzyme. Explain.

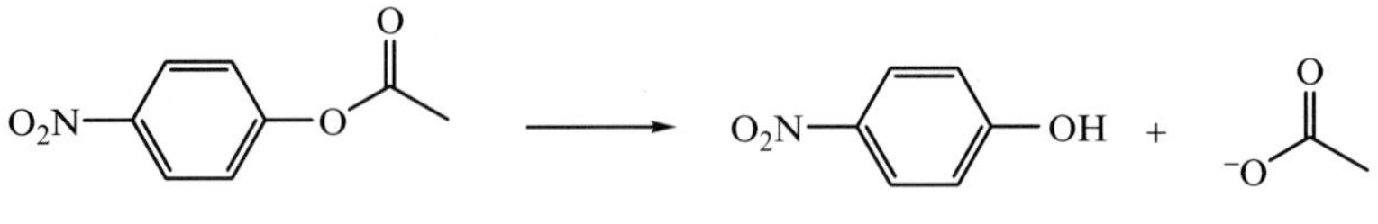

(2) Acetylcholinesterase is an esterase enzyme which catalyses the hydrolysis of acetylcholine (a neurotransmitter) to choline and acetate at nerve synaptic junctions. Inhibition of acetylcholinesterase is catastrophic and usually fatal. Given that the enzyme contains a serine catalytic triad, suggest a mechanism for this enzyme.

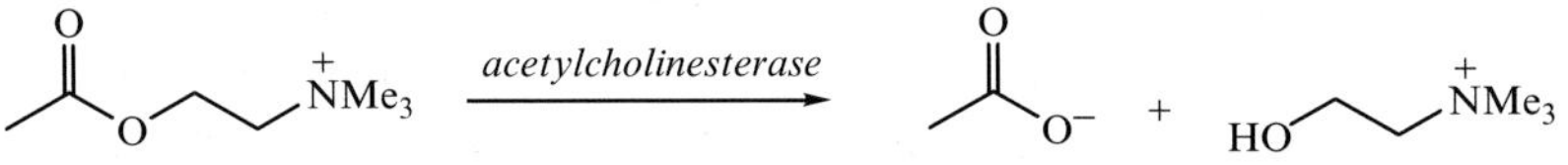

acetylcholine

Nerve gas sarin and insecticide parathion both act on this enzyme – suggest a common mechanism of action. Suggest reasons why parathion is not so toxic to humans (LD_{50} 6800 mg kg$^-$) as sarin.

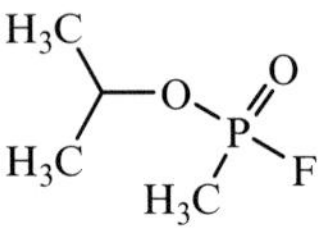

sarin 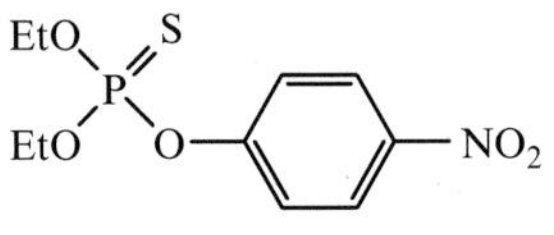   parathion

Pyridine aldoximine methiodide is an effective antidote for organophosphorus poisonings (used at the time of the Tokyo subway incident in March 1995). Suggest a mechanism for how it works.

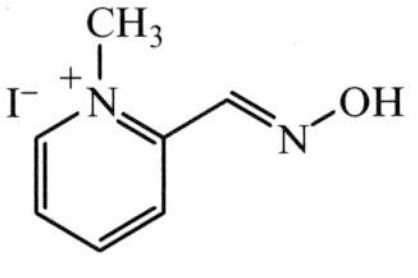

pyridine aldoximine
methiodide (PAM)

(3) Cysteine proteases are effectively inhibited by substrate analogues containing an aldehyde functional group in place of the amide targetted by the enzyme. Suggest a possible mechanism of inactivation.

(4) Acetyl CoA is biosynthesised from acetate by different pathways in bacteria versus higher organisms, as shown below. In the case of the bacterial pathway incubation of ^{18}O-labelled acetate yields one atom of ^{18}O in the phosphate product, whereas in higher organisms the same experiment yields one atom of ^{18}O in the adenosine monophosphate (AMP) product. Suggest intermediates and mechanisms for the two pathways.

Bacteria:

(1) acetate kinase
(2) phosphotransacetylase

$CH_3CO_2^- + ATP + CoASH \xrightarrow{\hspace{3cm}} CH_3COSCoA + ADP + P_i$

Higher organisms:

acetate thiokinase

$CH_3CO_2^- + ATP + CoASH \xrightarrow{\hspace{3cm}} CH_3COSCoA + AMP + PP_i$

(5) Glycogen is a mammalian polysaccharide consisting of repeating α-1,4-linked D-glucose units. It is stored in liver and muscle and is used as a carbohydrate energy source by the body, being converted to D-glucose when required by the pathway shown below. Suggest mechanisms for each of the enzymes on the pathway. What would be the medical consequences of a genetic defect in muscle glycogen phosphorylase?

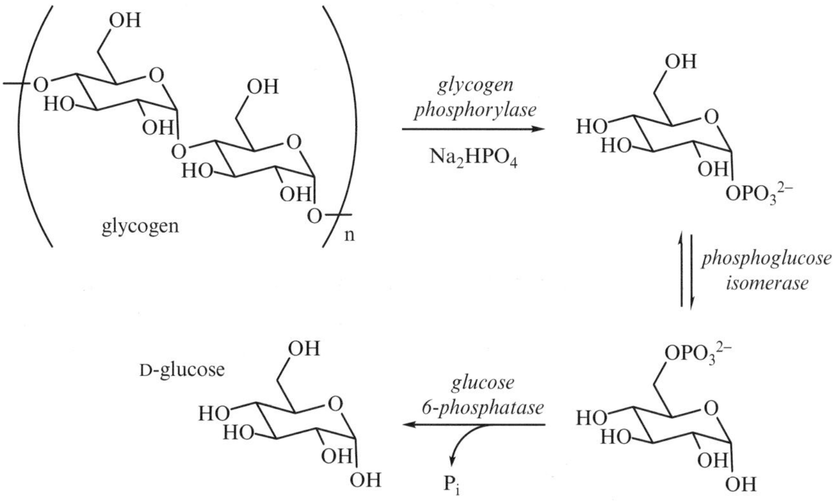

(6) Retaining glycosidases (like lysozyme) are irreversibly inhibited by substrate analogues containing a 2′-fluorine substituent, leading to covalent modification of the enzyme active site. Write a mechanism for this process, and explain why the covalent intermediate is not hydrolysed by the normal reaction pathway of the enzyme.

2′-fluoro substrate analogue

Further reading

General

R.H. Abeles, P.A. Frey & W.P. Jencks (1992) *Biochemistry*. Jones & Bartlett, Boston.
C.T. Walsh (1979) *Enzymatic Reaction Mechanisms*. Freeman, San Francisco.
C.H. Wong & G.M. Whitesides (1994) *Enzymes in Synthetic Organic Chemistry*. Pergamon, Oxford.

Serine proteases

S. Brenner (1988) The molecular evolution of genes and proteins: a tale of two serines. *Nature*, **334**, 528–30.
C.S. Craik, S. Roczniak, C. Largman & W.J. Rutter (1987) The catalytic role of the active site aspartic acid in serine proteases. *Science*, **237**, 909–13.
J. Kraut (1977) Serine proteases: structure and mechanism of catalysis. *Annu. Rev. Biochem.*, **46**, 331–58.
K. Kreutter, A.C.U. Steinmetz, T.C. Liang, M. Prorok, R.H. Abeles & D. Ringe (1994) Three-dimensional structure of chymotrypsin inactivated with (2*S*)-N-acetyl-L-Ala-L-Phe α-chloroethane: implication for the mechanism of inactivation of serine proteases by chloroketones. *Biochemistry*, **33**, 13792–800.
M. Prorok, A. Albeck, B.M. Foxman & R.H. Abeles (1994) Chloroketone hydrolysis by chymotrypsin and N-MeHis57-chymotrypsin: implications for the mechanism of chymotrypsin inactivation by chloroketones. *Biochemistry*, **33**, 9784–90.
S. Sprang, T. Standing, R.J. Fletterick, R.M. Stroud, J. Finer-Moore, N.H. Xuong, R. Hamlin, W.J. Rutter, & C.S. Craik (1987) The three-dimensional structure of Asn[102] mutant of trypsin: role of Asp[102] in serine protease catalysis. *Science*, **237**, 905–9.

Cysteine proteases

F.A. Johnson, S.D. Lewis & J.A. Shafer (1981) Determination of a low pK_a for histidine-159 in the S-methylthio derivative of papain by proton NMR spectroscopy. *Biochemistry*, **20**, 44–8.
S.D. Lewis, F.A. Johnson & J.A. Schafer (1981) Effect of cysteine-25 on the ionization of histidine-159 in papain as determined by proton NMR spectroscopy. *Biochemistry*, **20**, 48–51.

J.P.G. Malthouse, M.P. Gamesik, A.S.F. Boyd, N.E. Mackenzie & A.I. Scott (1982) Cryoenzymology of proteases: NMR detection of a productive thioacyl derivative of papain at subzero temperature. *J. Am. Chem. Soc.*, **104**, 6811–13.

E. Shaw (1990) Cysteinyl proteases and their selective inactivation. *Adv. Enzymol.*, **63**, 271–348.

T. Vernet, D.C. Tessier, J. Chatellier, C. Plouffe, T.S. Lee, D.Y. Thomas, A.C. Storer & R. Ménard (1995) Structural and functional roles of asparagine-175 in the cysteine protease papain. *J. Biol. Chem.*, **270**, 16645–52.

Metalloproteases

P.A. Bartlett & C.K. Marlowe (1987) Possible role for water dissociation in the slow binding of phosphorus-containing transition-state-analogue inhibitors of thermolysin. *Biochemistry*, **26**, 8553–61.

B.M. Britt & W.L. Peticolas (1992) Raman spectral evidence for an anhydride intermediate in the catalysis of ester hydrolysis by carboxypeptidase A. *J. Am. Chem. Soc.*, **114**, 5295–303.

D.W. Christianson & W.N. Lipscomb (1989) Carboxypeptidase A. *Acc. Chem. Res.*, **22**, 62–9.

H.M. Holden, D.E. Tronrud, A.F. Monzingo, L.M. Weaver & B.W. Matthews (1987) Slow- and fast-binding inhibitors of thermolysin display different modes of binding: crystallographic analysis of extended phosphonamidate transition state analogues. *Biochemistry*, **26**, 8542–53.

B.W. Matthews (1988) Structural basis of the action of thermolysin and related zinc proteases. *Acc. Chem. Res.*, **21**, 333–40.

M.E. Sander & H. Witzel (1985) Direct chemical evidence for the mixed anhydride intermediate of carboxypeptidase A in ester and peptide hydrolysis. *Biochem. Biophys. Res. Commun.*, **132**, 681–7.

HIV protease

J. Erickson, D.J. Neidhart, J. VanDrie, D.J. Kempf, X.C. Wang, D.W. Norbeck, J.J. Plattner, J.W. Rittenhouse, M. Turon, N. Wideburg, W.E. Kohlbrenner, R. Simmer, R. Helfrich, D.A. Paul & M. Knigge (1990) Design, activity, and 2.8 Å crystal structure of a C_2 symmetric inhibitor complexed to HIV-1 protease. *Science*, **249**, 527–33.

L.J. Hyland, T.A. Tomaszek, Jr., G.D. Roberts, S.A. Carr, V.W. Magaard, H.L. Bryan, S.A. Fakhoury, M.L. Moore, M.G. Minnich, J.S. Culp, R.L. DesJarlais & T.D. Meek (1991) Human immunodeficiency virus-1 protease. 1. Initial velocity studies and kinetic characterization of reaction intermediates by [18]O isotope exchange. *Biochemistry*, **30**, 8441–53.

M. Jaskolski, A.G. Tomasselli, T.K. Sawyer, D.G. Staples, R.L. Heinrikson, J. Schneider, S.B.H. Kent & A. Wlodawer (1991) Structure at 2.5 Å resolution of chemically synthesised HIV type 1 protease complexed with a hydroxyethylene-based inhibitor. *Biochemistry*, **30**, 1600–609.

A. Wlodawer & J.W. Erickson (1993) Structure-based inhibitors of HIV-1 protease. *Annu. Rev. Biochem.*, **62**, 543–86.

Phosphoryl transfer

A.C. Hengge (2002) Isotope effects in the study of phosphoryl and sulfuryl transfer reactions. *Acc. Chem. Res.*, **35**, 105–12.

J.R. Knowles (1980) Enzyme-catalysed phosphoryl transfer reactions. *Annu. Rev. Biochem.*, **49**, 877–920.

Glycosyl transfer

M.L. Sinnott (1990) Catalytic mechanisms of enzymic glycosyl transfer. *Chem. Rev.*, **90**, 1171–202.
D.J. Vocadlo, G.J. Davies, R. Laine & S.G. Withers (2001) Catalysis by hen egg-white lysozyme proceeds via a covalent intermediate. *Nature*, **412**, 835–8.
D.L. Zechel & S.G. Withers (2000) Glycosidase mechanisms: anatomy of a finely tuned catalyst. *Acc. Chem. Res.*, **33**, 11–18.

One-carbon transfers

R.G. Matthews & J.T. Drummond (1990) Providing one-carbon units for biological methylations: mechanistic studies on serine hydroxymethyltransferase, methylene FH$_4$ reductase and methylene FH$_4$-homocysteine methyltransferase. *Chem. Rev.*, **90**, 1275–90.

6　Enzymatic Redox Chemistry

6.1　Introduction

Oxidation and reduction involves the transfer of electrons between one chemical species and another. Electron transfer processes are widespread in biological systems and underpin the production of biochemical energy in all cells. The ultimate source of energy for all life on Earth is sunlight, which is utilised by plants for the process of photosynthesis. Photosynthesis involves a series of high-energy electron transfer processes, converting sunlight energy into high-energy reducing equivalents, which are then used to drive biochemical processes. These biochemical processes lead ultimately to the fixation of carbon dioxide and the production of oxygen. Oxygen in turn serves a vital role for mammalian cellular metabolism as an electron acceptor.

Cells contain a number of intermediate electron carriers which serve as cofactors for enzymatic redox processes. In this chapter we shall examine some of these redox cofactors, and we shall also examine metallo-enzymes which utilise redox-active metal ions to harness the oxidising power of molecular oxygen.

First of all we need to define a scale to measure the effectiveness of these different electron carriers as oxidising or reducing agents. How do we measure quantitatively whether something is a strong or weak oxidising/reducing agent? We can measure the strength of an oxidising agent electrochemically by dissolving it in water and measuring the voltage required to reduce it (i.e. add electrons) to a stable reduced form. The voltage measured under standard conditions with respect to a reference half-cell is known as the *redox potential*. The reference half-cell is the reaction $2H^+ + 2e^- \rightarrow H_2$, whose redox potential is $-0.42\,V$ at pH 7.0. A strong oxidising agent will be reduced very readily, corresponding to a strongly positive redox potential; whilst a weak oxidising agent will be reduced much less readily, corresponding to a less positive or a negative redox potential. A scale showing the redox potential of some important biological electron carriers is shown in Figure 6.1. One obvious point is that the strongest available oxidising agent is oxygen, hence the large number of enzymes which use molecular oxygen either as a substrate (oxygenases) or as an electron acceptor (oxidases).

We can use redox potentials to work out whether a particular redox reaction will be thermodynamically favourable. For example, the enzyme lactate dehydrogenase catalyses the reduction of pyruvate to lactate, using nicotinamide adenine dinucleotide (NADH) as a redox cofactor. NAD^+ (the oxidised form of NADH) has a redox potential of $-0.32\,V$, whereas pyruvate has a redox

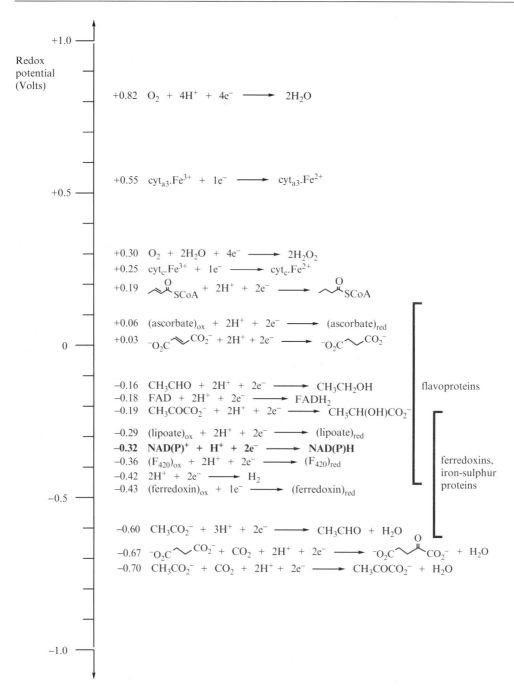

Figure 6.1 Some biologically important redox potentials.

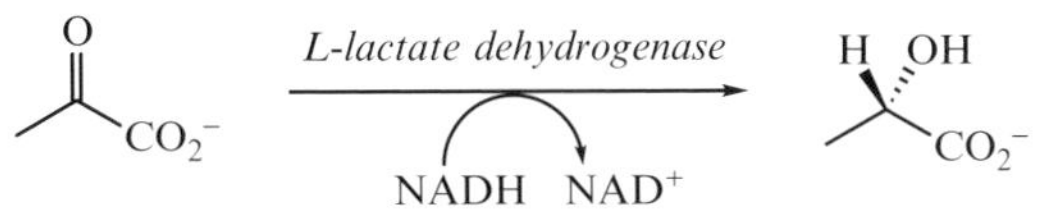

$$(1) \qquad NAD^+ + H^+ + 2e^- \longrightarrow NADH \qquad -0.32\ V$$

$$(2) \qquad CH_3COCO_2H + 2H^+ + 2e^- \longrightarrow CH_3CH(OH)CO_2H \qquad -0.19\ V$$

$$(2) - (1) \qquad CH_3COCO_2H + NADH \longrightarrow CH_3CH(OH)CO_2H + NAD^+ \qquad +0.13\ V$$

Figure 6.2 Redox potentials in the lactate dehydrogenase reaction.

potential of $-0.19\,V$. We simply subtract the redox potentials as shown in Figure 6.2, giving an overall redox potential difference for the pyruvate-to-lactate reaction of $+0.13\,V$. If the redox potential difference is above zero, then the reaction is thermodynamically favourable. The more positive the redox potential difference is, the more favourable the reaction is and the greater the equilibrium constant for the reaction (the equilibrium constant can be calculated from the redox potential difference using the Nernst equation – see physical chemistry texts). Remember that a highly positive redox potential means a strong oxidising agent, whilst a highly negative redox potential means a strong reducing agent.

One final point is that an enzyme cannot change the equilibrium constant of a chemical reaction that is thermodynamically unfavourable, however it can modify the redox potential of a cofactor *bound at its active site* by selectively stabilising either the oxidised or the reduced form of the cofactor. For example, the redox potential for free oxidised riboflavin is $-0.20\,V$, whereas redox potentials of between -0.45 and $+0.15\,V$ have been measured for flavo-enzymes in general. Just as with amino acid side chain pK_a values, enzymes are able to modify chemical reactivity at their active sites using subtle changes in micro-environment.

6.2   Nicotinamide adenine dinucleotide-dependent dehydrogenases

The first class of redox enzymes that we shall meet are the dehydrogenases (Table 6.1). These enzymes transfer two hydrogen atoms from a reduced substrate, which is usually an alcohol, to an electron acceptor. The electron acceptor in these enzymes is the coenzyme NAD, whose oxidised and reduced forms are shown in Figure 6.3.

The redox-active part of this coenzyme is the nicotinamide heterocyclic ring. In the oxidised form NAD^+ this is a pyridinium salt, which is reduced to a 1,4-dihydro-pyridine in the reduced form NADH. A phosphorylated version of the cofactor is also found, bearing a $2'$-phosphate on the adenosine portion of the structure, which is written as $NADP^+$ in the oxidised form and NADPH

Table 6.1 Classes of redox enzymes.

Transformation	Enzyme class	Redox cofactors
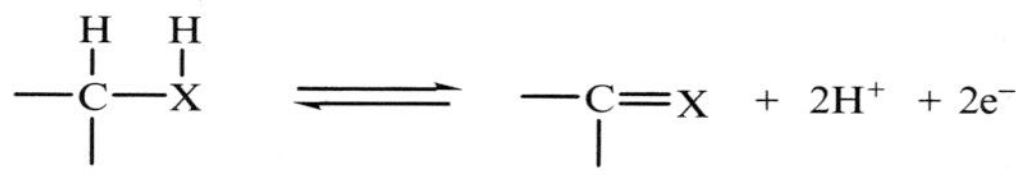 (H, H–X on C) $-\overset{H}{\underset{H}{C}}-X \;\rightleftharpoons\; -\overset{}{\underset{}{C}}=X \;+\; 2H^+ + 2e^-$	Dehydrogenases	(1) NAD^+ (X = O, NH) (2) flavin/$1e^-$ acceptors (X = O, NH, CHR)
	Oxidases	Flavin / O_2 (X = NH)
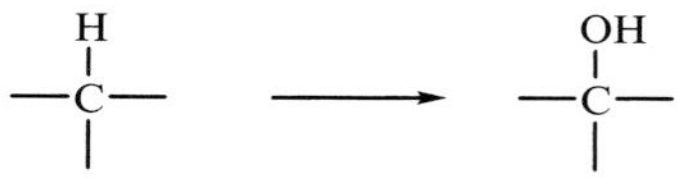 $-\overset{H}{\underset{}{C}}- \;\rightarrow\; -\overset{OH}{\underset{}{C}}-$	Mono-oxygenases, hydroxylases	Aromatic C–H: (1) Flavin/O_2/NADPH (2) Pterin/O_2/NADPH Aliphatic or aromatic C–H: (1) P_{450}/O_2/NADH (2) Non-haem metal/O_2/e^- donor (3) Fe^{2+}/O_2/α-ketoglutarate
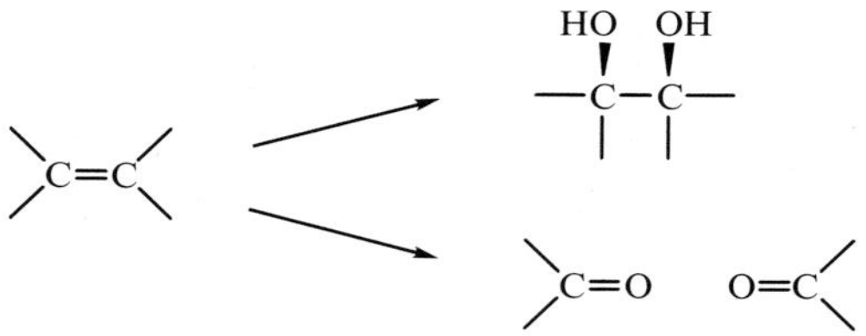 $C=C \;\rightarrow\; \overset{HO\ \ OH}{-\overset{}{C}-\overset{}{C}-}$	Dioxygenases (dihydroxylating)	Fe^{2+}/O_2/flavin/NADH
$C=C \;\rightarrow\; C=O \quad O=C$	Dioxygenases (C–C cleaving)	Fe^{2+} or Fe^{3+}/O_2
$A\overset{H}{\underset{H}{\big\langle}} \;\rightarrow\; A \;+\; 2H^+ + 2e^-$	Peroxidases	(1) P_{450}/H_2O_2 (2) Non-haem metal/H_2O_2
$H–H \;\rightarrow\; 2H^+ + 2e^-$	Hydrogenases	Ni^{2+}/factor F_{420}

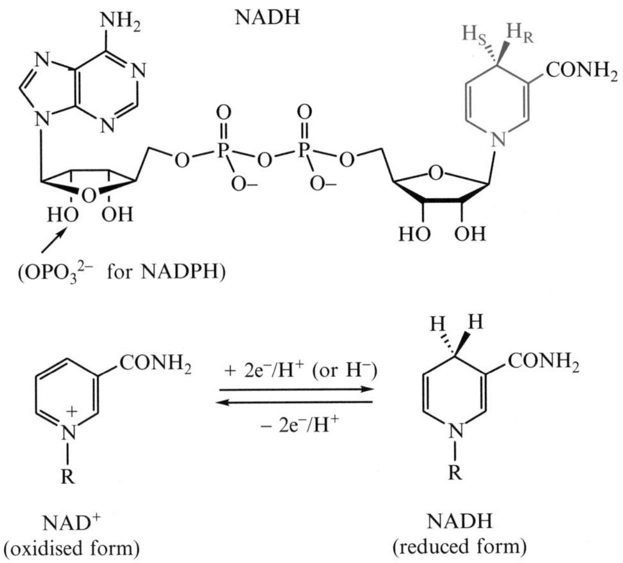

Figure 6.3 Structures of NAD$^+$ and NADH.

in the reduced form. Dehydrogenase enzymes which utilise this cofactor are usually specific either for NAD$^+$/NADH or for NADP$^+$/NADPH. NAD$^+$ and NADH tend to be utilised for catabolic processes (which break down cellular metabolites into smaller pieces), whereas NADP$^+$ and NADPH tend to be used for biosynthetic processes.

The redox potential for NAD$^+$ is -0.32 V, consequently NADH is the most powerful of the commonly available biological reducing agents. NAD$^+$ is, therefore, a relatively weak oxidising agent. NADH is most commonly used for the reduction of ketones to alcohols, although it is also sometimes used for reduction of the carbon–carbon double bond of an α,β-unsaturated carbonyl compound, as shown in Figure 6.4.

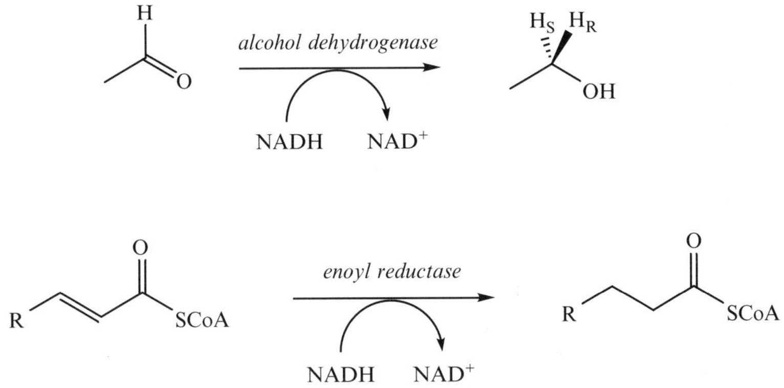

Figure 6.4 Examples of NAD$^+$-dependent dehydrogenases.

An important point is that NADH is a stoichiometric reagent which is bound non-covalently by the enzyme at the start of the reduction reaction and released as NAD$^+$ at the end of the reaction. The concentration of NADH in solution, therefore, decreases as an enzyme-catalysed reduction proceeds, which allows the reaction to be monitored by UV spectroscopy, since NADH has a strong absorption at 340 nm ($\varepsilon = 6.3 \times 10^3$ M^{-1} cm^{-1}).

We have already met the example of horse liver alcohol dehydrogenase, which catalyses the NAD$^+$-dependent oxidation of ethanol to acetaldehyde (see Section 4.4). In a classic series of experiments by Westheimer, this enzyme was shown to be entirely stereospecific for the removal of prochiral hydrogen atoms in both the substrate and the cofactor. These experiments are illustrated in Figure 6.5.

Incubation of enzyme with CH$_3$CD$_2$OH followed by re-isolation of oxidised NAD$^+$ revealed no deuterium incorporation into the oxidised cofactor, demonstrating that the deuterium atom transferred to the cofactor was stereospecifically removed in the reverse reaction. Isolation of the deuterium-containing reduced cofactor (NADD) followed by chemical conversion to a substance of established configuration revealed that the reduced cofactor had the 4R stereochemistry. Incubation of acetaldehyde and [4R-^{2}H]-NADD with enzyme for prolonged reaction times gave no exchange of the deuterium label into acetaldehyde, indicating that the enzyme was also stereoselective for the removal of a hydrogen atom from C-1 of ethanol.

Incubation with specifically deuteriated substrates subsequently revealed that the enzyme removes specifically the *proR* hydrogen, which is transferred to the C-4 *proR* position of NAD$^+$. The mechanism is thought to proceed via direct hydride transfer, as shown in Figure 6.6. The active site of horse liver alcohol dehydrogenase contains a Zn^{2+} cofactor which activates the substrate for nucleophilic attack. Protonation of the reduced carbonyl group is carried out by a dyad of Ser-48 and His-51, as illustrated in Figure 6.7.

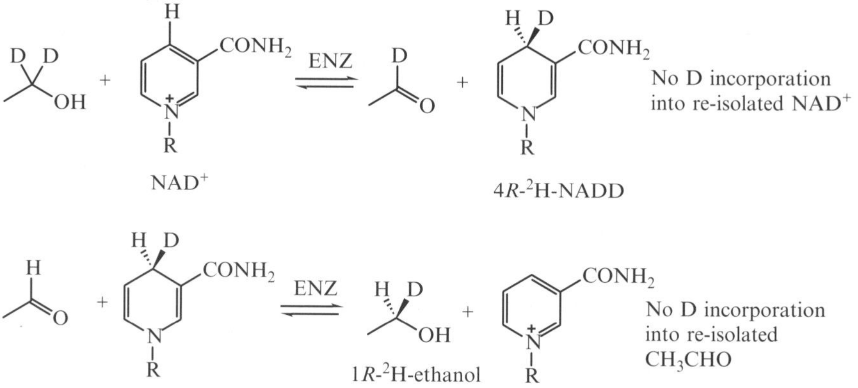

Figure 6.5 Stereospecificity of alcohol dehydrogenase. NADD, deuterium-containing reduced NADH.

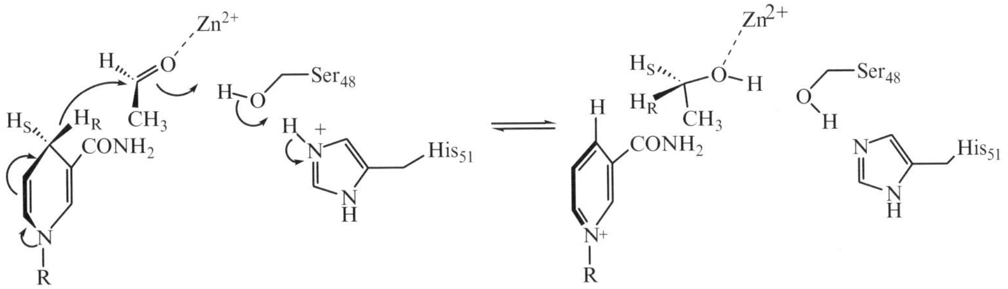

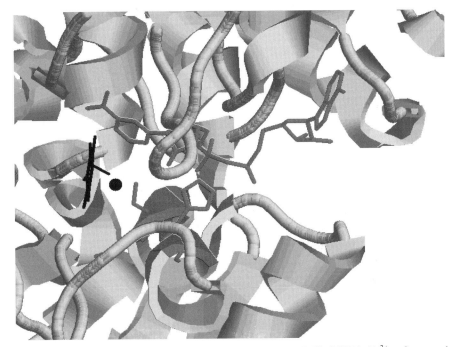

Figure 6.6 Mechanism for alcohol dehydrogenase.

Figure 6.7 Active site of horse liver alcohol dehydrogenase (PDB file 1HLD). Zn^{2+} cofactor and pentafluorobenzyl alcohol substrate shown in black; NAD^+ cofactor, Ser-48, and His-51 shown in red.

Although hydride transfer might at first sight seem unlikely in aqueous solution, note that this transfer is occurring at close proximity within the confines of an enzyme active site. There is also precedent for organic reactions involving hydride transfer occurring in aqueous solution, such as the Cannizzaro reaction shown in Figure 6.8.

Deuterium-labelling studies carried out on alcohol dehydrogenase have established that the same hydrogen atom abstracted from ethanol is transferred to the nicotinamide ring. It has also been established, using kinetic isotope effect studies, that there is a single transition state for the enzyme-bound

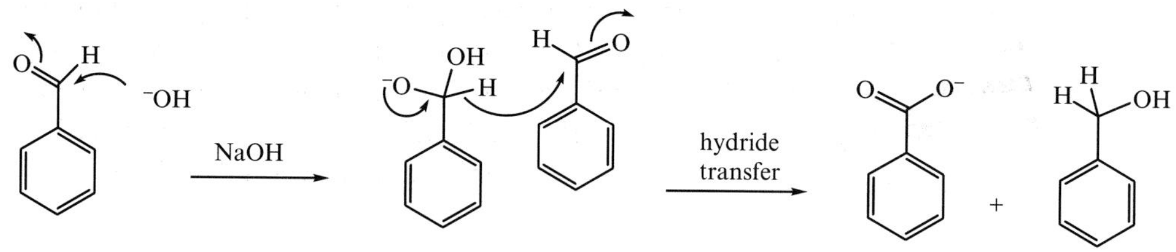

Figure 6.8 The Cannizzaro reaction.

reaction. Although radical intermediates are in theory possible, attempts to identify such intermediates have been unsuccessful, so hydride transfer via a single transition state is the accepted mechanism.

Dehydrogenase enzymes are each specific for one of the enantiotopic C-4 protons of NADH: some are specific for the 4R hydrogen; others are specific for the 4S hydrogen. Stereospecificity of an NAD-dependent dehydrogenase can be readily determined as shown in Figure 6.9. [1-^{2}H$_2$]-Ethanol is incubated with alcohol dehydrogenase and the [4R-^{2}H]-NADD product isolated. This is then incubated with the dehydrogenase of unknown stereospecificity, and the incorporation of ^{1}H or ^{2}H into the reduced product is monitored. The experiment can be done more sensitively using a ^{3}H label in a similar way.

While this stereospecificity is largely due to the orientation of the cofactor in the enzyme active site, there are indications that a further stereoelectronic effect may also be important. Model studies using dihydropyridine rings contained in molecules in which the two C-4 hydrogens are diastereotopic have shown that the environments of the two hydrogens are quite different (see Figure 6.10). This

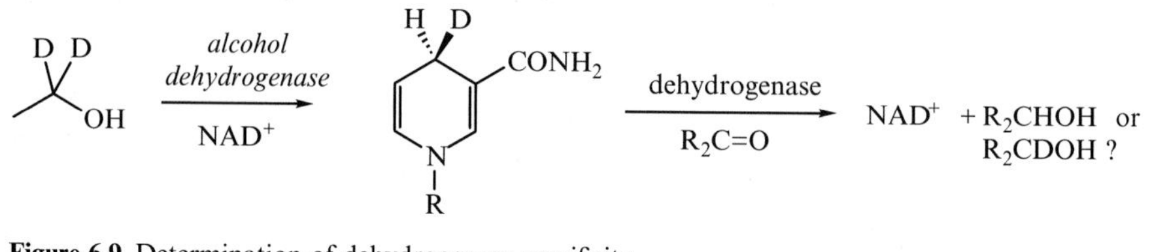

Figure 6.9 Determination of dehydrogenase specificity.

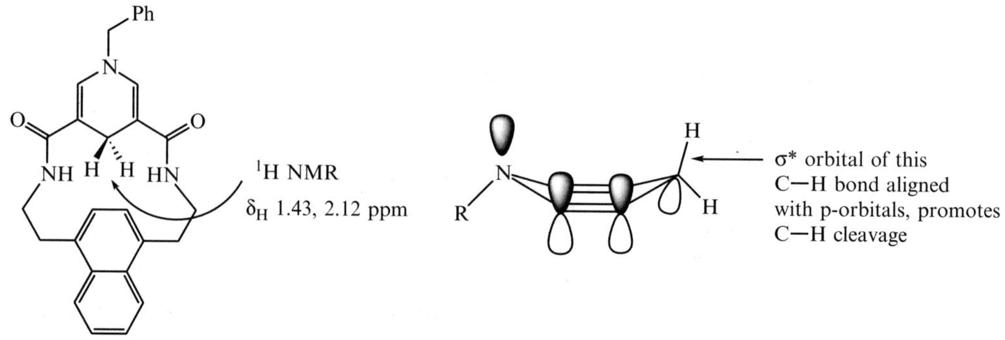

Figure 6.10 Puckering of the dihydropyridine ring of NADH.

can be explained by a puckering of the dihydropyridine ring, causing one of the two hydrogens to adopt a pseudo-axial orientation. This C−H bond is aligned more favourably with the *p*-orbitals of the dihydropyridine ring and is, therefore, stereoelectronically better aligned for hydride transfer. It seems likely that dehydrogenase enzymes also use this stereoelectronic effect to assist the reaction.

6.3 Flavin-dependent dehydrogenases and oxidases

Riboflavin was first isolated from egg white in 1933 as a vitamin whose deficiency causes skin lesions and dermatitis. The most striking property of riboflavin is its strong yellow–green fluorescence, a property conveyed onto those enzymes which bind this cofactor. The structure of riboflavin consists of an isoalloxidine heterocyclic ring system, which is responsible for its redox activity. Attached to N-10 is a ribitol side chain, which can be phosphorylated in the case of flavin mononucleotide (FMN) or attached through a diphosphate linkage to adenosine in flavin adenine dinucleotide (FAD), as shown in Figure 6.11.

The first important difference between NAD and flavin is that enzymes which use flavin bind it very tightly, sometimes covalently (attached to Cys or His through C-8a), such that it is not released during the enzymatic reaction but remains bound to the enzyme throughout. Consequently the active form of the cofactor must be regenerated at the end of each catalytic cycle by external redox reagents. The next important difference is that flavin can exist either as oxidised FAD, or reduced $FADH_2$, or as an intermediate semiquinone radical species $FADH^{\bullet}$, as shown in Figure 6.12. Therefore flavin is able to carry out

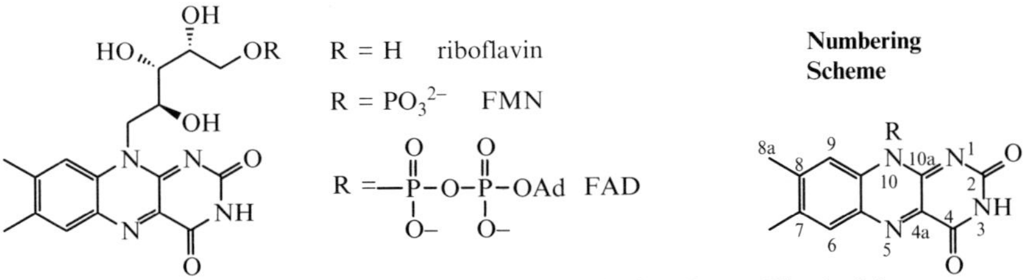

- occasionally covalently attached to Cys or His via C-8a.

Figure 6.11 Structures of flavin redox cofactors.

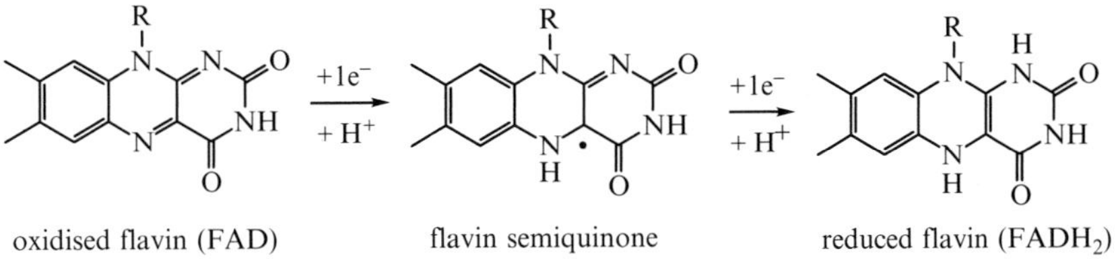

oxidised flavin (FAD) flavin semiquinone reduced flavin ($FADH_2$)

Figure 6.12 Redox states of riboflavin.

one-electron transfer reactions, whereas NAD is restricted to two-electron hydride transfers. This seemingly minor point has far-reaching consequences, since it allows flavin to react with the most powerful oxidising agent in biological systems: molecular oxygen.

In the reactions of flavin-dependent dehydrogenases and oxidases, a pair of hydrogen atoms is transferred from the substrate to the flavin nucleus, generating reduced $FADH_2$ (or $FMNH_2$). A few examples of reactions catalysed by these enzymes are shown in Figure 6.13. Since oxidised FAD is required for the next catalytic cycle, the enzyme-bound $FADH_2$ must be oxidised *in situ*. In the flavin-dependent dehydrogenases this is done by external oxidants, which *in vivo* are electron carriers such as cytochromes. *In vitro* the reduced flavin can be oxidised by chemical oxidants such as benzoquinone.

In the case of the flavin-dependent oxidases the regeneration of oxidised flavin is carried out by molecular oxygen, which is reduced to hydrogen peroxide. Since the ground state of molecular oxygen contains two unpaired electrons (in its $\pi_{2px,y}$ molecular orbitals) it is spin-forbidden to react with species containing paired electrons. However, reduced flavin is able to transfer a single electron to dioxygen to give superoxide and flavin semiquinone. Recombination of superoxide with the flavin semiquinone followed by fragmentation of the peroxy adduct generates oxidised flavin and hydrogen peroxide as shown in Figure 6.14.

How does the flavin nucleus carry out the dehydrogenation reaction? A number of possible mechanisms have been proposed for flavin-catalysed dehydrogenation, which have been much debated in the literature. The four main possibilities are illustrated in Figure 6.15: hydride transfer from the

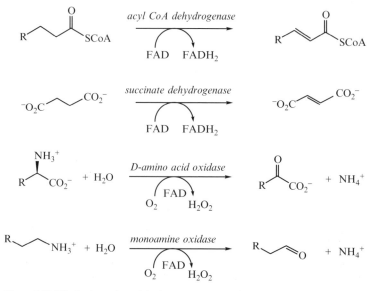

Figure 6.13 Flavin-dependent dehydrogenases and oxidases.

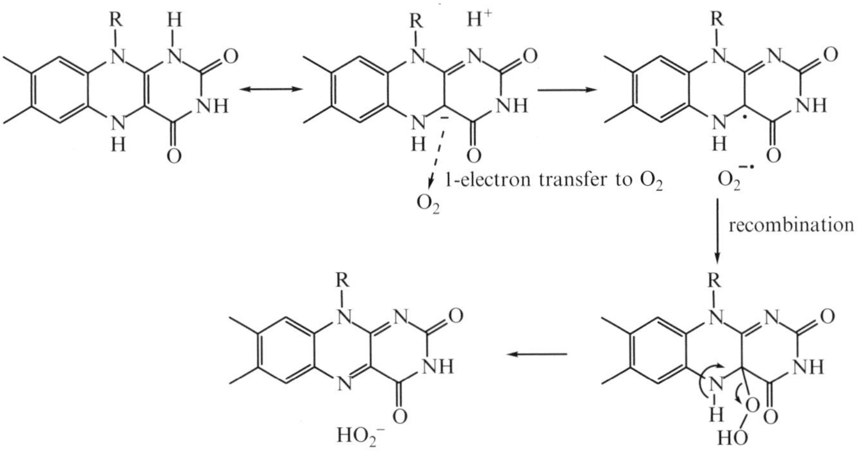

Figure 6.14 Regeneration of oxidised flavin by molecular oxygen.

substrate to oxidised flavin (a); nucleophilic attack by the substrate on the flavin nucleus, either by a heteroatom at C-4a (b), or by a substrate carbanion at N-5 (c); radical mechanisms involving single electron transfer (d).

Nucleophilic attack on the flavin nucleus (Figure 6.15b) is well precedented in flavin model systems, but there is no firm evidence for the existence of covalent substrate–flavin adducts in flavin-dependent dehydrogenases. We shall examine the available evidence for the enzymes acyl coenzyme A (CoA) dehydrogenase and monoamine oxidase.

There is good evidence in the reaction of acyl CoA dehydrogenase that the first step of the mechanism involves deprotonation of the substrate adjacent to the thioester carbonyl, since the enzyme catalyses the exchange of a C_α proton with ^2H_2O. However, there is also evidence for an α-radical intermediate, since the enzyme is inactivated by a substrate analogue containing a β-cyclopropyl group. This inhibitor is a metabolite of hypoglycin A, which is the causative agent of Jamaican vomiting sickness. The proposed mechanism of inactivation is shown in Figure 6.16. Upon formation of the α-radical a rapid opening of the strained cyclopropyl ring takes place, giving a rearranged radical intermediate which recombines with the flavin semiquinone irreversibly.

A possible mechanism for acyl CoA dehydrogenase consistent with both observations is shown in Figure 6.17. Deprotonation to form an α-carbanion is followed by a one-electron transfer from the substrate to FAD, generating a substrate radical and the flavin semiquinone. The reaction can be completed by abstraction of H$^\bullet$ at the β-position by the flavin radical, generating reduced $FADH_2$. In support of the final proposed step in the mechanism there are isolated examples of flavo-enzymes which catalyse hydrogen atom transfer from $FADH_2$ to the β-position of α,β-unsaturated carbonyl substrates (see Problem 4).

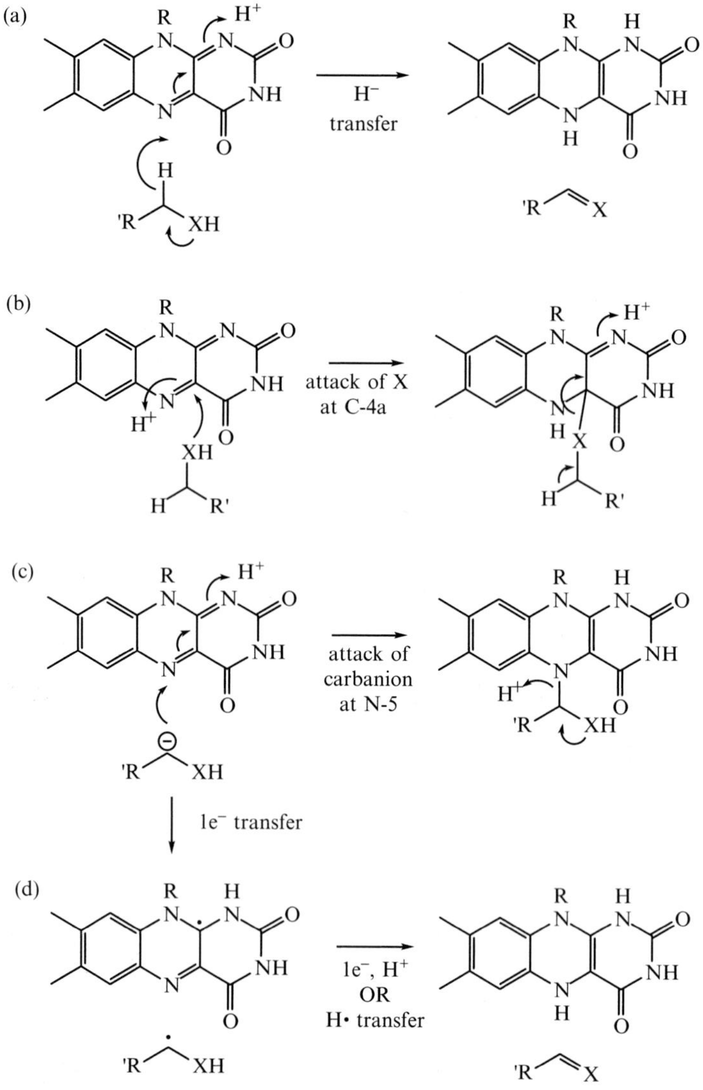

Figure 6.15 Possible mechanisms for flavin-catalysed dehydrogenation. (a) Hydride transfer from the substrate to oxidised flavin; nucleophilic attack by the substrate on the flavin nucleus by a heteroatom at C-4a (b); by a substrate carbanion at N-5 (c); (d) radical mechanism involving single electron transfer.

The monoamine oxidase reaction is a very important medicinal target for the mammalian nervous system, since it is responsible for the deamination of several neuro-active primary amines. Inhibitors of monoamine oxidase therefore have potential clinical applications as antidepressants. Monoamine oxidase is rapidly inactivated by *trans*-2-phenylcyclopropylamine (marketed as the antidepressant Tranylcypromine) via a radical mechanism shown in Figure 6.18. This behaviour indicates that an amine radical cation intermediate is

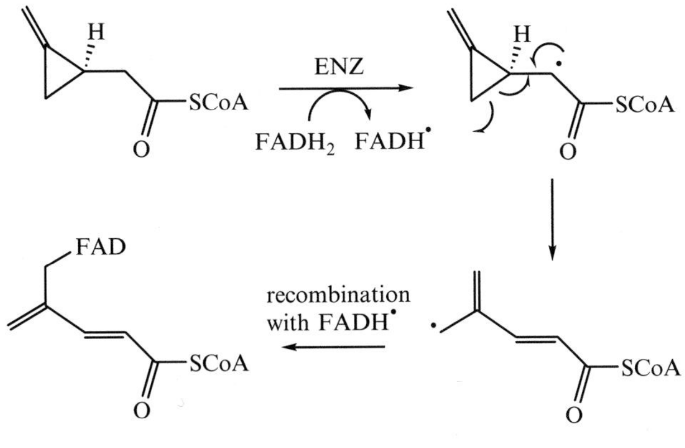

Figure 6.16 Inactivation of acyl CoA dehydrogenase by hypoglycin A.

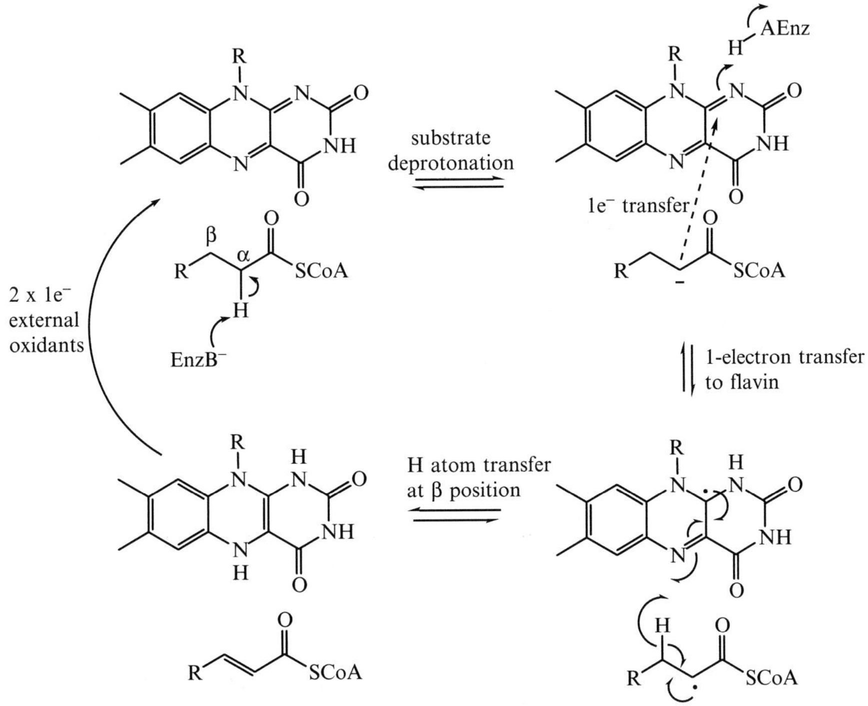

Figure 6.17 Probable mechanism for acyl CoA dehydrogenase.

formed in the mechanism via single electron transfer to FAD. This amine radical cation undergoes a further one-electron oxidation to give an iminium ion, which is then hydrolysed to give the corresponding aldehyde, as shown in Figure 6.18. The second one-electron transfer could proceed either by hydrogen atom transfer (i.e. loss of H• from the α-position, route *a*) or by loss of a proton from the α-carbon followed by single electron transfer (route *b*). There is evidence from further 'radical trap' inhibitors for the existence of an α-radical intermediate, consistent with route *b*.

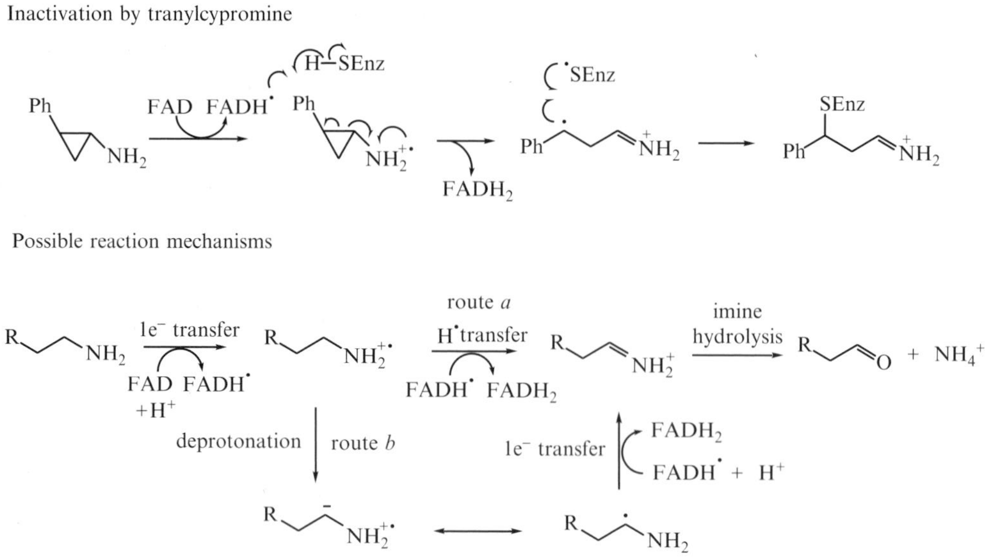

Inactivation by tranylcypromine

Possible reaction mechanisms

Figure 6.18 Monoamine oxidase inactivation and reaction mechanisms.

Thus, both the flavin-dependent dehydrogenases and oxidases appear to follow radical mechanisms involving single electron transfers to flavin. In the case of the amine oxidases, there is good chemical precedent for this type of mechanism in the form of electrochemical single electron oxidation of amines to the corresponding imines. Similar radical mechanisms can be written for succinate dehydrogenase and D-amino acid oxidase. Note that in acyl CoA dehydrogenase deprotonation occurs adjacent to a thioester carbonyl to form a stabilised enolate anion, whereas in succinate dehydrogenase deprotonation would occur adjacent to a carboxylate anion. As discussed in Section 7.3, a proton adjacent to an ester or a carboxylic acid is much less acidic than a proton adjacent to a ketone or thioester. The mechanism by which this problem is resolved is not well understood, but may involve protonation of the carboxylate and/or electrostatic stabilisation of the carbanion intermediate. This issue is also relevant to the cofactor-independent amino acid racemases which will be discussed in Section 10.2.

6.4   Flavin-dependent mono-oxygenases

The ability of flavin to react with molecular oxygen, seen above with the oxidases, also makes possible a number of *mono-oxygenase* reactions, in which one atom from molecular oxygen is incorporated into the product. The most common flavin-dependent mono-oxygenases are the phenolic hydroxylases, in which a hydroxyl group is inserted into the *ortho-* or *para-* position of a phenol (or aniline). As in the flavin-dependent oxidases, oxygen reacts with reduced FADH$_2$ to form a peroxy flavin adduct, which then acts as an electrophilic species for attack of the phenol. The mechanism for *para*-hydroxybenzoate hydroxylase is shown in Figure 6.19. The structure of *p*-hydroxybenzoate hydroxylase is shown in Figure 6.20.

In this case FADH$_2$ is generated from FAD by reduction with NADPH. Presumably the mechanism for this reduction involves hydride transfer to either N-1 or N-5 of the flavin nucleus. Thus flavin has the ability to accept two electrons from NADPH, but then transfer them via two one-electron transfers. It is apparent from this mechanism why hydroxylation occurs *ortho*- or *para*- to a phenol, because the phenolic hydroxyl group activates the *ortho*- and *para*-positions for nucleophilic attack.

There is also at least one example of an enzyme in which the flavin hydroperoxide intermediate can act as a nucleophile, rather than an electrophile. This is the enzyme cyclohexanone mono-oxygenase, which catalyses the oxidation of cyclohexanone to the corresponding seven-membered lactone. This reaction is analogous to the well-studied Baeyer–Villiger oxidation which uses a peracid to achieve the same transformation. A likely mechanism for this reaction is shown in Figure 6.21. Again NADPH is used to generate reduced flavin for reaction with oxygen.

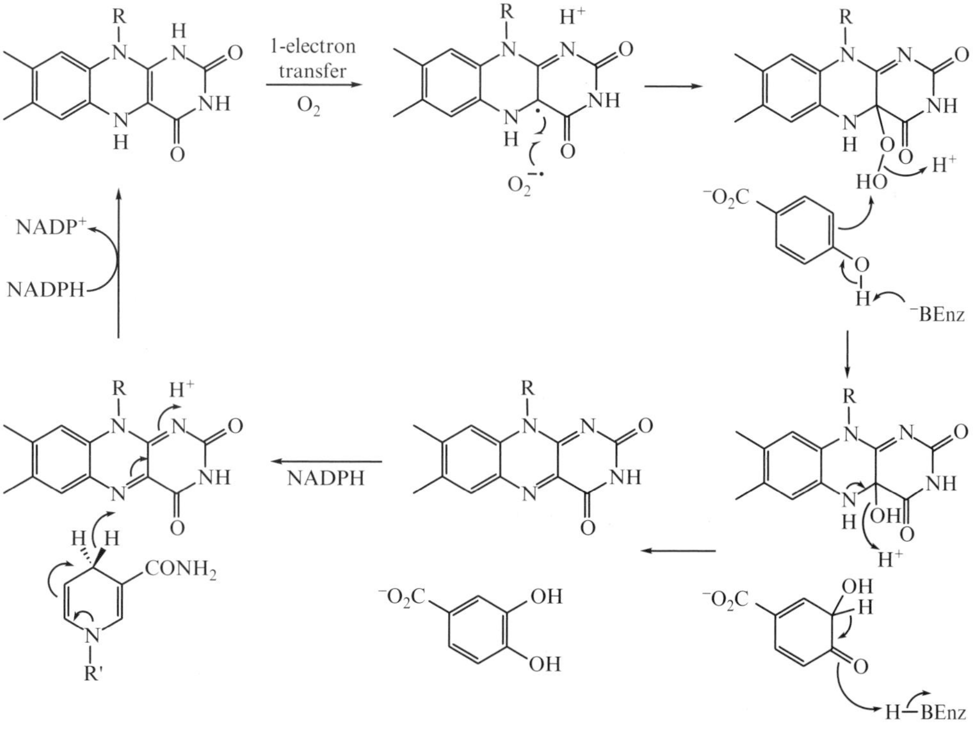

Figure 6.19 Mechanism for *p*-hydroxybenzoate hydroxylase.

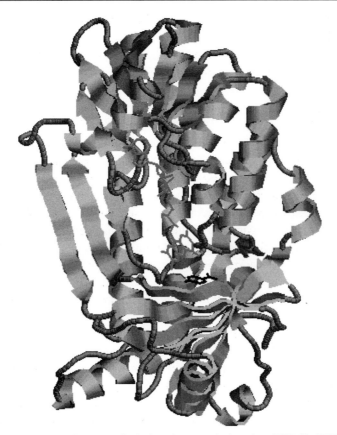

Figure 6.20 Structure of *p*-hydroxybenzoate hydroxylase (PDB file 1IUT). FAD cofactor shown in red; *p*-aminobenzoate substrate analogue shown in black.

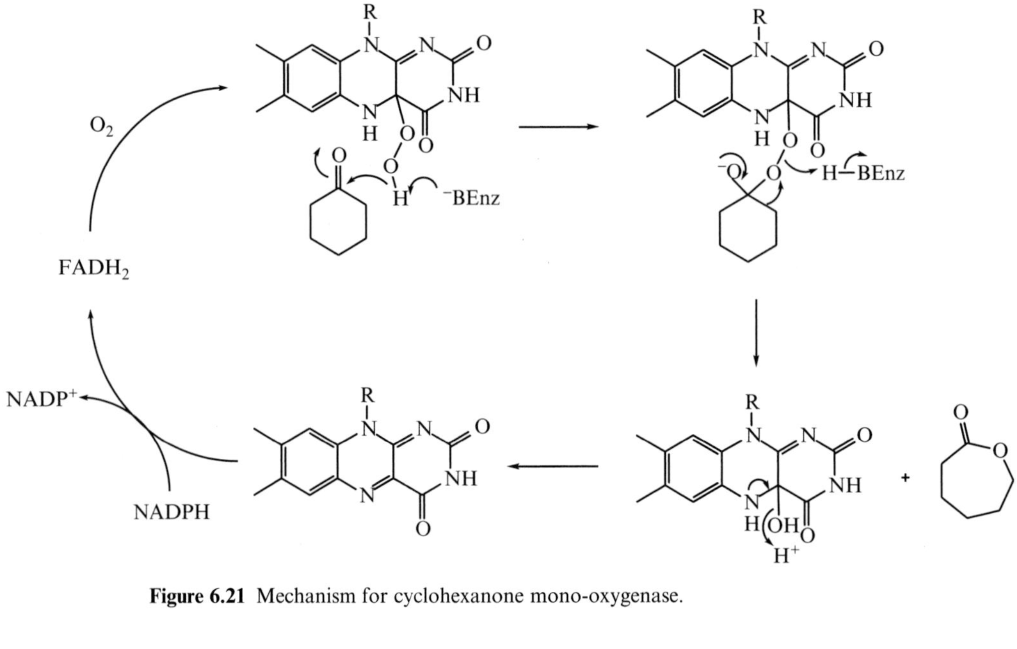

Figure 6.21 Mechanism for cyclohexanone mono-oxygenase.

Cyclohexanone mono-oxygenase is the enzyme used by *Pseudomonas* for the biodegradation of aliphatic hydrocarbons containing cyclohexane rings. The lactone product is subsequently hydrolysed and broken down into small fragments which can be utilised for growth.

6.5   CASE STUDY: Glutathione and trypanothione reductases

Glutathione is a tripeptide L-Glu-γ-L-Cys-Gly which is found at concentrations of about 1 mM in all mammalian cells. Its function is to protect the cell against 'oxidative stress', or the presence of activated oxygen species which would otherwise have harmful effects upon the cell. In the course of its reaction with these activated oxygen species, the thiol side chain of glutathione (GSH) is converted into an oxidised disulphide (GSSG). Reduced glutathione is regenerated in the cell by the enzyme glutathione reductase (see Figure 6.22), which uses NADPH as reducing equivalent and contains one equivalent of tightly bound FAD.

Glutathione reductase is a dimer of identical 50-kDa protein sub-units. Each sub-unit contains FAD- and NADPH-binding domains which are composed of a βαβαβ secondary structural motif common to many nucleotide-binding proteins. The active site of the enzyme lies at the interface of the two protein sub-units. On one face of the flavin cofactor is the NADPH cofactor responsible for generation of reduced $FADH_2$, as shown in Figure 6.23a. On the other face of the flavin cofactor are two active site cysteine residues (Cys-46 and Cys-41) which form an active site disulphide at the start of the catalytic cycle.

Figure 6.22 Glutathione reductase.

(a)

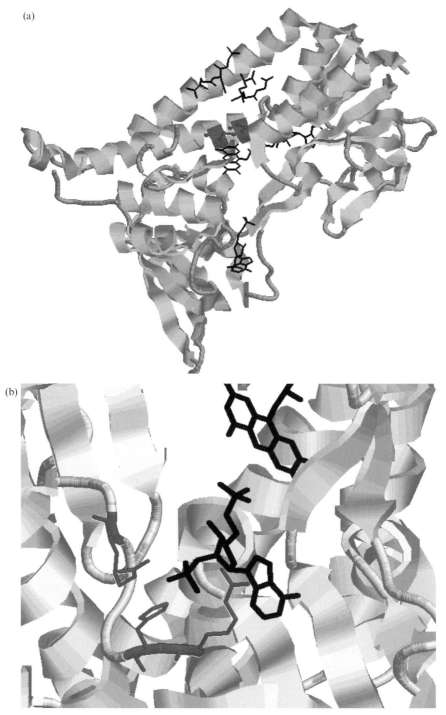

(b)

Figure 6.23 Structure of human glutathione reductase (PDB file 1GRA). (a) Overall structure; shown in black are reduced glutathione substrate (top), FAD cofactor (centre) and nicotinamide diphosphate (bottom); Cys-41 and Cys-46 shown in red. (b) Binding of 2′-phosphate of NADP (black) by Arg-218, His-219 and Arg-224 (shown in red).

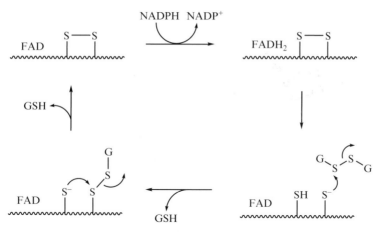

Figure 6.24 Mechanism for glutathione reductase.

The mechanism of the glutathione reductase catalytic cycle is shown in Figure 6.24. NADPH reduces the bound flavin to $FADH_2$, which in turn reduces the active site disulphide into two reduced cysteine residues. Attack of a free cysteine thiol onto the disulphide linkage of oxidised glutathione generates one equivalent of reduced glutathione. Attack of the second free cysteine thiol generates a second equivalent of reduced glutathione and regenerates the active site disulphide.

A high-resolution X-ray crystal structure for the human glutathione reductase has made possible extensive protein engineering studies on the glutathione reductase active site. Close examination of the NADPH binding site reveals that the 2'-phosphate of NADPH is bound by three positively charged residues: Arg-218, His-219 and Arg-224 (see Figure 6.23b). Arg-218 and Arg-224 are strictly conserved in the amino acid sequences of other flavoprotein disulphide oxido-reductases that use NADPH. However, the amino acid sequence of NADH-specific lipoamide dehydrogenase contains methionine and proline, respectively, at these positions, which are incapable of forming electrostatic interactions.

Mutation of either of these two arginine residues in the *Escherichia coli* glutathione reductase enzyme (to methionine and leucine, respectively) gave mutant enzymes whose k_{cat}/K_M values for NADPH were reduced by approximately 100-fold, whilst a mutant enzyme containing both mutations had a 500-fold reduced k_{cat}/K_M. Mutation of four additional residues identified in the NADPH binding site to the corresponding residues in the NADH-specific lipoamide dehydrogenase gave a mutant enzyme with an eight-fold preference for NADH over NADPH. The wild type enzyme in contrast has a 2000-fold preference for NADPH over NADH. This type of study shows that enzyme characteristics such as cofactor specificity can in principle be rationally modified using protein engineering.

A closely related enzyme trypanothione reductase (TR) has been found in *Trypanosoma* and *Leishmania* parasites which cause human diseases such as sleeping sickness and Chagas' disease. These parasites use a modified form of glutathione called trypanothione in which the C-terminal glycine carboxylates are connected by a spermidine linker, as shown in Figure 6.25.

Examination of the amino acid sequence of the *Trypanosoma congolense* enzyme revealed that three of the amino acid residues involved in substrate binding in human glutathione reductase are modified in the parasite enzyme: Arg-347 (found as alanine in TR), Ala-34 (found as glutamine in TR) and Arg-37 (found as tryptophan in TR). The location of these residues is shown in Figure 6.26. Mutation of these three residues in the parasite TR to the

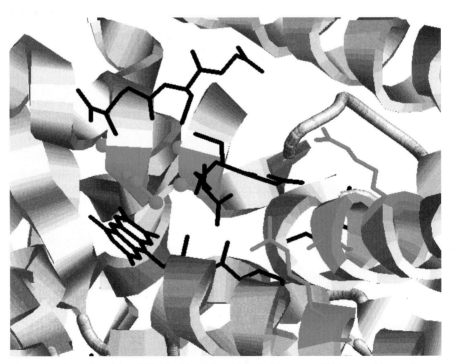

oxidised trypanothione   reduced trypanothione

Figure 6.25 Trypanothione reductase.

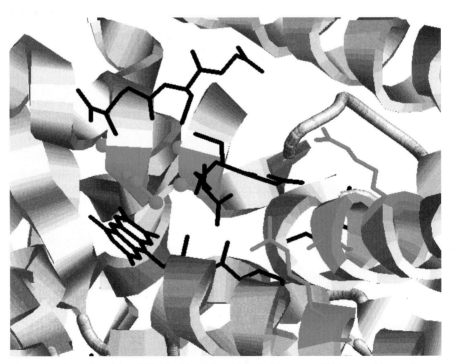

Figure 6.26 Location of Arg-347, Ala-34 and Arg-37 (red, right hand side) in glutathione reductase active site, in relation to reduced glutathione (black), FAD (black, top) and Cys 41/46 (red, ball-and-stick). Picture prepared using RASMOL.

corresponding residues in the human glutathione reductase gave a mutant enzyme with 10^3-fold lower trypanothione reductase activity and 10^4-fold higher glutathione reductase activity. The specificity of the parasite enzyme for trypanothione over glutathione offers the potential for selective anti-parasite activity via inhibition of trypanothione reductase.

6.6   Deazaflavins and pterins

Nicotinamide and riboflavin are by far the most common carbon-based redox cofactors used by enzymes, but there are several other heterocyclic redox cofactors used by particular enzymes.

Factor F_{420} was isolated in 1978 from methanogenic bacteria (strictly anaerobic bacteria which ferment acetate to methane and carbon dioxide) in yields of up to 100 mg per kg of cells. It was found to have a structure similar to that of riboflavin, except that N-5 is replaced by a carbon atom. The discovery of this deazaflavin prompted an investigation into the properties of other deaza-analogues of riboflavin, which are shown in Figure 6.27. Both factor F_{420} and 5-deazaflavin have much lower redox potentials than riboflavin itself, but

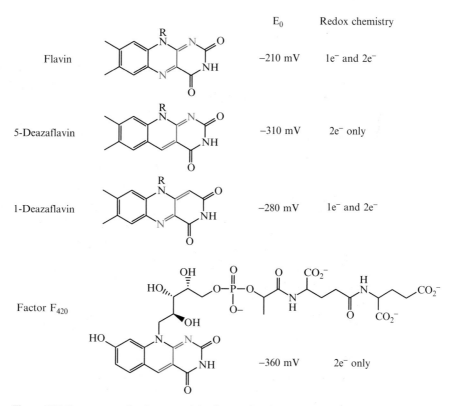

Figure 6.27 Structures and redox potentials of natural and synthetic deazaflavins.

neither has a stable semiquinone form, so are capable of only two-electron transfers. In these respects their chemistry is more similar to nicotinamide than riboflavin. 1-Deazaflavin on the other hand behaves more like riboflavin in terms of its redox potential and its ability to carry out one-electron redox chemistry.

Factor F_{420} is used as a cofactor in a Ni^{2+}-dependent hydrogenase enzyme found in methanogenic bacteria which uses hydrogen gas to reduce carbon dioxide to methane. Its role appears to be as one component of a complex chain of electron carriers in this multi-enzyme complex. The redox potential of F_{420} is ideally suited for its role in this enzyme, since at $-0.36\,V$ it is higher than the redox potential for hydrogen ($-0.42\,V$) but lower than the redox potentials of other redox cofactors such as NADH and flavin. Therefore F_{420} is able to accept electrons from hydrogen and transfer them to NAD^+ or FAD.

The pterin cofactor is used in a number of redox enzymes, in particular a small family of mono-oxygenase enzymes which hydroxylate aromatic rings. Phenylalanine hydroxylase catalyses the conversion of L-phenylalanine to L-tyrosine, using tetrahydropterin as a cofactor. The enzyme incorporates one atom of oxygen from dioxygen into the product, similar to the flavin-dependent mono-oxygenases. However, unlike the flavin-dependent mono-oxygenases, there is no hydroxyl group present in the *ortho-* or *para-* positions of the substrate.

Conversion of phenylalanine labelled with deuterium at C-4 of the ring by phenylalanine hydroxylase gives 3-^2H-tyrosine, indicating that a 1,2-shift (known historically as the 'NIH shift', since this startling result was discovered at the National Institute of Health research laboratories) is taking place during the reaction. It is likely that a pterin hydroperoxide is formed upon reaction with dioxygen, as found with flavin. The mammalian phenylalanine hydroxylase requires iron(II) for activity, and it is believed that a high-valent iron-oxo species is formed which carries out substrate hydroxylation. The NIH shift could result upon formation either of an epoxide intermediate or a carbonium ion intermediate, shown in Figure 6.28. Although rearrangement is observed with the $[4\text{-}^2H]$ substrate, there is no kinetic isotope effect observed with this substrate, implying that the rearrangement occurs after the rate-determining step of the reaction.

6.7   Iron–sulphur clusters

We have seen in the case of flavin how single electron transfer is an important process in biological systems. The most common type of one-electron carrier found in biological systems is the family of iron–sulphur clusters. They are inorganic clusters of general formula $(FeS)_n$, where n is commonly 2 or 4. The [2Fe2S] and [4Fe4S] clusters shown in Figure 6.29 are commonly found in biological electron carriers known as ferredoxins, and are also found in a number of redox enzymes. They have the ability to accept a single electron from a single electron donor such as flavin, and transfer the single electron to

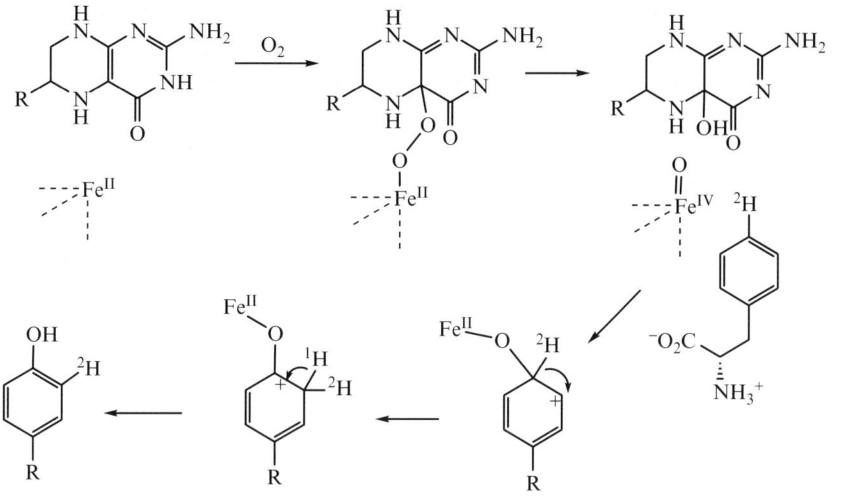

Figure 6.28 Mechanism for phenylalanine hydroxylase involving iron(IV)-oxo species.

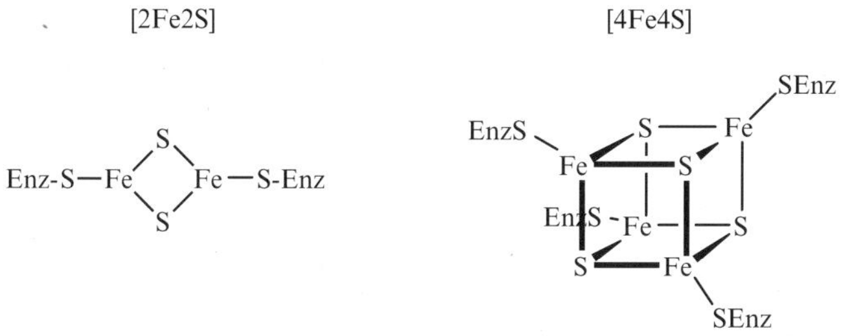

Figure 6.29 Iron–sulphur clusters.

another electron carrier or to the active site of a redox enzyme. We shall meet one such example in Section 6.10. Redox potentials for ferredoxins are typically in the range -0.2 to $-0.6\,V$ (see Figure 6.1), implying that these are strongly reducing biological redox agents.

6.8   Metal-dependent mono-oxygenases

Enzymatic hydroxylation of organic molecules is one of the remarkable examples of enzymatic reactions for which there is little precedent in organic chemistry. We have seen above how flavin is able to utilise dioxygen to carry out the hydroxylation of a phenolic aromatic ring. However, more remarkably there is a class of mammalian enzymes that are able to catalyse the specific hydroxylation of unactivated alkanes. These enzymes are known as the P_{450} mono-oxygenases, due to the presence of a haem cofactor which upon treatment with carbon monoxide gives a characteristic UV absorbance at 450 nm. At the centre of the haem cofactor is an iron centre which in the resting enzyme is in the $+3$ oxidation state, but which is reduced to the $+2$ oxidation state upon substrate binding. This reduction is carried out by a reductase sub-unit of the enzyme which contains a flavin cofactor, itself reduced by NADPH.

The mechanism of hydroxylation by P_{450} enzymes shown in Figure 6.30 is explained as follows. Just as reduced flavin is able to donate a single electron to dioxygen in the case of the flavin-dependent mono-oxygenases, so the reduced iron(II) is able to donate a single electron to dioxygen, forming iron(III)-superoxide. Transfer of a second electron from the flavin reductase sub-unit via the iron centre generates iron(III)-peroxide. Protonation of the peroxide generates a good leaving group for cleavage of the O—O bond, giving formally an iron(V) oxo species, which is believed to exist as an iron(IV) oxo/porphyrin radical cation. The mechanism of substrate hydroxylation by this reactive intermediate has been much debated in the literature. One possible mechanism, shown in Figure 6.30, is the abstraction of $H^\bullet$ from the substrate, forming a substrate radical species and iron(IV)-hydroxide. Homolytic cleavage of the Fe—O bond and transfer of $HO^\bullet$ to the substrate radical gives the hydroxylated product and regenerates iron(III). The lack of evidence for a substrate radical intermediate from radical trapping experiments has led to the more recent proposal of a concerted mechanism involving oxygen insertion into the C—H bond, illustrated in Figure 6.31.

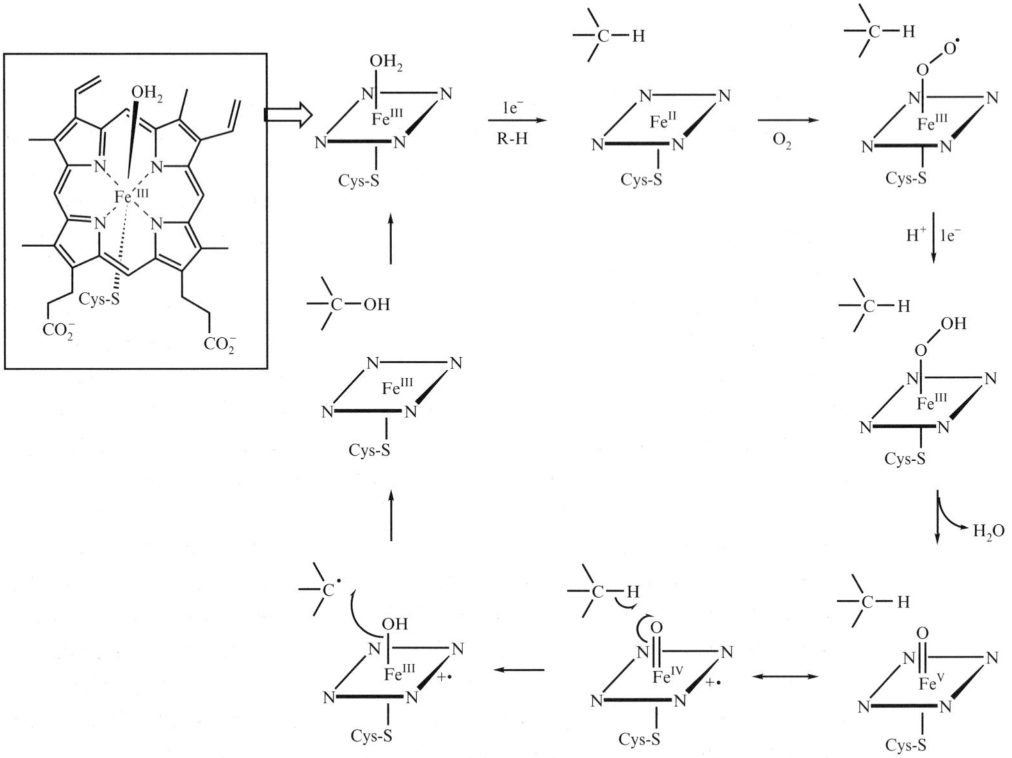

Figure 6.30 Mechanism for P_{450}-dependent hydroxylation.

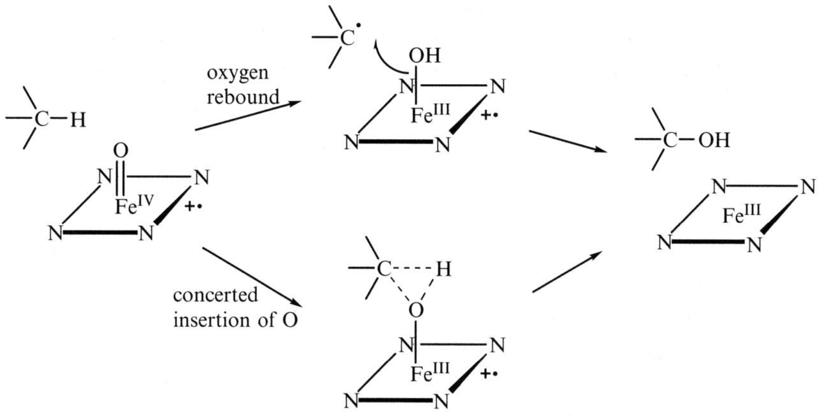

Figure 6.31 Mechanisms for oxygen insertion in P_{450} mono-oxygenases.

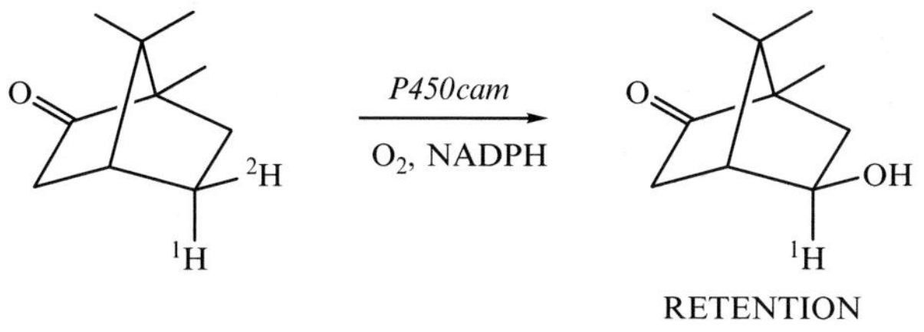

Figure 6.32 Stereochemistry of P450cam.

The best characterised enzyme of this class is P450cam, which catalyses the stereospecific hydroxylation of camphor. This reaction has been shown by deuterium labelling to proceed with retention of stereochemistry at the position of hydroxylation, as shown in Figure 6.32. Retention of stereochemistry is generally found in other P_{450}-dependent hydroxylases. The structure of P450cam is shown in Figure 6.33.

P_{450}-dependent enzymes also catalyse other types of oxidative reactions: one common example is demethylation. This reaction can be rationalised as shown in Figure 6.34 by hydroxylation of the terminal methyl group as above, generating a labile hemi-acetal which breaks down, liberating the free alcohol and formaldehyde.

Other haem-dependent enzymes utilise hydrogen peroxide instead of dioxygen to access the iron(III)-peroxide intermediate directly – these enzymes are known as peroxidases. An important function of peroxidases in plants, where they are widely found, is to initiate lignin formation by abstraction of $H^\bullet$, as will be discussed in Section 7.10. Finally, the chloroperoxidases are a family of haem-dependent enzymes found in marine organisms which catalyse the formation of carbon–chlorine bonds in organic molecules. These enzymes also use hydrogen peroxide to access the iron(III)-peroxide intermediate, which reacts with a chloride ion to generate an electrophilic hypochlorite species (i.e. a 'Cl^+' equivalent). Formation of a carbon–chlorine bond with a phenol substrate is shown in Figure 6.35.

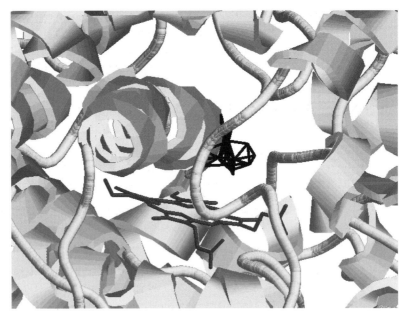

Figure 6.33 Active site of P450cam (PDB file 1AKD). Haem cofactor shown in red; camphor substrate shown in black (superposition of several bound conformations).

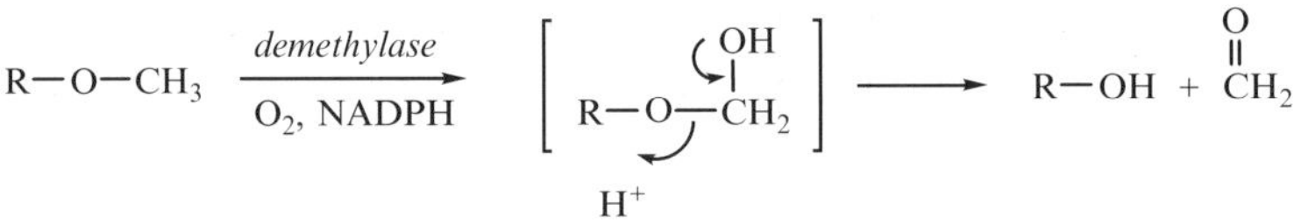

$$R{-}O{-}CH_3 \xrightarrow[\text{O}_2, \text{ NADPH}]{\textit{demethylase}} \left[\begin{array}{c} \text{OH} \\ R{-}O{-}CH_2 \end{array} \right] \longrightarrow R{-}OH \ + \ \overset{\text{O}}{\underset{}{CH_2}}$$

$$H^+$$

Figure 6.34 P_{450} demethylation reactions.

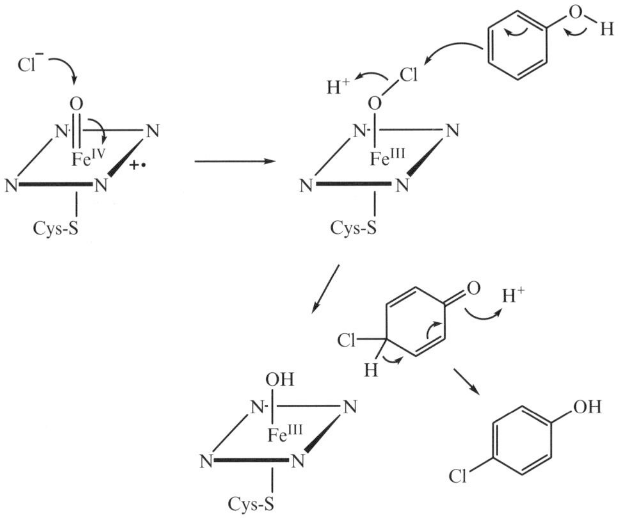

Figure 6.35 Mechanism for chloroperoxidase.

Iron is the only metal utilised in haem-dependent mono-oxygenases. However, metal-dependent mono-oxygenases are found which involve other redox-active metal ions such as copper, which can access the +1 and +2 oxidation states, and hence also has the ability to interact with molecular oxygen.

6.9   α-Ketoglutarate-dependent dioxygenases

Dioxygenases are enzymes that incorporate both atoms of dioxygen into the product(s) of their enzymatic reactions. The first class of dioxygenases that we shall consider catalyse hydroxylation reactions that at first glance seem similar to the P_{450}-dependent mono-oxygenases. However, there are two differences:

(1) these enzymes use α-ketoglutarate as a co-substrate, which they convert into succinate and carbon dioxide;
(2) they use non-haem iron rather than the haem cofactor.

As in the haem enzymes, the non-haem iron(II) can donate one electron to oxygen, forming iron(III)-superoxide which can react with the carbonyl group of α-ketoglutarate, forming a cyclic peroxy intermediate. Decarboxylation of this intermediate releases succinate and produces an iron(IV)-oxo species. This species can carry out the hydroxylation reaction via a radical mechanism shown in Figure 6.36, which is similar to the P_{450} mono-oxygenase mechanism.

An important member of this family of enzymes is prolyl hydroxylase, which catalyses the hydroxylation of proline amino acid residues in developing

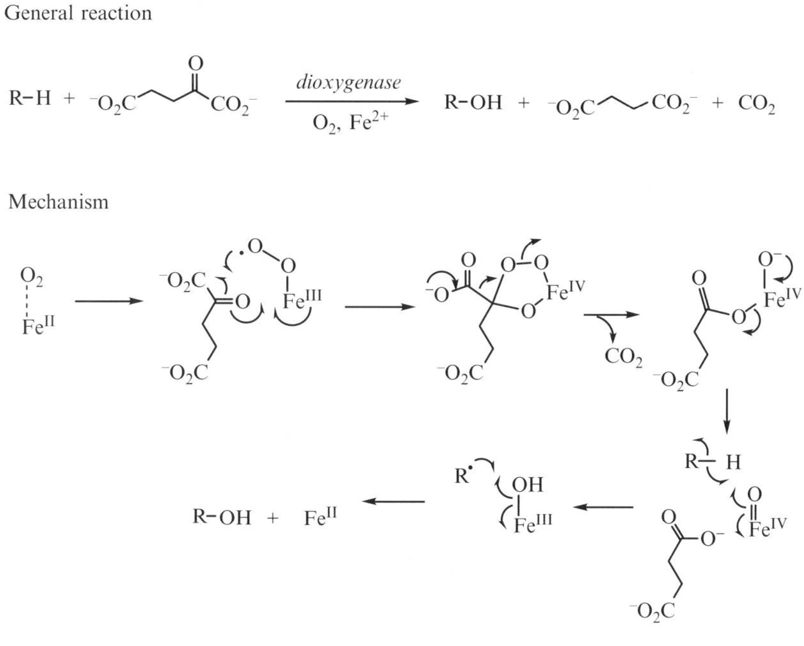

General reaction

Mechanism

Figure 6.36 α-Ketoglutarate-dependent iron(II) dioxygenases.

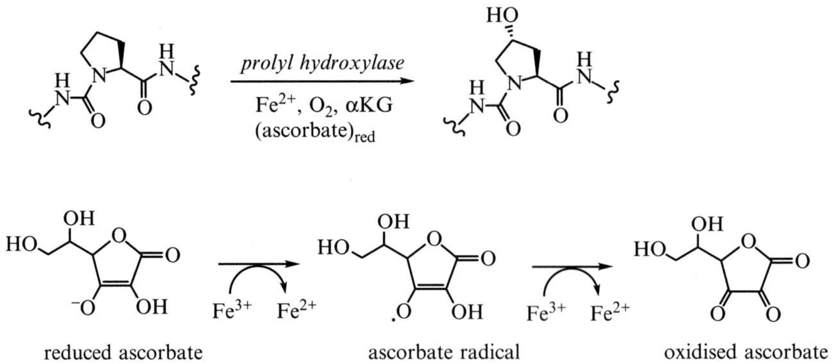

Figure 6.37 Prolyl hydroxylase and the role of ascorbate as a reducing agent.

collagen fibres to 4-hydroxyproline. Collagen is a very important structural protein found in skin, teeth, nail and hair, and the presence of hydroxyproline is necessary for maintenance of the tertiary structure of the protein. This enzyme also uses ascorbic acid – vitamin C – as a cofactor, and this is one of the major functions of vitamin C in the body. The overall reaction does not apparently require any further redox equivalents, but ascorbic acid is required in non-stoichiometric amounts for full activity. The probable role of ascorbic acid is to maintain the iron cofactor in the reduced iron(II) state in cases where the catalytic cycle is not completed. For example, if superoxide is released from the enzyme active site an inactive iron(III) form of the enzyme is generated. Ascorbic acid has the ability to reduce iron(III) to iron(II), generating a stable ascorbate radical shown in Figure 6.37. Thus in the absence of vitamin C the body's machinery for making collagen is impaired, leading to the symptoms of the dietary deficiency disease scurvy.

6.10   Non-haem iron-dependent dioxygenases

There are a number of further non-haem iron-dependent dioxygenase enzymes, of which I shall introduce two: the catechol dioxygenases and the aromatic

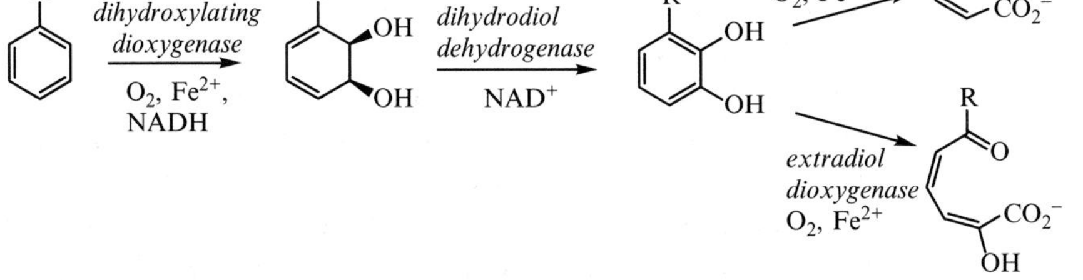

Figure 6.38 Aromatic hydroxylation/cleavage reactions.

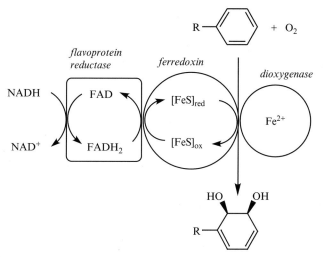

Figure 6.39 Electron transfer in dihydroxylating dioxygenases.

dihydroxylating dioxygenases. Both these classes of enzyme are involved in the bacterial degradation of aromatic compounds in the environment. As shown in Figure 6.38, the dihydroxylating dioxygenases catalyse the oxidative conversion of an aromatic substrate into the corresponding *cis*-dihydrodiol. The *cis*-diol is then oxidised by a class of NAD^+-dependent dehydrogenases into the corresponding aromatic diol, or catechol. Oxidative cleavage of the aromatic ring is then carried out in one of two ways. Cleavage of the carbon–carbon bond between the two hydroxyl groups and insertion of both atoms of dioxygen is catalysed by non-haem iron(III)-dependent intradiol dioxygenases. Alternatively, cleavage of a carbon–carbon bond adjacent to the two hydroxyl groups and insertion of two oxygen atoms can be catalysed by non-haem iron(II)-dependent extradiol dioxygenases.

The dihydroxylating dioxygenases are multi-component enzymes whose sub-units are involved in electron transfer. Two electrons are transferred by a flavoprotein reductase sub-unit from NADH to $FADH_2$. The two electrons are then transferred singly to iron–sulphur clusters contained in a ferredoxin sub-unit. These electrons are then transferred to the active site of the dioxygenase sub-unit, as shown in Figure 6.39.

The mechanisms of these non-haem iron-dependent dioxygenases are not well understood, but all presumably involve activation of dioxygen by electron transfer from the iron cofactor. In the case of the dihydroxylating dioxygenases, formation of a four-membered dioxetane intermediate has been proposed, followed by reduction to generate the *cis*-diol product, as shown in Figure 6.40. This mechanism would explain the *cis*-stereochemistry of the product and the incorporation of ^{18}O into both hydroxyl groups from $^{18}O_2$. However, the formation of dioxetane intermediates from ground state oxygen is endothermic and therefore apparently unlikely.

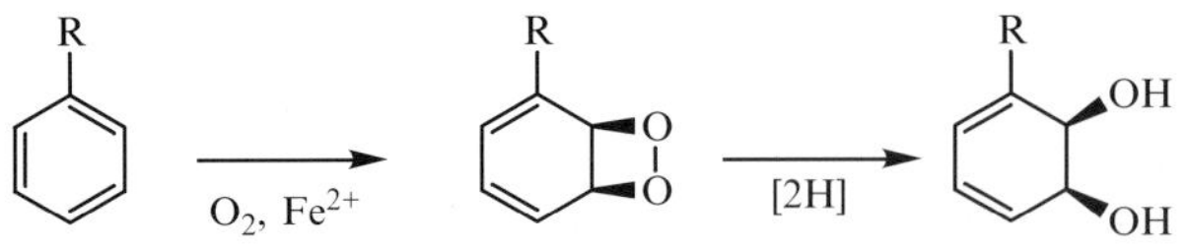

Figure 6.40 Dioxetane mechanism for dihydroxylating dioxygenases.

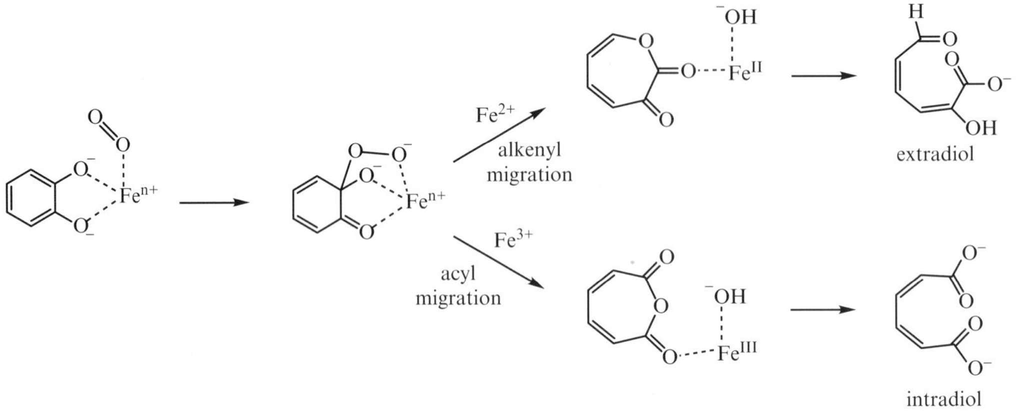

Figure 6.41 Possible mechanisms for catechol dioxygenases.

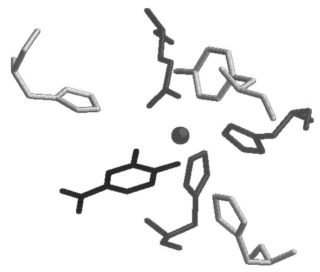

Figure 6.42 Active site of protocatechuate 4,5-dioxygenase (PDB file 1B4U), showing the tridentate His, His, Glu motif responsible for binding the non-haem iron(II) cofactor. This motif is found in many non-haem iron-dependent oxygenases. The bound substrate is shown in black. Picture prepared using RASMOL.

Dioxetane intermediates were also proposed for the intradiol and extradiol catechol dioxygenases, based on $^{18}O_2$ incorporation experiments. However, recent evidence suggests that 1,2-rearrangements (similar to the Baeyer–Villiger oxidation of ketones) of peroxy intermediates are taking place in these reactions leading to anhydride and lactone intermediates respectively, as shown in Figure 6.41. The non-haem iron(II) centre of the extradiol catechol dioxygenases, the dihydroxylating dioxygenases, and the α-ketoglutarate dioxygenases, is ligated by a common motif of two histidine residues and one carboxylate group from a glutamic acid or aspartic acid residue. This tridentate motif seems to be especially used in Nature for activation of dioxygen and organic substrates by non-haem iron(II). It is illustrated in Figure 6.42 in the case of protocatechuate 4,5-dioxygenase.

Many of the redox reactions in the second half of this chapter have little precedent in organic chemistry, although in some cases inorganic model complexes have been prepared which mimic the action of metallo-enzymes. Thus, utilising only a small selection of organic redox coenzymes and metal redox cofactors as electron carriers, an extraordinary range of enzyme-catalysed oxidation/reduction chemistry is possible.

Problems

(1) Work out the redox potential differences for the following enzymatic reactions, using the data in Figure 6.1. In example (c) what can you deduce about the redox potential of the enzyme-bound flavin?

(2) In the enoyl reductase reaction illustrated in Problem 1(b), incubation of 4S-^{2}H-NADPD with enzyme and crotonyl CoA gives no incorporation of deuterium in the butyryl CoA product. Incubation of 4R-^{2}H-NADPD with enzyme and crotonyl CoA gives 3R-^{2}H-butyryl CoA. Incubation of enzyme, NADH and crotonyl CoA in ^2H_2O gives 2S-^{2}H-butyryl

CoA. Explain these results, and write a mechanism for the enzymatic reaction.

(3) The enzyme which catalyses the conversion below has been purified and requires a catalytic amount of NAD^+ for activity. Given that H^* is transferred intact as shown, suggest a mechanism for the enzymatic reaction.

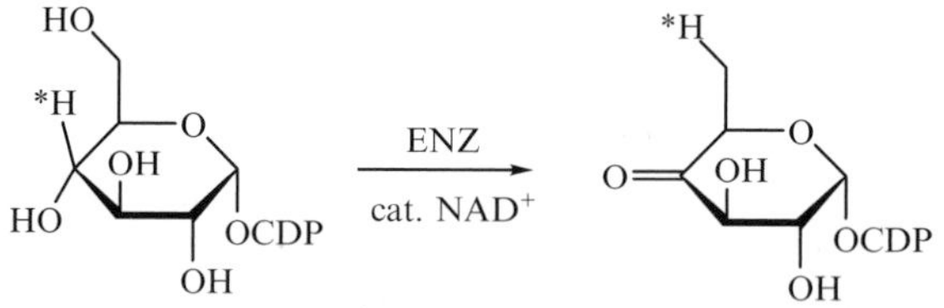

(4) The reductase enzyme which catalyses the reaction shown below contains a stoichiometric amount of tightly bound FAD, which is reducible during catalytic turnover, and utilises NADPH as a reducing equivalent. Incubation of $4S$-^{2}H-NADPD with enzyme and substrate gives product containing one atom of deuterium β to the carboxylate. Incubation with unlabelled NADPH in ^2H_2O gives product containing one atom of deuterium α to the carboxylate. Suggest a mechanism for the enzymatic reaction consistent with these data.

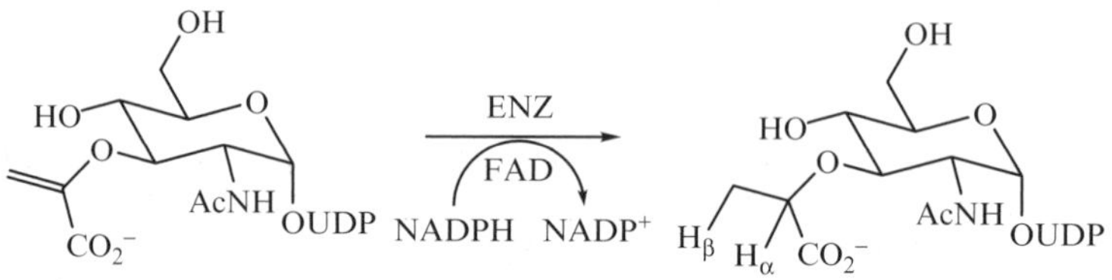

(5) A flavo-enzyme has been purified which catalyses the conversion of *para*-nitrophenol into hydroquinone and nitrite. The enzyme contains tightly bound FAD, and each catalytic cycle consumes one equivalent of dioxygen and *two* equivalents of NADH. Suggest a mechanism.

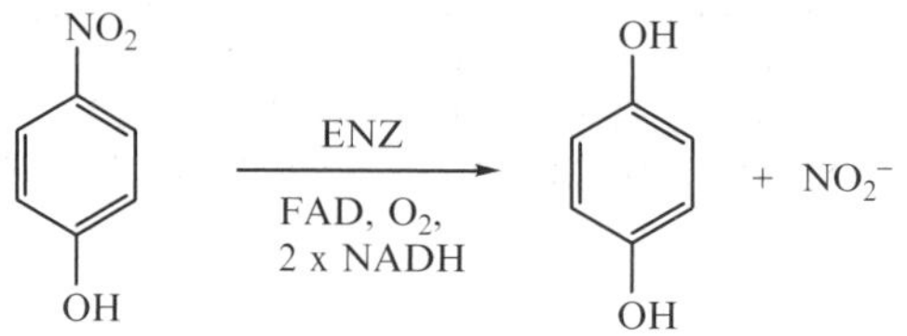

(6) Thymine hydroxylase is an α-ketoglutarate-dependent iron(II) dioxygenase which catalyses three successive oxidations of thymine, as shown below.

(a) Write a mechanism for the first oxidation.

(b) 5-Ethynyluracil is a potent inhibitor of this enzyme. Inactivation results in covalent modification of the enzyme with a stoichiometry of one adduct/enzyme sub-unit, and also generates a byproduct 5-carboxyuracil. Suggest a possible mechanism of inactivation which would account for these observations.

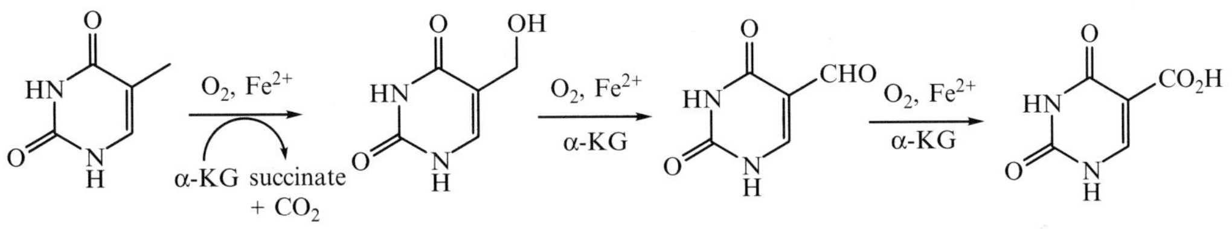

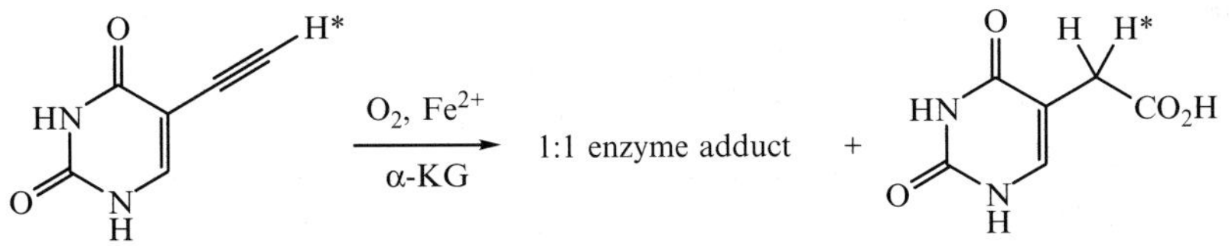

Further reading

General

R.H. Abeles, P.A. Frey & W.P. Jencks (1992) *Biochemistry*. Jones & Bartlett, Boston.

I. Bertini, H.B. Gray, S.J. Lippard, & J.S. Valentine (1994) *Bio-inorganic Chemistry*. University Science Books, Mill Valley, California.

C.T. Walsh (1979) *Enzymatic Reaction Mechanisms*. Freeman, San Francisco.

C.H. Wong & G.M. Whitesides (1994) *Applications for Organic Synthesis: Enzymes in Synthetic Organic Chemistry*. Pergamon, Oxford.

NAD-dependent dehydrogenases

J. Everse & N.O. Kaplan (1973) Lactate dehydrogenases: structure and function. *Adv. Enzymol.*, **37**, 61–134.

F.A. Loewus, F.H. Westheimer & B. Vennesland (1953) Enzymatic synthesis of the enantiomorphs of ethanol-1-*d*. *J. Am. Chem. Soc.*, **75**, 5018-23.

F.H. Westheimer, H.F. Fisher, E.E. Conn & B. Vennesland (1951) The enzymatic transfer of hydrogen from alcohol to DPN. *J. Am. Chem. Soc.*, **73**, 2403.

NADH models

O. Almarsson, R. Karaman & T.C. Bruice (1992) Kinetic importance of conformations of NAD in the reactions of dehydrogenase enzymes. *J. Am. Chem. Soc.*, **114**, 8702–4.
F. Rob, H.J. van Ramesdonk, W. von Gerresheim, P. Bosma, J.J. Scheele & J.W. Verhoeven (1984) Diastereo-differentiating hydride transfer in bridged NAD(H) models. *J. Am. Chem. Soc.*, **106**, 3826–32.

Flavin-dependent enzymes

T. Bruice (1980) Mechanisms of flavin catalysis. *Acc. Chem. Res.*, **13**, 256–62.
P.F. Fitzpatrick (2001) Substrate dehydrogenation by flavoproteins. *Acc. Chem. Res.*, **34**, 299–307.
S. Ghisla, C. Thorpe & V. Massey (1984) Mechanistic studies with general acyl-CoA dehydrogenase and butyryl-CoA dehydrogenase: evidence for the transfer of the β-hydrogen to the flavin N(5)-position as a hydride. *Biochemistry*, **23**, 3154–61.
M. Lai, D. Li, E. Oh & H. Liu (1993) Inactivation of medium-chain acyl-CoA dehydrogenase by a metabolite of hypoglycine. *J. Am. Chem. Soc.*, **115**, 1619–28.
V. Massey (1994) Activation of molecular oxygen by flavins and flavoproteins. *J. Biol. Chem.*, **269**, 22459–62.
G.E. Schulz, R.H. Schirmer, W. Sachsenheimer & E.F. Pai (1978) The structure of the flavoenzyme glutathione reductase. *Nature* **273**, 120–24.
N.S. Scrutton, A. Berry, & R.N. Perham (1990) Redesign of the coenzyme specificity of a dehydrogenase by protein engineering. *Nature*, **343**, 38–43.
R.B. Silverman (1995) Radical ideas about monoamine oxidase. *Acc. Chem. Res.*, **28**, 335–42.
F.X. Sullivan, S.B. Sobolov, M. Bradley & C.T. Walsh (1991) Mutational analysis of parasite trypanothione reductase: acquisition of glutathione reductase activity in a triple mutant. *Biochemistry*, **30**, 2761–7.
C. Walsh, (1980) Flavin coenzymes: at the crossroads of biological redox chemistry. *Acc. Chem. Res.*, **13**, 148–55.
C.T. Walsh (1986) Naturally occurring 5-deazaflavin coenzymes: biological redox roles. *Acc. Chem. Res.*, **19**, 216–21.

Pterin-dependent mono-oxygenases

S.J. Benkovic (1980) On the mechanism of action of folate- and biopterin-requiring enzymes. *Annu. Rev. Biochem.*, **49**, 227–52.
G.R. Moran, A. Derecskei-Kovacs, P.J. Hillas, and P.F. Fitzpatrick (2000) On the catalytic mechanism of tryptophan hydroxylase. *J. Am. Chem. Soc.*, **122**, 4535–41.

Iron–sulphur clusters

R.H. Holm, S. Ciurli & J.A. Weigel (1990) Subsite-specific structures and reactions in native and synthetic [4Fe-4*S*] cubane-type clusters. *Prog. Inorg. Chem.*, **38**, 1–74.
W.V. Sweeney & J.C. Rabinowitz, (1980) Proteins containing [4Fe-4*S*] clusters. *Annu. Rev. Biochem.*, **49**, 139–62.

Haem-dependent oxygenases

M. Akhtar & J.N. Wright (1991) A unified mechanistic view of oxidative reactions catalysed by P_{450} and related iron-containing enzymes. *Nat. Prod. Reports*, **8**, 527–52.

M. Newcomb & P.H. Toy (2000) Hypersensitive radical probes and the mechanisms of cytochrome P450-catalysed hydroxylation reactions. *Acc. Chem. Res.*, **33**, 449–55.

M. Sono, M.P. Roach, E.D. Coulter & J.H. Dawson (1996) Heme-containing oxygenases. *Chem. Rev.*, **96**, 2841–88.

Non-haem iron-dependent oxygenases

G.J. Cardinale & S. Udenfriend (1974) Prolyl hydroxylase. *Adv. Enzymol.*, **41**, 245–300.

A.L. Feig & S.J. Lippard (1994) Reactions of non-heme iron (II) centers with dioxygen in biology and chemistry. *Chem. Rev.*, **94**, 759–805.

K.I. Kivirikko, R. Myllyla & T. Pihlajaniemi (1989) Protein hydroxylation: prolyl 4-hydroxylase, an enzyme with four cosubstrates and a multifunctional subunit. *FASEB J.*, **3**, 1609–17.

L. Que Jr & R.Y.N. Ho (1996) Dioxygen activation by enzymes with mononuclear non-heme iron active sites. *Chem. Rev.*, **96**, 2607–24.

J. Sanvoisin, G.J. Langley & T.D.H. Bugg (1995) Mechanism of extradiol catechol dioxygenases: evidence for a lactone intermediate in the 2,3-dihydroxyphenylpropionate 1,2-dioxygenase reaction. *J. Am. Chem. Soc.*, **117**, 7836–7.

T.D.H. Bugg (2003) Dioxygenase enzymes: catalytic mechanisms and chemical models. *Tetrahedron*, **59**, 7075–7101.

7 Enzymatic Carbon–Carbon Bond Formation

7.1 Introduction

The formation of carbon–carbon bonds is central to *biosynthesis*, which is the assembly of carbon-based compounds within living cells. Most of these compounds are *primary metabolites* – molecules such as amino acids, carbohydrates and nucleic acids which are necessary to support life. Many organisms also produce *secondary metabolites* – molecules whose presence is not essential for the survival of the cell, but which often have other biological properties such as defence against micro-organisms or communication with other organisms. In this chapter we shall analyse the types of enzymatic reactions used in the assembly of the carbon skeletons of primary and secondary metabolites, and also carbon–carbon cleavage reactions involved in their degradation.

A carbon–carbon bond consists of a pair of electrons contained within a filled molecular orbital. Formation of a carbon–carbon bond can be achieved either by donation of a pair of electrons from one carbon atom to an empty orbital on another carbon atom, or by the combination of two single electron species. Examples of these processes are illustrated in Figure 7.1, involving:

(1) nucleophilic attack of a carbanion onto an electron-accepting carbonyl group;
(2) electrophilic attack of a carbocation onto an electron-rich alkene;
(3) recombination of two phenoxy radicals.

We shall meet examples of each of these types of carbon–carbon formation reactions in this chapter.

One general point to note is that carbanions and carbocations are high-energy species which can usually only be generated under strenuous reaction conditions in organic chemistry. How then are they generated by enzymes that work at neutral pH with relatively weak acidic and basic groups? The answer is that carbanion and carbocation intermediates in enzyme-catalysed reactions must be highly stabilised, either by neighbouring groups in the substrate molecule, or by the enzyme active site, using the type of enzyme–substrate interactions mentioned in Section 2.7. This stabilisation will be explained where possible in the following examples; however, in some cases the means by which high-energy intermediates are stabilised is not fully understood.

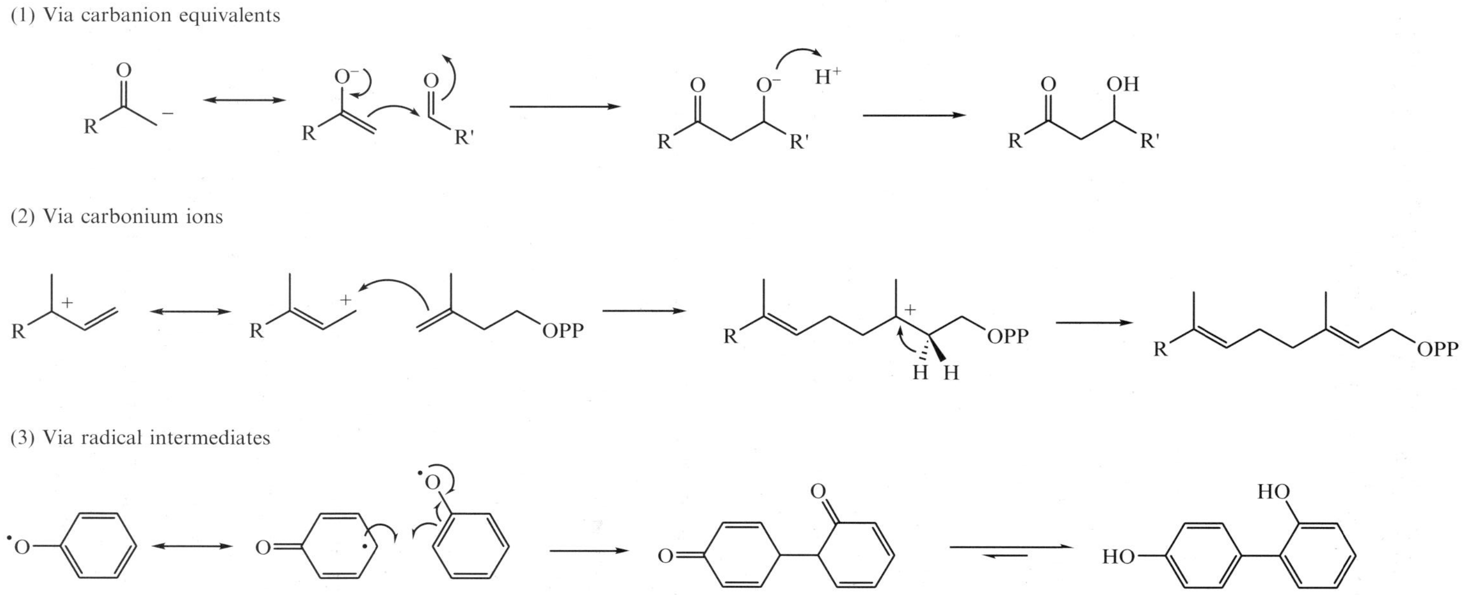

(1) Via carbanion equivalents

(2) Via carbonium ions

(3) Via radical intermediates

Figure 7.1 Formation of carbon–carbon bonds.

Figure 7.2 Aldol reaction.

Carbon–carbon bond formation via carbanion equivalents

7.2   Aldolases

The aldol reaction involves the condensation of two carbonyl compounds via an enolate intermediate. The reaction is illustrated in Figure 7.2 for the case of the self-condensation of acetone in alkaline aqueous solution. A wide range of aldol reactions occur in biological systems that are used for the formation of carbon–carbon bonds. Cleavage of carbon–carbon bonds by the reverse reaction is also found.

The enzymes which catalyse these aldol reactions are known as aldolases, and they are divided into two families based on their mechanism of action. The class I aldolases function by formation of an imine linkage between one carbonyl reagent and the ε-amino group of an active site lysine residue, followed by deprotonation of the adjacent carbon to generate an enamine intermediate. Enamines are well-known enolate equivalents in synthetic organic chemistry: they can be formed under mild conditions by condensation of the carbonyl compound with a primary or secondary amine, and they react with carbonyl compounds also under mild conditions.

The class II aldolases do not proceed through enamine intermediates but instead use a metal ion to assist catalysis. The metal ion is usually a divalent metal ion such as Mg^{2+}, Mn^{2+} or Zn^{2+}. The two classes are exemplified by the enzyme fructose-1,6-bisphosphate aldolase, which is found in mammals as a class I enzyme, and in bacteria as a class II enzyme.

CASE STUDY: Fructose-1,6-bisphosphate aldolase

Fructose-1,6-bisphosphate aldolase catalyses the reversible reaction of dihydroxyacetone phosphate (DHAP) with glyceraldehyde 3-phosphate (G3P) to give fructose-1,6-bisphosphate, shown in Figure 7.3.

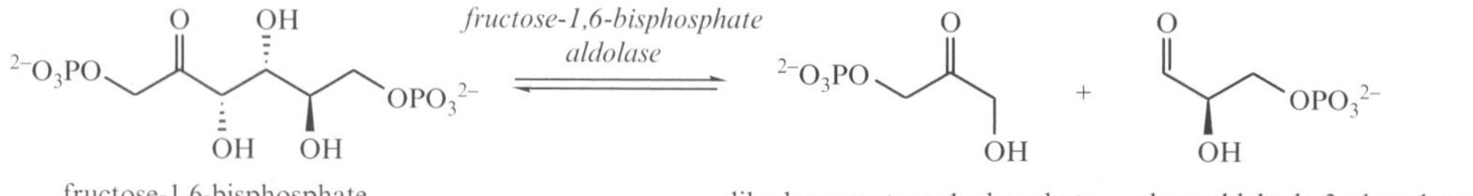

Figure 7.3 Reaction catalysed by fructose-1,6-bisphosphate aldolase.

The reaction catalysed by the class I enzyme from rabbit muscle has been shown to proceed via formation of an imine linkage between the ε-amino group of an active site lysine residue and the C-2 carbonyl of DHAP. The imine linkage can be reduced by sodium borohydride to give an irreversibly inactivated secondary amine. Incubation of enzyme with radiolabelled G3P and sodium borohydride followed by proteolytic digestion of the inactivated enzyme has identified the imine-forming active site residue as Lys-229.

Incubation of enzyme and DHAP in ^2H_2O (in the absence of G3P) leads to stereospecific ^2H exchange at C-3 to give 3S-[3-^2H]-DHAP. This indicates that the *proS* proton at C-3 is specifically removed following imine formation, giving an enamine intermediate. In order to generate the observed stereochemistry at C-3 and C-4 of the product the enamine intermediate must react from its *si*-face with the *si*-face of the aldehyde group of G3P (see Figure 7.5 later). The resulting imine intermediate is hydrolysed to yield the product fructose-1,6-bisphosphate.

An X-ray crystal structure was determined for rabbit muscle fructose-1,6-bisphosphate aldolase in 1987. The tertiary structure of the protein subunit is an αβ barrel consisting of 9 parallel β-sheets interconnected by α-helices. The active site lies in the centre of the barrel formed by the β-sheets, as shown in Figure 7.4. Close to the imine-forming Lys-229 in the centre of the active site cavity are three active site residues which participate in catalysis: Glu-187, Lys-146 and Asp-33. Site-directed mutagenesis has been used to replace each of these residues with a non-functional side chain. Replacement of Glu-187 by alanine dramatically lowers the rate of Schiff base formation, indicating that Glu-187 participates in acid–base catalysis during Schiff base formation. Replacement of Asp-33 by alanine dramatically reduces the rate of C−C bond formation, consistent with a role as a catalytic base in C−C bond cleavage. Mutation of Lys-146 to alanine gives a mutant enzyme which is $> 10^6$-fold less active, whilst a histidine mutant enzyme is only 2000-fold less active than the wild type enzyme, suggesting that Lys-146 fulfils an acid–base function.

The most plausible catalytic mechanism is shown in Figure 7.5. Schiff base formation with Lys-229 is assisted by nearby Glu-187 and Lys-146. Asp-33 acts as the base for enamine formation, and protonates the carbonyl oxygen of G3P on the same face.

The fructose-1,6-bisphosphate aldolase enzyme in yeast and bacteria is a class II enzyme, dependent upon Zn^{2+} for activity. Studies of the yeast aldolase by infrared spectroscopy have shown that the carbonyl stretching frequency of G3P is shifted from $1730\,cm^{-1}$ in aqueous solution to $1706\,cm^{-1}$ when bound to the enzyme, whereas the stretching frequency of the DHAP is unchanged upon binding. These data indicate that the aldehyde of G3P is polarised when bound to the active site. The X-ray crystal structure of the *Escherichia coli* aldolase has been solved, revealing that the ketone and hydroxyl groups of DHAP are chelated to the Zn^{2+} cofactor, which is ligated by

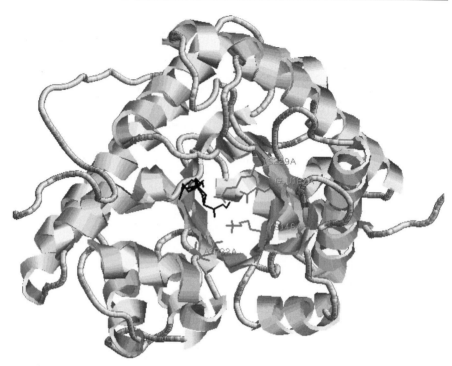

Figure 7.4 Structure of rabbit muscle fructose-1,6-bisphosphate aldolase (PDB file 1ADO), a class I aldolase. Catalytic residues Lys-229, Glu-187, Lys-146 and Asp-33 shown in red. Bound substrate analogue shown in black. Picture prepared using RASMOL.

three histidine residues, as shown in Figure 7.6. The *proS* hydrogen is removed by an active site base, which appears to be Glu-182, to give a dienolate intermediate, as shown in Figure 7.7. Reaction of the *si*-face of the enolate with the *si*-face of G3P forms the C−C bond, with Asp-109 protonating the aldehyde carbonyl of G3P.

Fructose-1,6-bisphosphate aldolase has been used for enantioselective carbon–carbon bond formation in organic synthesis. The class I enzyme is highly selective for the DHAP substrate, but can react with a wide range of aldehyde substrates. Reaction with racemic 2-hydroxy-3-azido-propionalde-hyde on a 1–10 mmol scale yields a single enantiomer of the aldol product containing three chiral centres. Biotransformation of the same substrates with aldolase enzymes of different stereospecificity yields diastereomeric products as shown in Figure 7.8. Dephosphorylation of the products followed by reduction of the azido group generates the corresponding amines, which undergo stereospecific reductive amination reactions with the free keto group. The cyclic amine products can be readily converted into a range of aza-sugar analogues which are of considerable interest as glycosidase inhibitors.

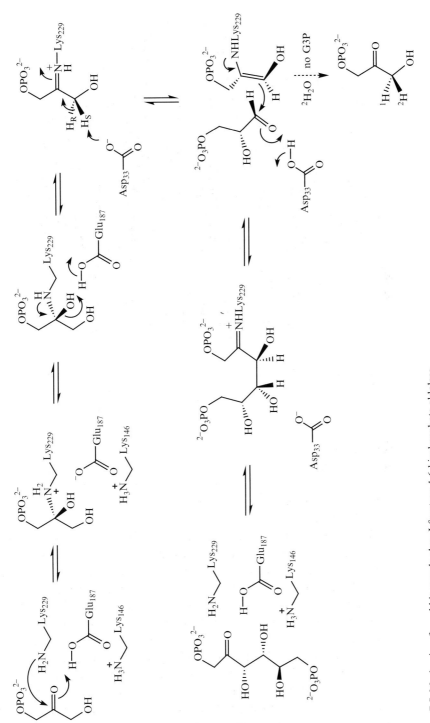

Figure 7.5 Mechanism for rabbit muscle class I fructose-1,6-bisphosphate aldolase.

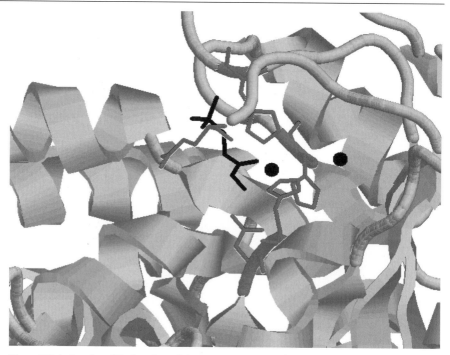

Figure 7.6 Active site of *Escherichia coli* fructose-1,6-bisphosphate aldolase (PDB file 1B57), a class II aldolase. Bound substrate analogue phosphoglycolohydroxamate and Zn²⁺ cofactor shown in black. Catalytic residues Asp-109 (bottom) and Glu-182 (top left), and Zn²⁺ ligands shown in red.

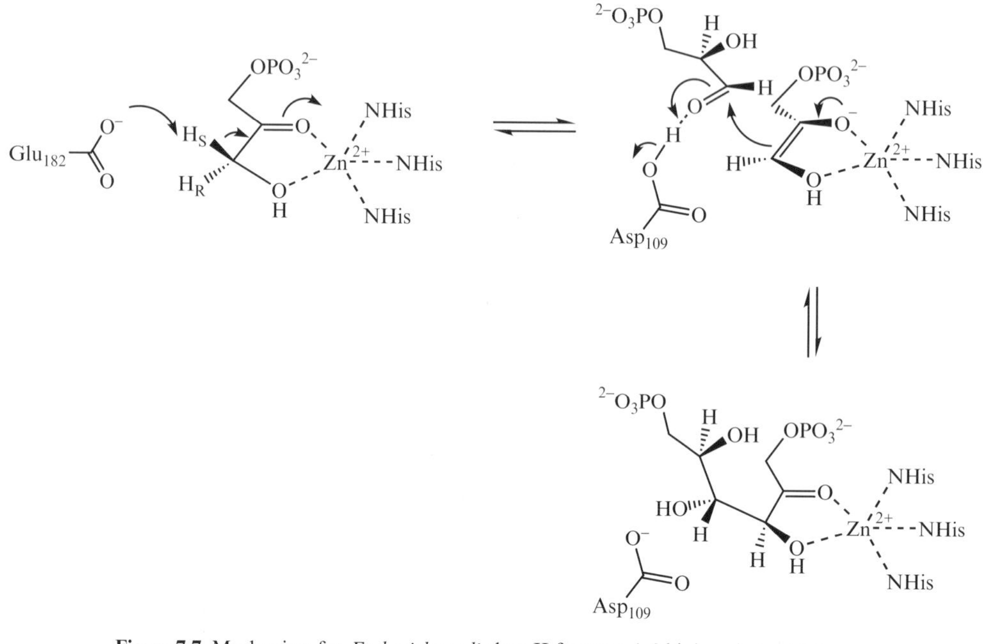

Figure 7.7 Mechanism for *Eschericha coli* class II fructose-1,6-biphosphate aldolase.

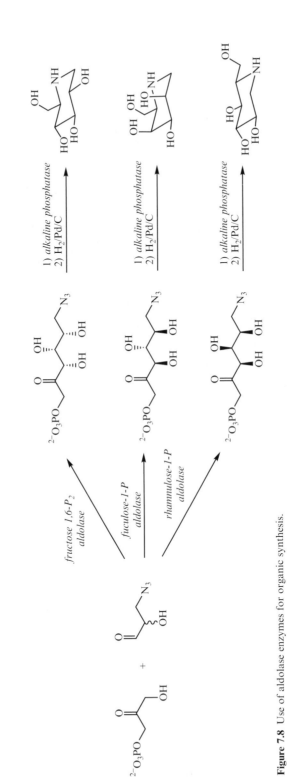

Figure 7.8 Use of aldolase enzymes for organic synthesis.

7.3   Claisen enzymes

In the presence of alkoxide ions, carboxylic esters react via an ester enolate intermediate to form β-keto-esters. This reaction, illustrated in Figure 7.9 for ethyl butyrate, is known as the Claisen ester condensation. The Claisen ester condensation requires more vigorous reaction conditions than the aldol condensation, since the C−H proton adjacent to an ester is significantly less acidic than the C−H proton adjacent to a ketone.

Claisen reactions also occur in biological systems, using thioesters rather than oxygen esters. Since the sulphur atom of a thioester is larger than oxygen and utilises $3p$ valence electrons rather than $2p$ electrons, there is consequently much less overlap between the sulphur lone pair of electrons and the carbonyl group than in the oxygen equivalent. Thus the carbonyl group of a thioester behaves more like a ketone in terms of reactivity than an ester. Shown in Figure 7.10 are pK_a values for formation of enolate or enolate equivalents for a selection of ketone and ester systems. The first set of data highlight the dramatic increase in acidity of an iminium salt compared to a ketone, which rationalises the use of enamine intermediates in the class I aldolases. From the second set of data we can see that the pK_a of a β-keto-thioester is similar to that of a β-diketone, and significantly less than for a β-keto-ester.

The thioester used for biological Claisen ester reactions is the coenzyme A (CoA) ester that we encountered in Chapter 5 as an acyl transfer coenzyme. Acetyl Co A is the substrate for several types of carbon–carbon forming reactions, shown in Figure 7.11.

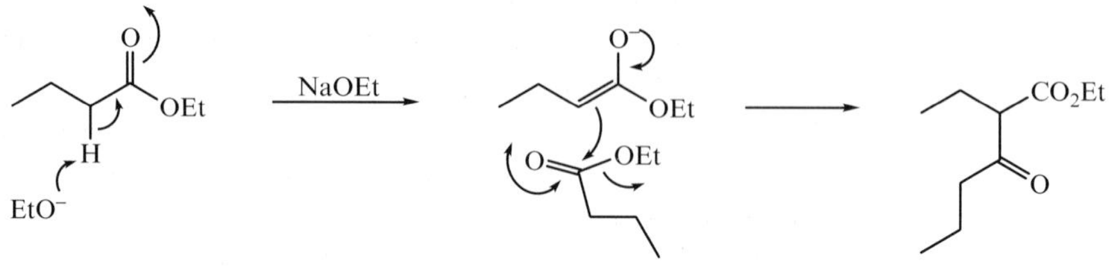

Figure 7.9 Example of a Claisen reaction.

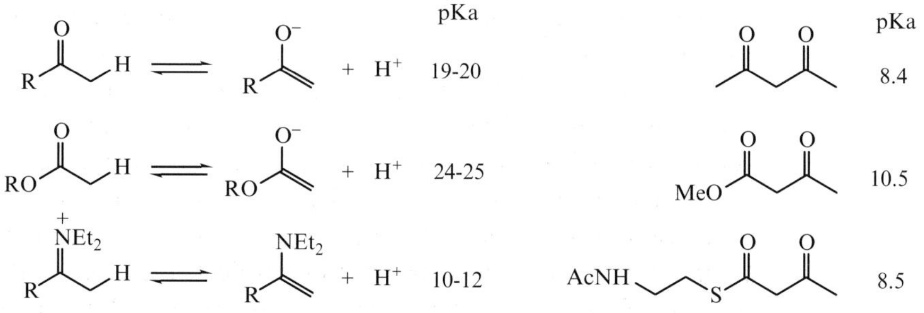

Figure 7.10 pK_a values for enolate formation.

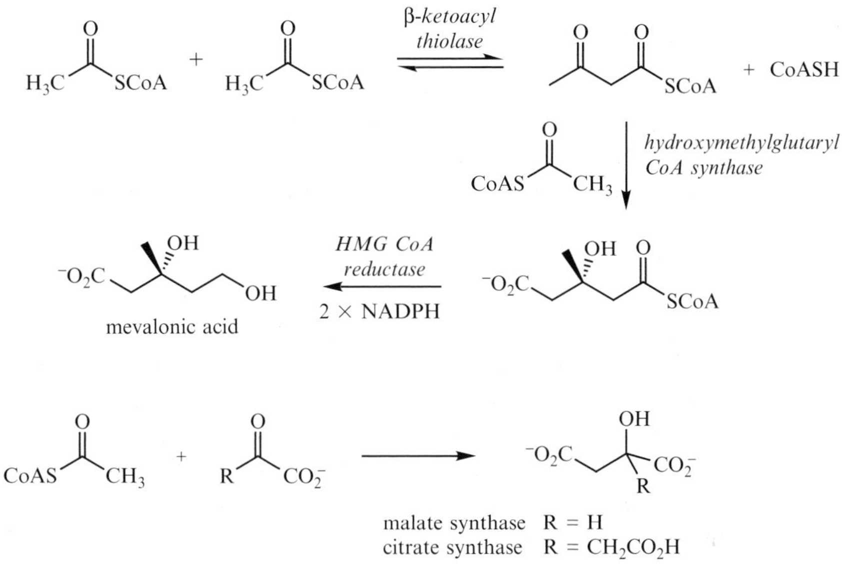

Figure 7.11 Reactions catalysed by Claisen enzymes, involving acetyl CoA.

Mevalonic acid is an important cellular precursor to terpene and steroid natural products. The biosynthesis of mevalonic acid involves two reactions in which carbon–carbon bonds are formed from the α-position of acetyl CoA. The first reaction is a condensation reaction with a second molecule of acetyl CoA to form 3-ketobutyryl CoA. Note that the good leaving group properties of the thiol group of CoA are also significant in this reaction. Reaction of the β-keto group with a further equivalent of acetyl CoA generates hydroxymethyl-glutaryl CoA, which is reduced by a nicotinamide adenine dinucleotide (NADPH)-dependent reductase to give mevalonic acid. Two further examples shown in Figure 7.11 are malate synthase and citrate synthase, which catalyse important metabolic reactions of acetyl CoA.

The mechanism of these 'Claisen enzymes' which react through the α-position of acetyl CoA could either proceed via formation of an α-carbanion intermediate, or by a concerted deprotonation/bond formation step. Incubation of acetyl CoA with malate synthase, or citrate synthase in the presence of ^3H_2O followed by re-isolation of substrate, gives no exchange of ^3H into acetyl CoA, ruling out the reversible formation of a carbanion intermediate. Stereochemical studies using chiral [2-^{2}H, ^{3}H]-labelled acetyl CoA illustrated in Figure 7.12, have established that the malate synthase reaction proceeds with overall inversion of stereochemistry, and with a kinetic isotope effect ($k_H/k_T = 2.7$).

These results suggest that there is a transient thioester enolate intermediate formed in the malate synthase reaction, which then reacts with the aldehyde carbonyl of glyoxylate. Formation of the thioester enolate intermediate would be thermodynamically unfavourable due to the high pK$_a$ of the α-proton;

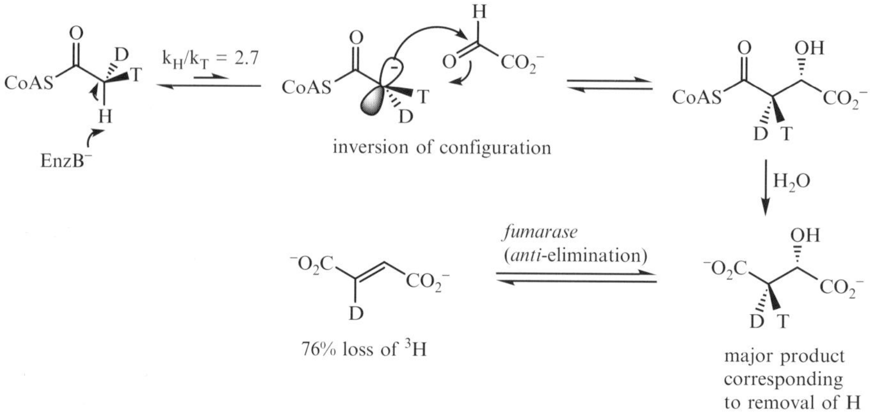

Figure 7.12 Stereochemistry of the malate synthase reaction.

however, the enzymatic reaction is made effectively irreversible by the subsequent hydrolysis of malyl CoA to malic acid. It is not known exactly how the formation of the thioester enolate (or enol) is made possible at the enzyme active site, but there must be significant stabilisation of this intermediate *in situ* by the enzyme.

7.4   Assembly of fatty acids and polyketides

Acetyl CoA is also used in the formation of the carbon chains of fatty acids and polyketide natural products in biological systems. Fatty acids are long-chain carboxylic acids, such as stearic acid ($C_{17}H_{35}CO_2H$), which are found widely as triglyceride esters in the lipid component of living cells. Polyketide natural products are assembled from a polyketide precursor containing ketone functional groups on alternate carbon atoms. In many cases these polyketide precursors are cyclised to form aromatic compounds, such as orsellinic acid shown in Figure 7.13.

There are many similarities between fatty acid biosynthesis and polyketide biosynthesis: they are both assembled from acyl CoA thioesters via a series of Claisen-type reactions by multi-enzyme synthase complexes (fatty acid synthase or polyketide synthase). The assembly is carried out via stepwise addition of two-carbon units onto a growing acyl chain which is attached to an acyl carrier protein (ACP) via a thioester linkage.

The first acetyl unit is transferred from acetyl CoA onto the acyl carrier protein, but thereafter the substrate for carbon–carbon bond formation is malonyl CoA, formed by carboxylation of acetyl CoA (described in Section 7.5). The malonyl group is transferred from malonyl CoA onto the acyl carrier protein by an acyltransferase (AT) activity (see Figure 7.14). Carbon–carbon bond formation is achieved by attack of the α-carbon of the malonyl-thioester

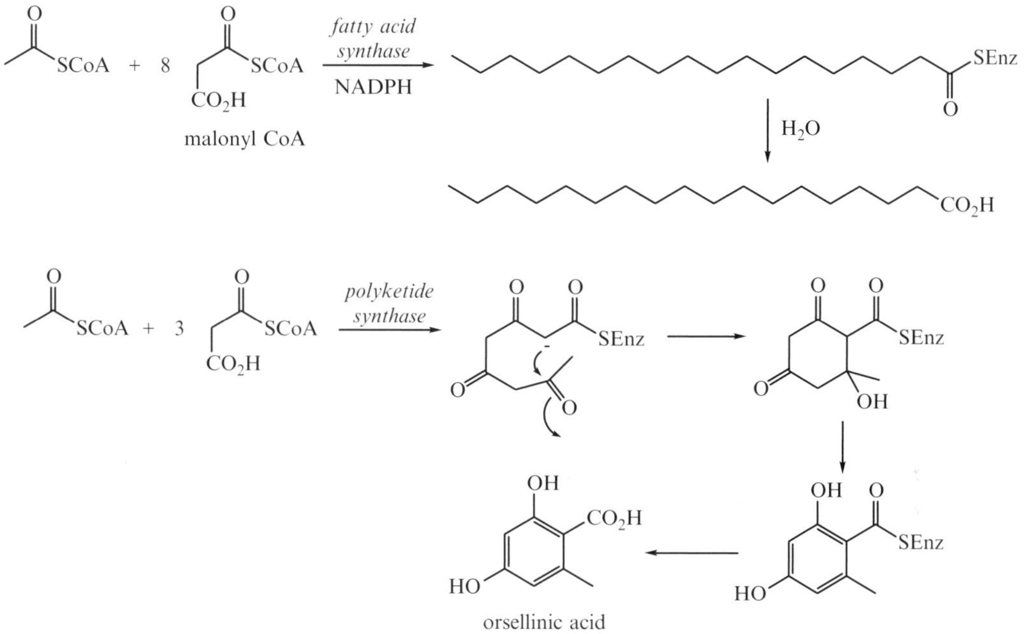

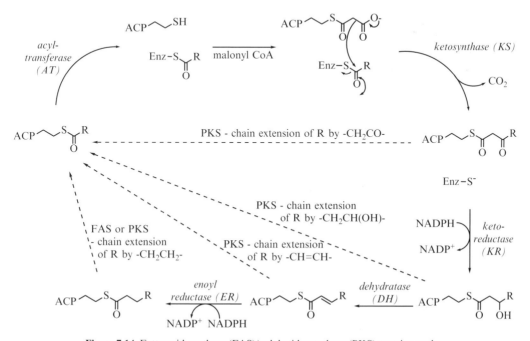

Figure 7.13 Assembly of fatty acids and polyketides.

Figure 7.14 Fatty acid synthase (FAS)/polyketide synthase (PKS) reaction cycle.

onto the acetyl-thioester intermediate, with decarboxylation of the malonyl group, by a ketosynthase (KS) activity. This reaction, producing a β-keto-thioester, is similar to that of the Claisen enzymes above, except that decarboxylation occurs at the same point.

The question which then arises is whether decarboxylation occurs before, after, or at the same time as carbon–carbon bond formation? Stereospecific labelling studies shown in Figure 7.15 have demonstrated that this reaction occurs with inversion of configuration at C-2 of the malonyl unit and with little or no hydrogen exchange at C-2. This implies that carbon–carbon bond formation and decarboxylation are, in fact, concerted.

In the case of fatty acid biosynthesis, the new β-keto-thioester is reduced to a β-hydroxy-thioester, eliminated to give an α,β-unsaturated thioester, and then further reduced to give a two-carbon-extended acyl chain, as shown in Figure 7.14. In the case of polyketide biosynthesis, each two-carbon unit can be processed as either the β-keto-thioester, the β-hydroxy-thioester, the α,β-unsaturated thioester or as the fully reduced thioester. Assembly of each polyketide is therefore controlled by the arrangement of processing enzyme activitites on the polyketide synthase multi-enzyme complex. How is this done?

Information regarding the molecular structure and organisation of polyketide synthases is now emerging from the cloning and sequencing of genes which encode these enzymes. The genes responsible for the biosynthesis of the polyketide antibiotic erythromycin have been identified and their nucleotide sequences determined. They encode three huge multi-functional polypeptides of size 300–500 kDa, illustrated in Figure 7.16. The enzyme activities responsible for processing of the growing polyketide chain have been identified by amino acid sequence alignments, and are found sequentially along the polypeptide chains. Remarkably, the arrangement of processing enzyme activities on the polyketide synthases matches the order of chemical steps required for biosynthesis of the polyketide precursor. It, therefore, appears that these multi-enzyme complexes function as molecular production lines built up of 'modules' of enzyme activities.

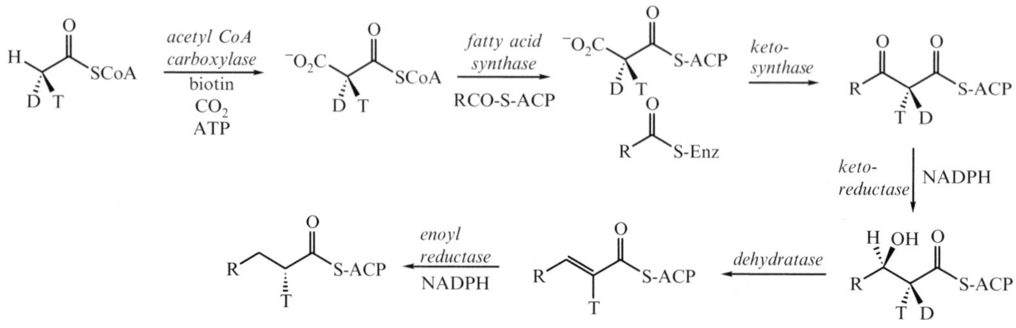

Figure 7.15 Stereochemistry of fatty acid biosynthesis.

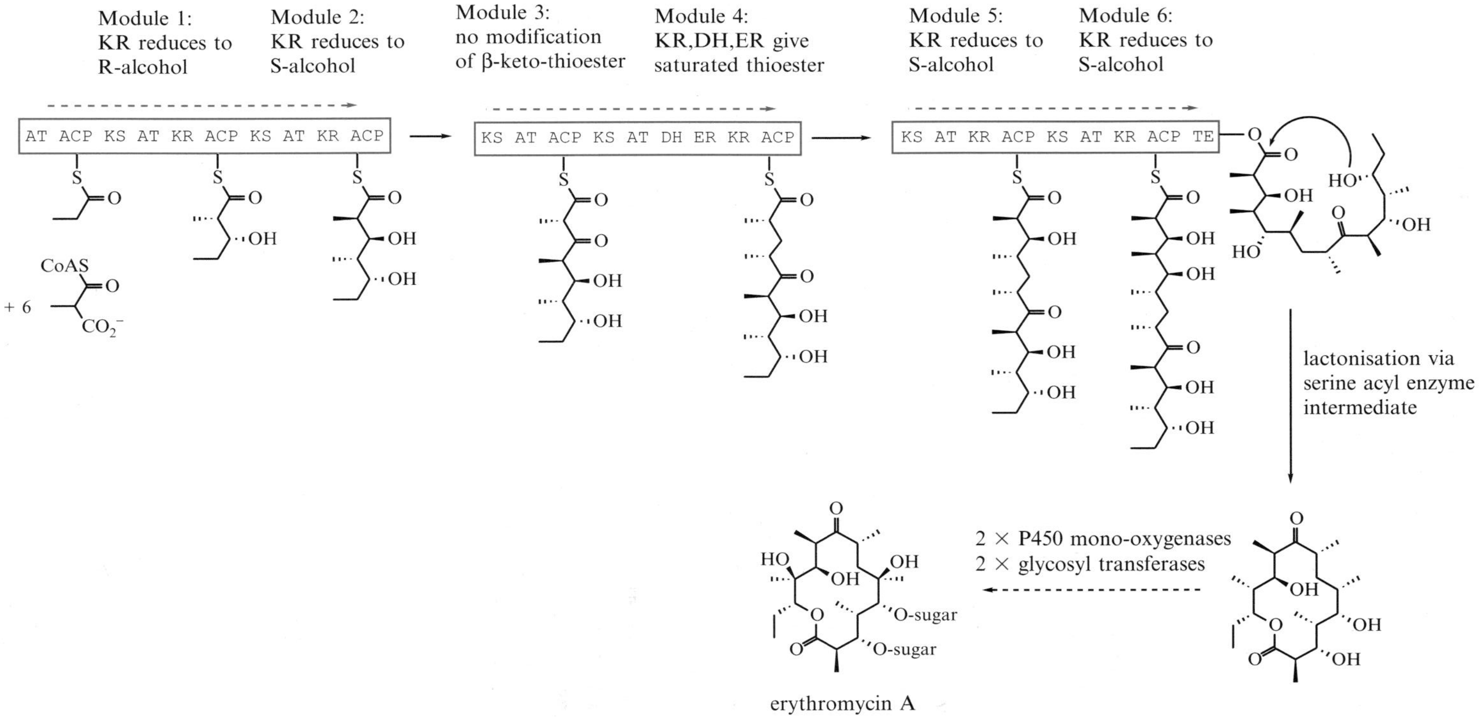

Figure 7.16 Assembly of erythromycin A via multi-enzyme polyketide synthases.

In the case of erythromycin the polyketide is assembled from the three-carbon unit of propionyl CoA, which is carboxylated to give methylmalonyl CoA. Acyl transfer and carbon–carbon bond formation takes place through the α-carbon of a methylmalonyl-thioester, in the same way as is shown in Figure 7.14, giving α-methyl-β-keto-thioesters at each stage. Each 'module' of enzyme activities contains the enzymes required for the assembly of a new β-keto-thioester and its subsequent modification. For example, the first module contains ketosynthase (KS) and acyltransferase (AT) activities to make the new β-keto-thioester, and a ketoreductase (KR) activity to reduce the β-keto-thioester to a β-R-hydroxy-thioester, and so on. Each of the multi-functional polyketide synthases contains two such 'modules' of enzymatic activities. At the end of the third polyketide synthase is a thioesterase (TE) activity which catalyses the intramolecular lactonisation via a serine acyl enzyme intermediate. Subsequent modification of the polyketide precursor to erythromycin A occurs by separate P_{450} mono-oxygenase and glycosyl transferase enzymes.

7.5   Carboxylases: use of biotin

We have already seen examples of nucleophilic attack of a carbanion equivalent onto aldehyde and ester electrophiles. There are finally a number of examples of nucleophilic attack of carbanion equivalents onto carbon dioxide to generate carboxylic acid products.

We have just seen that the carboxylation of acetyl CoA to give malonyl CoA is an important step in fatty acid and polyketide natural product biosynthesis. This step is catalysed by acetyl CoA carboxylase. This enzyme uses acetyl CoA and bicarbonate as substrates, but also requires adenosine triphosphate (ATP), which is converted to adenosine diphosphate (ADP) and inorganic phosphate (P_i), and the cofactor biotin. Biotin was first isolated from egg yolk in 1936, and was found to act as a vitamin whose deficiency causes dermatitis. Its structure is a bicyclic ring system containing a substituted urea functional group which is involved in its catalytic function. The biotin cofactor is covalently attached to the ε-amino side chain of an active site lysine residue.

How does such an apparently unreactive chemical structure serve to activate carbon dioxide for these carboxylation reactions, and what is the role of ATP in the reaction? These questions were answered by a series of experiments with isotopically labelled bicarbonate substrates. Bicarbonate is rapidly formed from carbon dioxide in aqueous solution and is the substrate for biotin-dependent carboxylases. Incubation of biotin-dependent β-methylcrotonyl-CoA carboxylase with ^{14}C-bicarbonate and ATP gave an intermediate ^{14}C-labelled enzyme species. Methylation with diazomethane followed by degradation of the enzyme structure revealed that the ^{14}C-label was covalently attached to the biotin cofactor, in the form of a carbon dioxide adduct onto N−1 of the cofactor. This

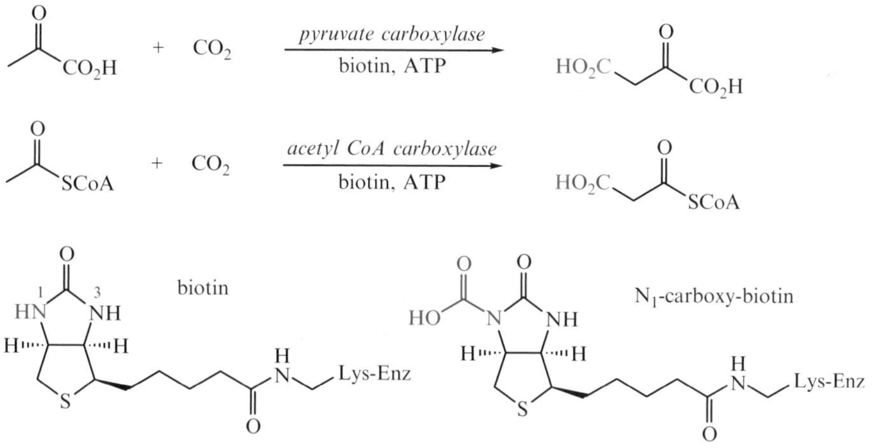

Figure 7.17 Biotin-dependent enzymes.

species N_1-carboxy-biotin, illustrated in Figure 7.17, has been shown to be chemically and kinetically competent as an intermediate in the carboxylation reaction.

Incubation of enzyme with ^{18}O-labelled bicarbonate led to the isolation of product containing two atoms of ^{18}O and inorganic phosphate containing one atom of ^{18}O. The transfer of ^{18}O to inorganic phosphate implies that ATP is used to activate bicarbonate by formation of an acyl phosphate intermediate, known as carboxyphosphate. It is thought that carboxyphosphate is then attacked by N-1 of the biotin cofactor. Since the NH group of amides and urea is normally a very unreactive nucleophile, it is thought that N-1 is deprotonated prior to attack on carboxyphosphate. The carboxy-biotin intermediate thus formed is then attacked by a deprotonated substrate, probably by initial decarboxylation to generate carbon dioxide, to form the carboxylated product and regenerate the biotin cofactor, as shown in Figure 7.18. It has been shown that the carboxylation of pyruvate to oxaloacetate catalysed by pyruvate carboxylase proceeds with retention of configuration at C-3, as shown in Figure 7.19. Other biotin-dependent carboxylases also proceed with retention of configuration.

7.6   Ribulose bisphosphate carboxylase/oxygenase (Rubisco)

All of the carbon-based moleules found in living systems are ultimately derived from the fixation of gaseous carbon dioxide by green plants during photosynthesis. Carbon dioxide is then regenerated from respiration of living organisms, from the decomposition of carbon-based material and dead organisms, and from the combustion of wood and fossil fuels. This global cycle of processes is known as the carbon cycle. The enzyme which is responsible for the fixation of carbon dioxide by green plants is an enzyme of primary importance to life on earth. This enzyme is **ribu**lose **bis**phosphate **c**arboxylate/**o**xygenase, commonly known as Rubisco.

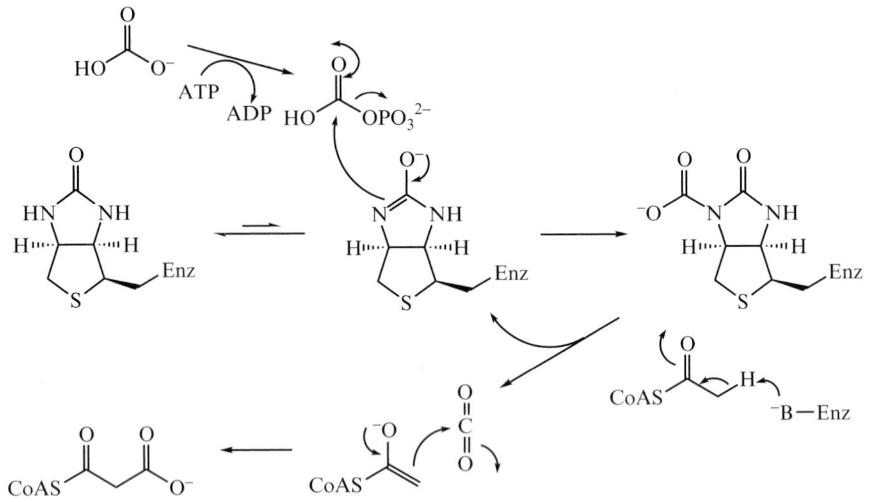

Figure 7.18 Mechanism for biotin-dependent carboxylation.

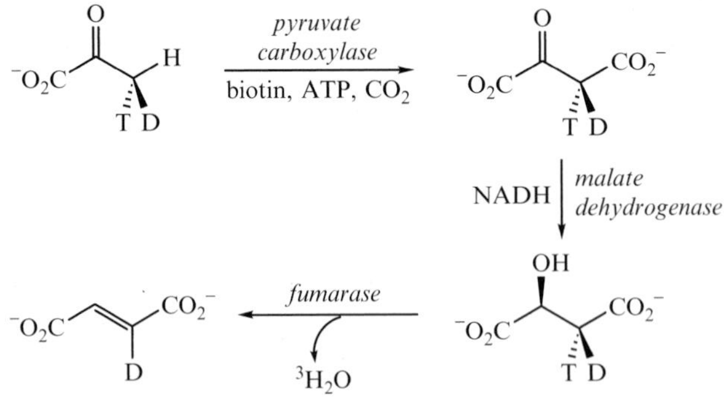

Figure 7.19 Stereochemistry of the pyruvate carboxylase reaction.

The reaction catalysed by Rubisco is the carboxylation and concomitant fragmentation of ribulose-1,5-bisphosphate to generate two molecules of 3-phosphoglycerate (see Figure 7.20). This is a step on the Calvin cycle of green plants, which transforms 3-phosphoglycerate via a series of aldolase and transketolase enzymes once again into ribulose-1,5-bisphosphate. Thus each cycle incorporates one equivalent of carbon dioxide into cellular carbon.

The enzymatic reaction has been shown to commence by deprotonation at C-3 of ribulose-1,5-bisphosphate, and enolisation of the C-2 ketone, since [3-^{3}H]-ribulose-1,5-bisphosphate rapidly exchanges ^{3}H with solvent in the presence of enzyme. The resulting enediol intermediate reacts with carbon dioxide to generate a carboxylated intermediate. Carbon–carbon bond cleavage is thought to occur by hydration of the C-3 ketone, followed by fragmentation of the C-2,C-3 bond to generate a carbanion product, which protonates to give

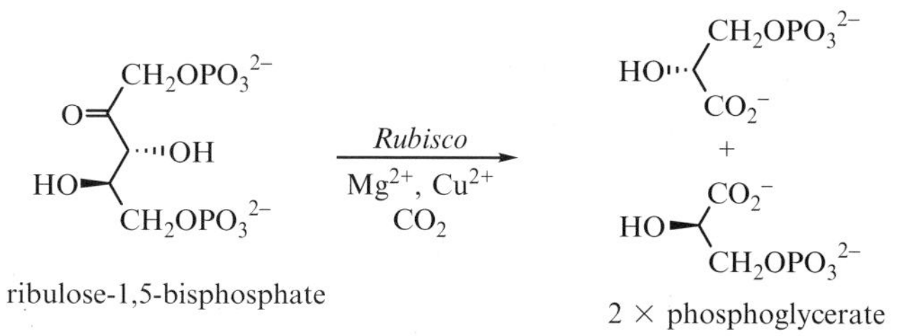

Figure 7.20 Reaction catalysed by ribulose-1,5-bisphosphate carboxylase.

the second molecule of 3-phosphoglycerate. The proposed mechanism is shown in Figure 7.21.

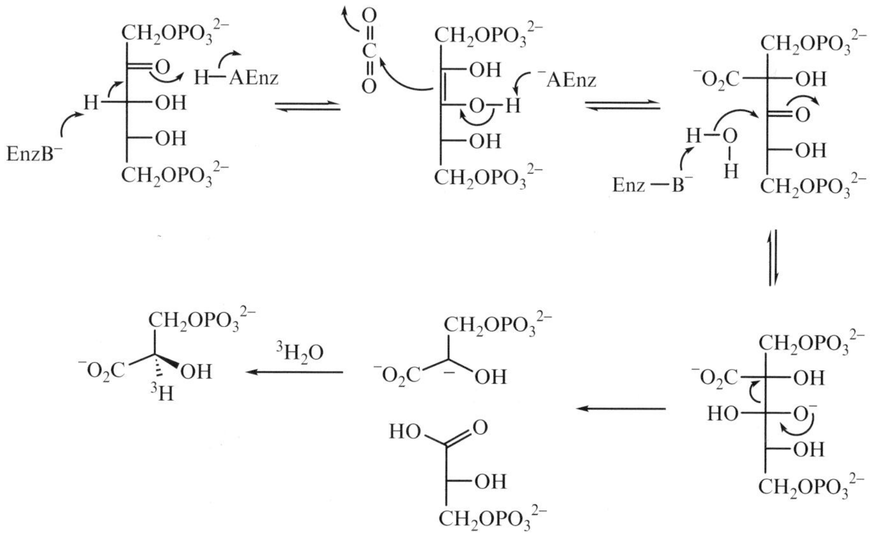

Figure 7.21 Mechanism for the ribulose-1,5-bisphosphate carboxylase reaction.

7.7 Vitamin K-dependent carboxylase

One other carboxylase enzyme worthy of note is the vitamin K-dependent carboxylase responsible for the activation of prothrombin via carboxylation of a series of glutamate amino acid residues. This step is involved in the calcium-dependent activation of platelets during the blood clotting response. Vitamin K is a fat-soluble naphthoquinone which can exist in either oxidised (quinone form) or reduced (hydroquinone form) states. The carboxylase enzyme is also an integral membrane protein which has only recently been purified. The carboxylation of prothrombin requires reduced vitamin K, carbon dioxide, oxygen and a carboxylation substrate (see Figure 7.22). Reduced

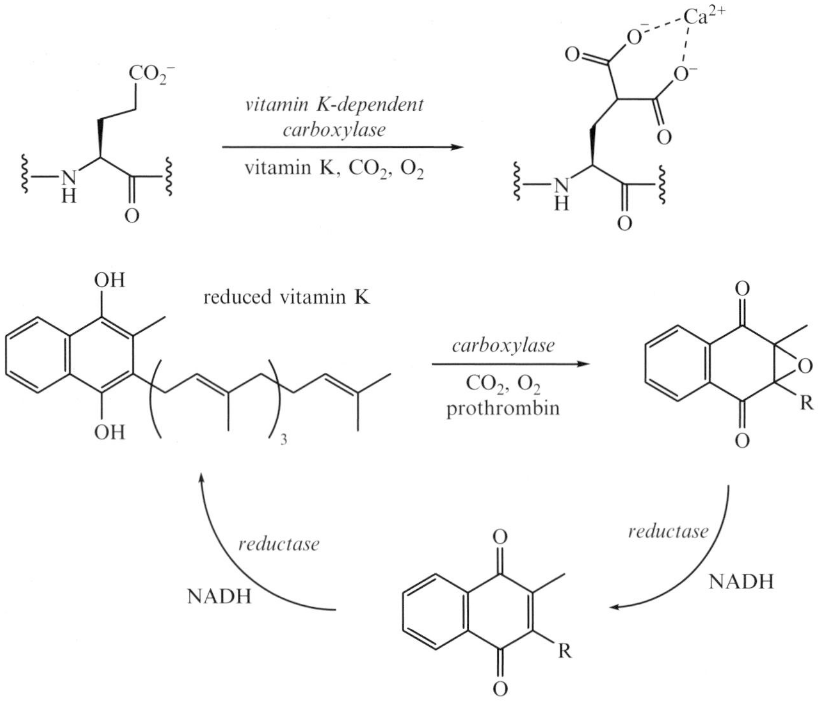

Figure 7.22 The vitamin K-dependent carboxylase reaction.

vitamin K is converted stoichiometrically into vitamin K epoxide, which is recycled via separate reductase enzymes back to reduced vitamin K.

A major clue to the role of vitamin K in the enzymatic reaction came from a chemical model reaction shown in Figure 7.23. In the presence of molecular oxygen and a crown ether, a naphthol analogue of reduced vitamin K was

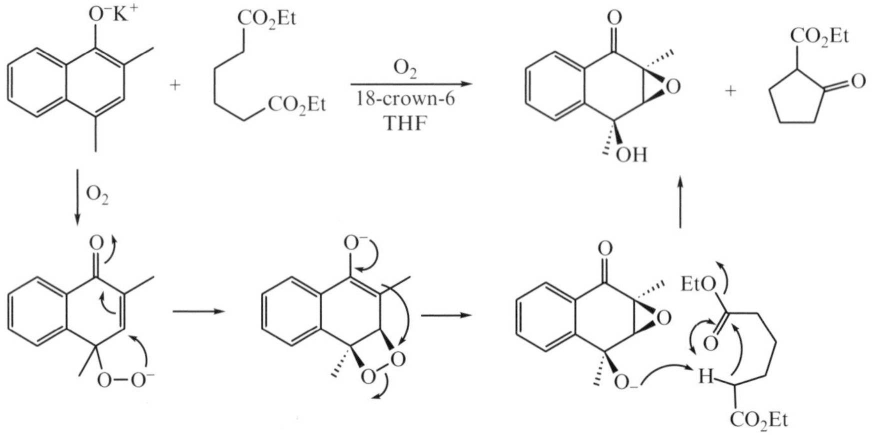

Figure 7.23 Model reaction for role of vitamin K.

found to act as a catalyst for the Dieckmann condensation of diethyl adipate, generating an epoxide byproduct. A mechanism was proposed for this reaction in which the naphthoxide salt and dioxygen react via a dioxetane intermediate to form a tertiary alkoxide species which acts as a base for the ester condensation reaction. Note that the naphthoxide salt itself is not a strong enough base to catalyse the reaction, but is converted into a much stronger base via reaction with oxygen.

In order to investigate the mechanism of the enzyme-catalysed carboxylation reaction, reduced vitamin K was incubated with carbon dioxide and enzyme in the presence of $^{18}O_2$. Complete incorporation of ^{18}O into the epoxide oxygen of vitamin K epoxide was observed, together with partial incorporation of ^{18}O (approx. 20%) into the ketone carbonyl, consistent with the dioxetane mechanism. It has been proposed that the hydrate alkoxide ion released from fragmentation of the dioxetane intermediate acts as a base to deprotonate the γ-H of a glutamyl substrate, followed by reaction with carbon dioxide, as shown in Figure 7.24. It has been shown that the C-4 *proS* hydrogen is abstracted, and subsequent reaction with carbon dioxide occurs with inversion of configuration. Note that it is remarkable that the hydrate alkoxide ion does not simply collapse to form the corresponding ketone: presumably the microenvironment of the enzyme active site disfavours the loss of hydroxide ion from the lower face of vitamin K.

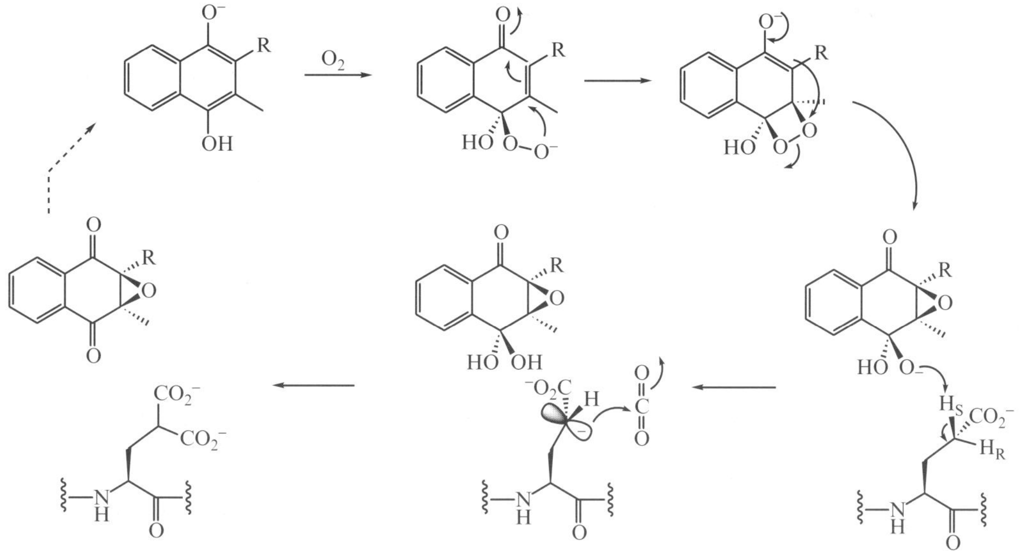

Figure 7.24 Mechanism for the vitamin K-dependent carboxylase.

7.8   Thiamine pyrophosphate-dependent enzymes

Decarboxylation reactions are also widely found in biological chemistry. In Chapter 3 we saw an example of the decarboxylation of a β-keto acid aceto-acetate by the action of acetoacetate decarboxylase, via an imine linkage. In general terms β-keto acids can be fairly readily decarboxylated since the β-keto group provides an electron sink for the decarboxylation reaction. No such electron sink exists for α-keto acids; however, Nature has found a way of decarboxylating α-keto acids, and in many cases forming a carbon–carbon bond at the same time, using the coenzyme thiamine pyrophosphate (TPP).

Thiamine pyrophosphate is formed by phosphorylation of the vitamin thiamine, lack of which causes the deficiency disease beri-beri. The structure of TPP consists of two heterocyclic rings: a substituted pyrimidine ring and a substituted thiazolium ring (see Figure 7.25). The thiazolium ring is responsible for the catalytic chemistry carried out by this coenzyme, due to two chemical properties:

(1) the acidity of the proton attached to the thiazolium ring;
(2) the presence of a carbon–nitrogen double bond that can act as an electron sink for decarboxylation.

The decarboxylation of pyruvic acid is carried out by several TPP-dependent enzymes, yielding acetaldehyde, acetic acid, acetyl CoA or α-hydroxyacetyl compounds. Each of these products is formed via closely related mechanisms.

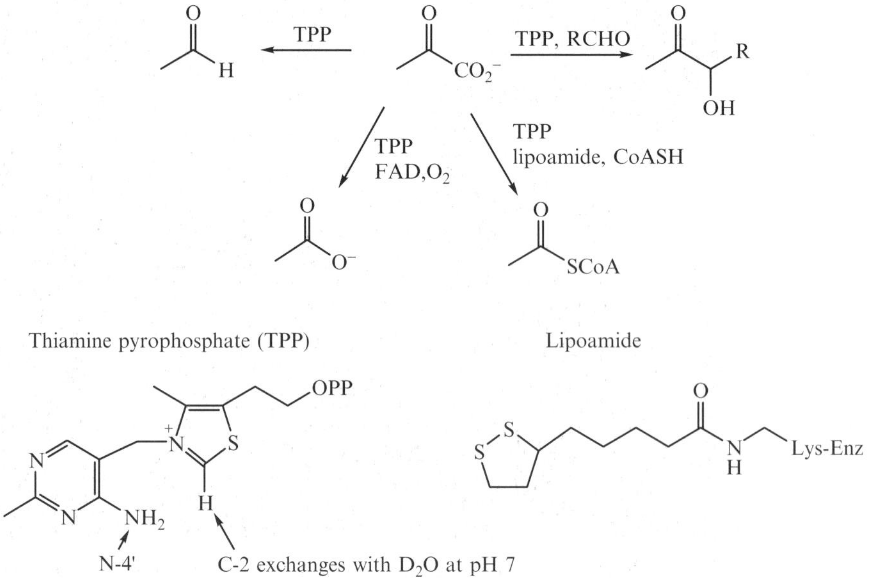

Figure 7.25 Thiamine pyrophosphate-dependent reactions of pyruvate.

The first step in each of the TPP-dependent reactions is deprotonation of the thiazolium ring to generate a carbanion ylid at C-2. In chemical models the pK_a of this proton is 17–19, yet when bound to the enzyme it is able to exchange with ^2H_2O, therefore the enzyme active site increases the acidity of this position dramatically. When bound to the active site, the TPP cofactor adopts a 'V' conformation (shown for pyruvate decarboxylase in Figure 7.26), which brings N-4' of the aminopyrimidine ring in close proximity to C-2 of the thiazolium ring. It is thought therefore that N-4' of an imine tautomer of the aminopyrimidine ring deprotonates C-2, to generate the C-2 carbanion.

The ylid carbanion then attacks the keto group of pyruvate to generate a covalently attached lactyl adduct (see Figure 7.27). Decarboxylation of the lactyl adduct then takes place, using the carbon–nitrogen double bond as an electron sink, forming an enamine intermediate. Protonation of this intermediate and fragmentation of the linkage with the coenzyme yields acetaldehyde and regenerates the coenzyme ylid. Alternatively, the enamine intermediate can react with an electrophile (E^+), generating after fragmentation of the linkage with the coenzyme, an acetyl-E species.

An important reaction in the cellular breakdown of D-glucose is the conversion of pyruvate to acetyl CoA, which is catalysed by the pyruvate

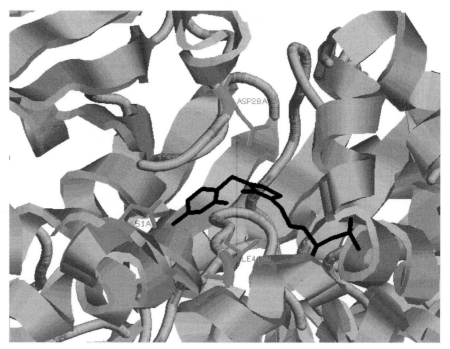

Figure 7.26 Active site of yeast pyruvate decarboxylase (PDB file 1QPB), showing the 'V' conformation of the bound TPP cofactor, which is shown in black. Three active site residues, Asp-28, Glu-51 and Ile-415, are shown in red.

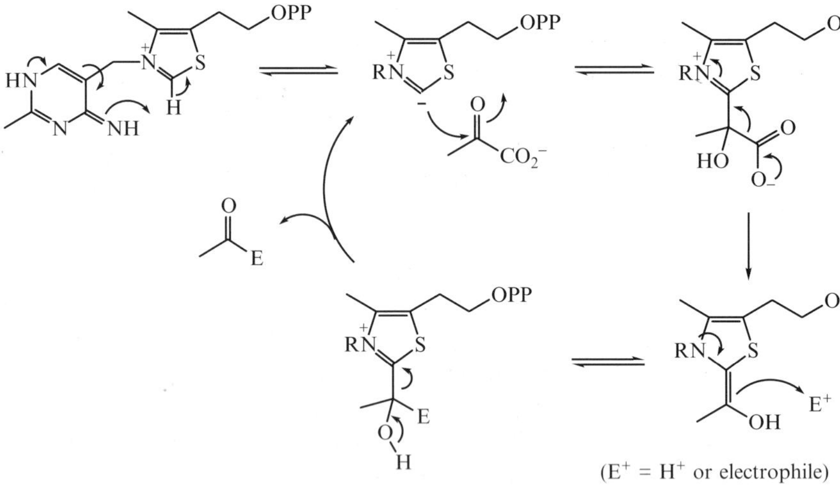

Figure 7.27 Mechanism for pyruvate decarboxylation.

dehydrogenase complex. This reaction follows the mechanism shown in Figure 7.27, but in this case the electrophile is a second cofactor present in the pyruvate dehydrogenase complex: lipoamide. The structure of lipoamide shown in Figure 7.28 is simply a five-membered ring containing a disulphide linkage attached via an acyl chain to the ε-amino group of a lysine residue. Upon reaction of the thiamine pyrophosphate–enamine intermediate with oxidised lipoamide and fragmentation of the linkage with thiamine pyrophosphate, an acetyl-lipoamide thioester is formed, as shown in Figure 7.28. Transfer of the acyl group to the thiol group of CoA generates acetyl CoA and reduced

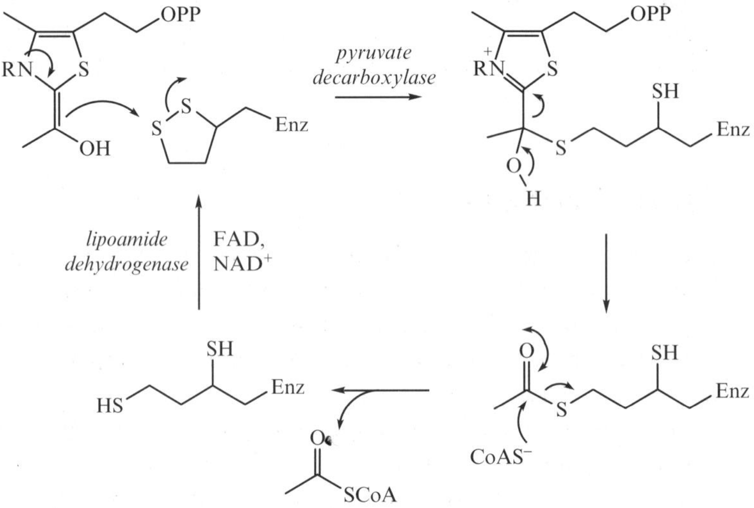

Figure 7.28 TPP-dependent production of acetyl CoA using lipoamide.

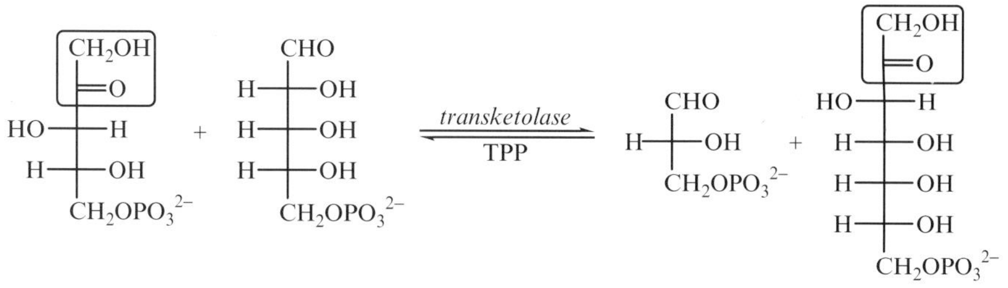

Figure 7.29 Example of a transketolase reaction.

lipoamide, which is recycled to oxidised lipoamide by flavin-dependent lipo-amide dehydrogenase (see Section 6.5).

Thiamine pyrophosphate is also used by several enzymes for carbon–carbon bond formation. Illustrated in Figure 7.29 is one example of the family of TPP-dependent transketolase enzymes which carry out a range of carbohydrate 2-hydroxyacetyl transfer reactions. A similar mechanism can be written for these reactions initiated by attack of the thiazolium ylid on the keto group, followed in this case not by decarboxylation but by carbon–carbon bond cleavage. The ability of these enzymes to form carbon–carbon bonds enantio-selectively is also being exploited for novel biotransformation reactions that are of use in organic synthesis.

Carbon–carbon bond formation via carbocation intermediates

Carbon–carbon bond formation via carbocation (or carbonium ion) intermediates is less widely found than via carbanion equivalents. However, there is one large class of biological reactions that involve highly stabilised carbocation ion intermediates: the conversion of allylic pyrophosphate metabolites into terpenoid natural products.

7.9   Terpene cyclases

Terpenes are a major class of natural products found widely in plants, but also including the steroid lipids and hormones found in animals. The common structural feature of the terpene natural products is the five-carbon isoprene unit. Some common examples shown in Figure 7.30 are plant natural products menthol, camphor and geraniol.

The biosynthesis of terpenoid natural products proceeds from a family of allylic pyrophosphates containing multiples of five carbon atoms. Two five-carbon units are joined together by the enzyme geranyl pyrophosphate synthase as shown in Figure 7.31. Loss of pyrophosphate from dimethylallyl

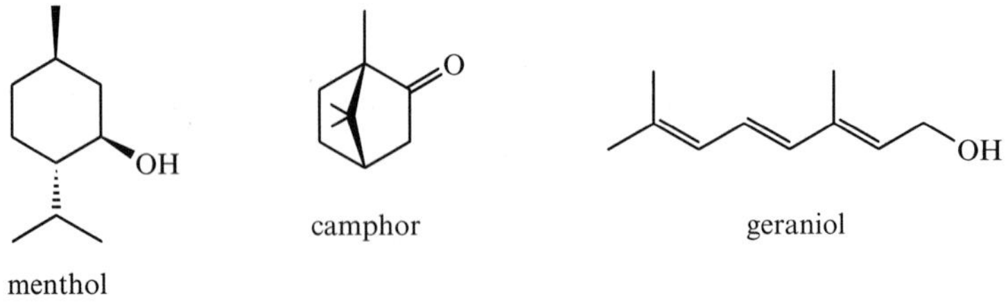

Figure 7.30 Terpenoid natural products.

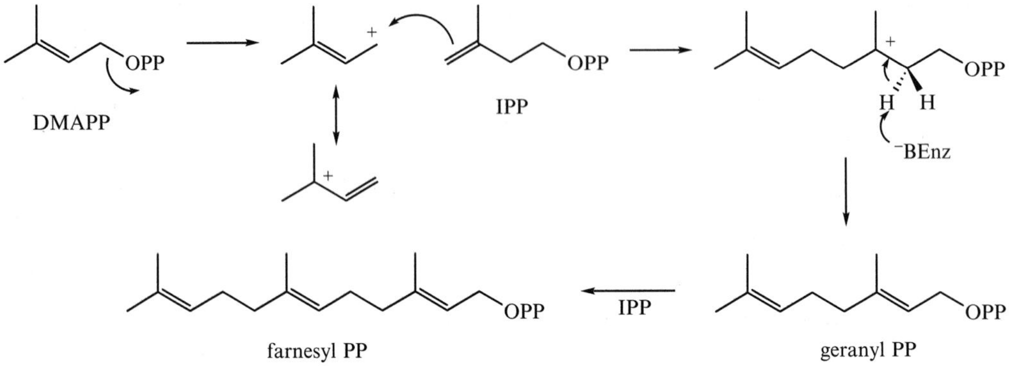

menthol

camphor

geraniol

DMAPP

IPP

IPP

farnesyl PP

geranyl PP

Figure 7.31 Biosynthesis of farnesyl pyrophosphate.

pyrophosphate (DMAPP) generates a stabilised allylic carbonium ion. Attack of the π-bond of isopentenyl pyrophosphate (IPP) generates a tertiary carbocation. Stereospecific loss of a proton generates the product geranyl pyrophosphate. Similar reactions with further units of IPP generate the C_{15} building block farnesyl pyrophosphate, the C_{20} geranylgeranyl pyrophosphate, and so on.

Geranyl pyrophosphate is converted to the family of C_{10} monoterpenes by the monoterpene cyclases. A 96-kDa cyclase enzyme responsible for the assembly of (+)-α-pinene has been purified from sage leaves, whilst a separate 55-kDa enzyme from the same source catalyses the production of the opposite enantiomer (−)-α-pinene. The 55-kDa cyclase II produces several other isomeric products. As well as (−)-α-pinene (26%) the enzyme produces (−)-β-pinene (21%), (−)-camphene (28%), myrcene (9%) and (−)-limonene (8%). These products can be rationalised by the mechanism shown in Figure 7.32. It has been shown that the first step in the assembly of these monoterpenes is a 1,3-migration of pyrophosphate to give linalyl pyrophosphate, which for cyclase II has the 3S stereochemistry. Loss of pyrophosphate gives an allylic cation, which is attacked by the second alkene to give a six-membered ring and a tertiary carbocation intermediate. Further ring closure to generate a four-membered ring gives another tertiary carbocation, which can deprotonate in either of two ways to give α-pinene or β-pinene. Limonene can be formed by

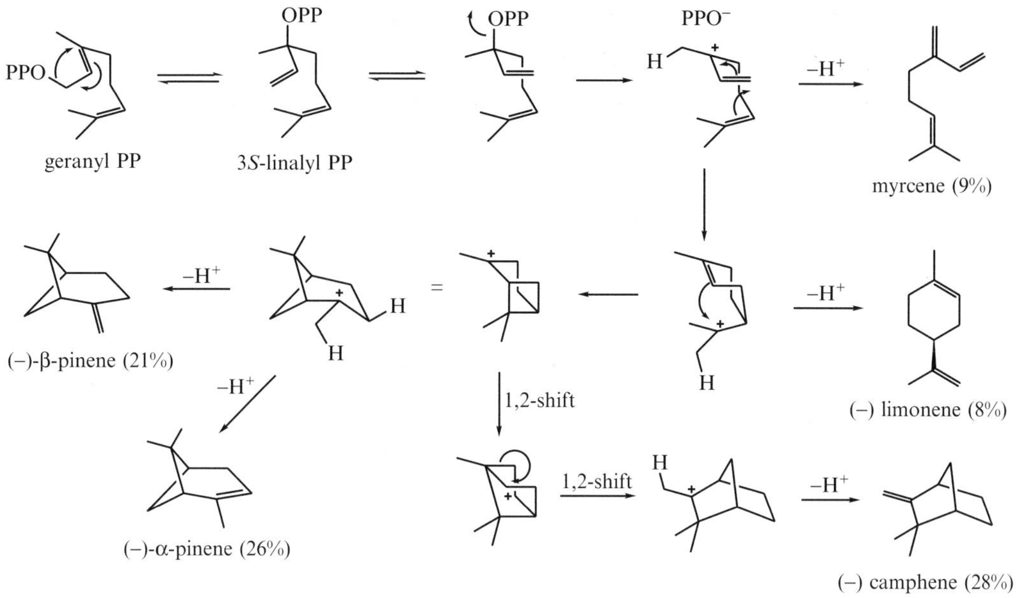

Figure 7.32 Mechanism for (−)-α-pinene synthase.

elimination from the monocyclic carbocation, whilst camphene is formed by two consecutive 1,2-alkyl migrations, followed by elimination.

Most enzymes are specific for the production of a single isomeric product, so in this case how can we be sure that all of these products are from one enzyme? Apart from demonstrating that there is only one homogeneous protein present, a chemical test can be made. If there is a common precursor to two or more products, then by introducing deuterium atoms at specific points in the substrate, the ratio of products should be affected by the presence of a deuterium isotope effect, which should favour one pathway over another. This test was applied to the (−)-pinene synthase reaction by incubation of the specifically deuterated substrate shown in Figure 7.33. At the point of divergence to α-pinene or β-pinene the presence of a terminal −CD$_3$ group would be expected to favour formation of α-pinene. This was indeed observed: the proportion of α-pinene in the product mixture rose from 26% to 38%, whilst the proportion of β-pinene dropped from 21% to 13%. The proportion of myrcene product also dropped significantly from 9% to 4%, also consistent with an isotope effect operating on the allylic carbocation intermediate.

Farnesyl pyrophosphate is cyclised to form the large family of sesquiterpene natural products by a further class of terpene cyclase enzymes, several of which have been purified to homogeneity. Pentalenene synthase catalyses the cyclisation of farnesyl pyrophosphate to give pentalenene, which is further processed in *Streptomyces* UC5319 to give the pentalenolactone family of antibiotics, as shown in Figure 7.34. The enzyme is a 41-kDa protein with a K$_M$ of 0.3 μM for

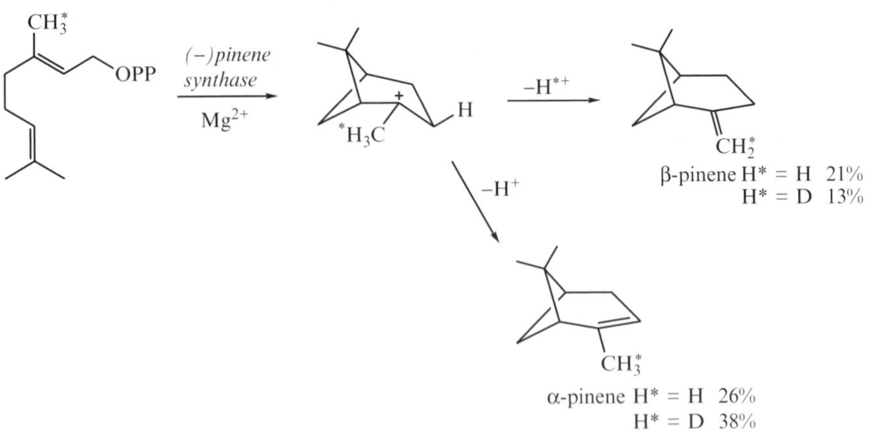

Figure 7.33 Kinetic isotope effect in the (−)-pinene synthase reaction.

farnesyl pyrophosphate and a k_{cat} of $0.3\,s^{-1}$, and an absolute requirement for Mg^{2+} ions. A series of isotopic labelling experiments have been carried out to support the mechanism of cyclisation shown in Figure 7.34. Cyclisation of farnesyl pyrophosphate is proposed to form an 11-membered intermediate, humulene, which is followed by a five-membered ring closure to form a bicyclic tertiary carbocation. 1,2-Hydride migration followed by a further five-membered ring closure gives a tricyclic carbocation, which upon elimination gives pentalenene.

Examination of the inferred amino acid sequence of pentalenene synthase revealed an amino acid sequence motif Asp–Asp–X–X–Asp (DDXXD) found in other pyrophosphate-utilising enzymes. It is believed that the Asp–Asp pair are involved in chelating the essential Mg^{2+} ion, which in turn chelates the pyrophosphate ion as shown in Figure 7.35. An indication of the tight binding of pyrophosphate is that this product inhibits the enzymatic reaction strongly ($K_i\,3\,\mu M$).

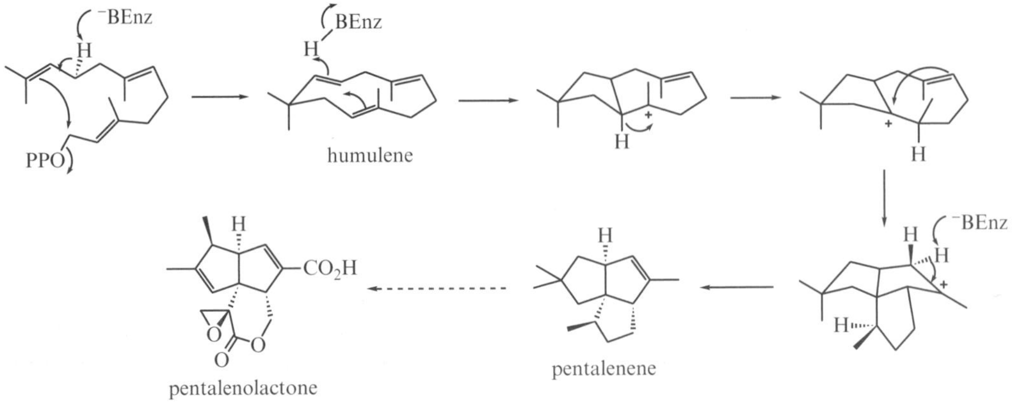

Figure 7.34 Mechanism for pentalenene synthase.

```
Pentalenene synthase            F  L  D  D  F  L  D
Trichodiene synthase            V  L  D  D  S  K  D
Aristolochene synthase          L  I  D  D  V  L  E
Casbene synthase                L  I  D  D  T  I  D
Limonene synthase               V  I  D  D  I  Y  D
Farnesyl PP synthase            V  Q  D  D  I  L  D
Geranylgeranyl PP synthase   I  A  D  D  Y  H  N


Consensus                                D  D  X  X (D)
```

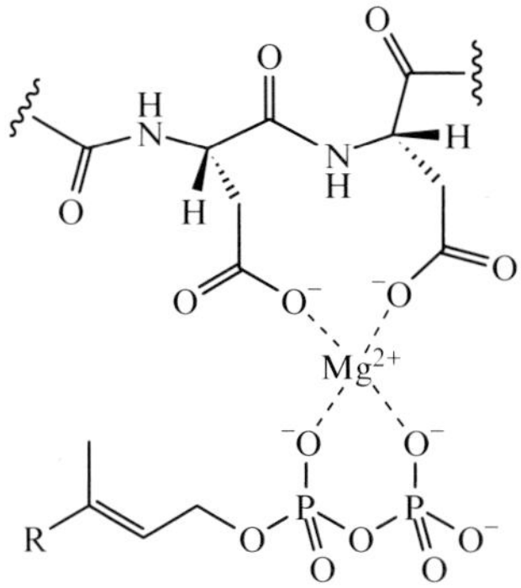

Figure 7.35 Mg^{2+}-pyrophosphate binding motif.

How do these cyclase enzymes control the precise regiochemistry and stereochemistry of these multi-step cyclisations? In 1997 the crystal structure of pentalenene synthase was solved. The active site consists of a conical hydrophobic cleft, flanked by six α-helices, lined with aromatic and non-polar residues. The DDXXD motif is located near the upper edge of the cavity, as shown in Figure 7.36. The identity of the catalytic base has not yet been identified, but replacement of Phe-76 or Phe-77 by alanine gave at least a 10-fold reduction in activity, suggesting that they may stabilise carbocationic intermediates through π-cation interactions. The presence of aromatic residues in key positions of the wall of terpene cyclase active sites suggests that the π-cation interaction, shown in Figure 7.37, may have an important role in the specific stabilisation of the carbocation intermediates.

Carbon–carbon bond formation via radical intermediates

In Chapter 6 we have seen how redox enzymes are able to generate radical intermediates in enzyme-catalysed reactions. In certain instances these radical intermediates are used for carbon–carbon bond formation reactions. The two examples which we shall examine are phenol radical couplings used in natural product biosynthesis and in lignin biosynthesis.

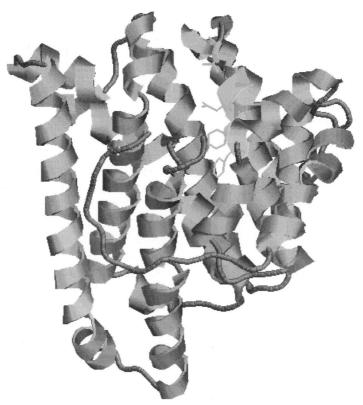

Figure 7.36 Structure of pentalanene synthase (PDB file 1PS1), showing the active site residues Asp-80, Asp-81, Asp-84 (DDXXD motif, top), Phe-76 and Phe-77 in red.

Figure 7.37 π-Cation interaction.

7.10   Phenolic radical couplings

Certain aromatic natural products are formed by the radical coupling of two phenol precursors in an enzyme-catalysed process. In some cases the enzymes are blue copper proteins (the characteristic colour due to the metal cofactor). For example, the phenol radical coupling of sulochrin shown in Figure 7.38 is catalysed by a 157-kDa oxidase enzyme containing six atoms of copper. The

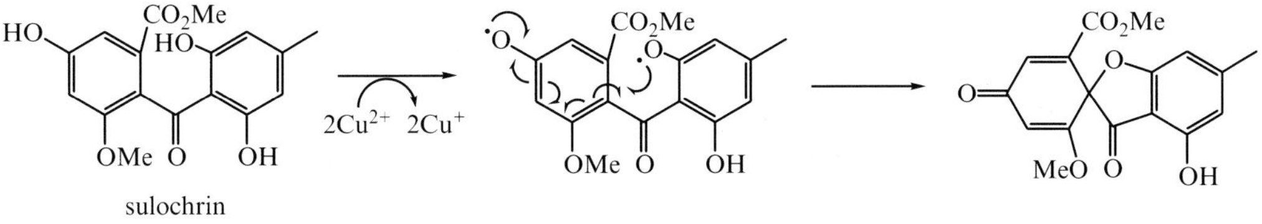

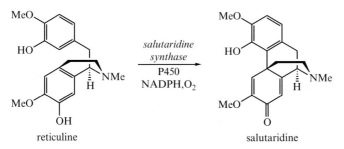

Figure 7.38 Reaction catalysed by sulochrin oxidase.

enzyme exhibits a deep blue colour (λ_{max} 605 nm) which disappears upon reduction with ascorbate under nitrogen. These observations suggest that there are active site Cu^{2+} metal ions which accept one electron from the phenol substrates to generate Cu^{+} intermediates. The phenoxy radicals thus generated are relatively stable, and react together as shown in Figure 7.38.

Oxidative phenol coupling is frequently found in pathways responsible for the biosynthesis of isoquinoline and indole alkaloids in plants. In several cases the enzymes responsible for the oxidative phenol coupling have been found to be cytochrome P_{450} enzymes (described in Section 6.8). One example is the enzyme salutaridine synthase, shown in Figure 7.39, involved in the biosynthesis of morphine in *Papaver somniferum*. In other cases, the enzymes responsible for oxidative phenol couplings have been found to be α-ketoglutarate-dependent dioxygenases (described in Section 6.9), or flavoprotein oxidases (described in Section 6.3).

Lignin is a complex aromatic polymeric material which is a major structural component of woody tissues in plants. In most trees that require a great deal of structural rigidity lignin constitutes 30–40% of the dry weight of the wood. It is highly cross-linked and exceptionally resistant to chemical degradation, making elucidation of its chemical structure very difficult. However, it is now known to be heterogeneous in structure, composed of phenylpropanoid (aromatic ring + three-carbon alkyl chain) units linked together by a variety of types of carbon–carbon linkage. Shown in Figure 7.40 are a selection of structural components found in lignin. Although structurally diverse, the common feature of these components is that they can all be formed via radical couplings.

Figure 7.39 Reaction catalysed by salutaridine synthase.

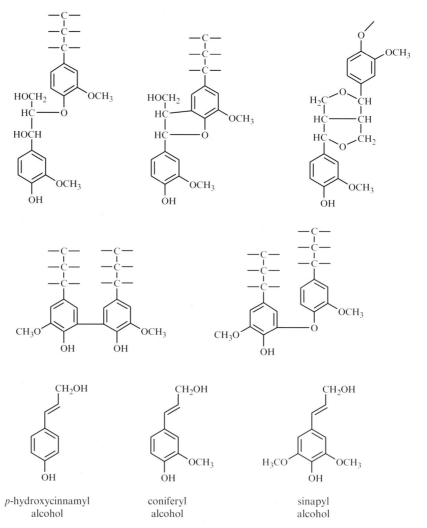

Figure 7.40 Structural components of lignin.

The biosynthesis of lignin starts from three cinnamyl alcohol precursors: *p*-hydroxycinnamyl alcohol, coniferyl alcohol and sinapyl alcohol, shown in Figure 7.40. The ratio of these precursors used for lignin assembly is species-dependent. They are each activated by formation of a phenoxy radical at the C-4 hydroxyl group by a family of peroxidase, phenolase and tyrosinase enzymes widely found in plants. As shown in Figure 7.41, the phenoxy radical can exist in a number of resonance structures in which radical character is found, for example, on oxygen, at the C-5 position and at the β-position of the side chain. Carbon–carbon bond formation can then take place with the α,β-double bond of another molecule, generating a new radical species which can

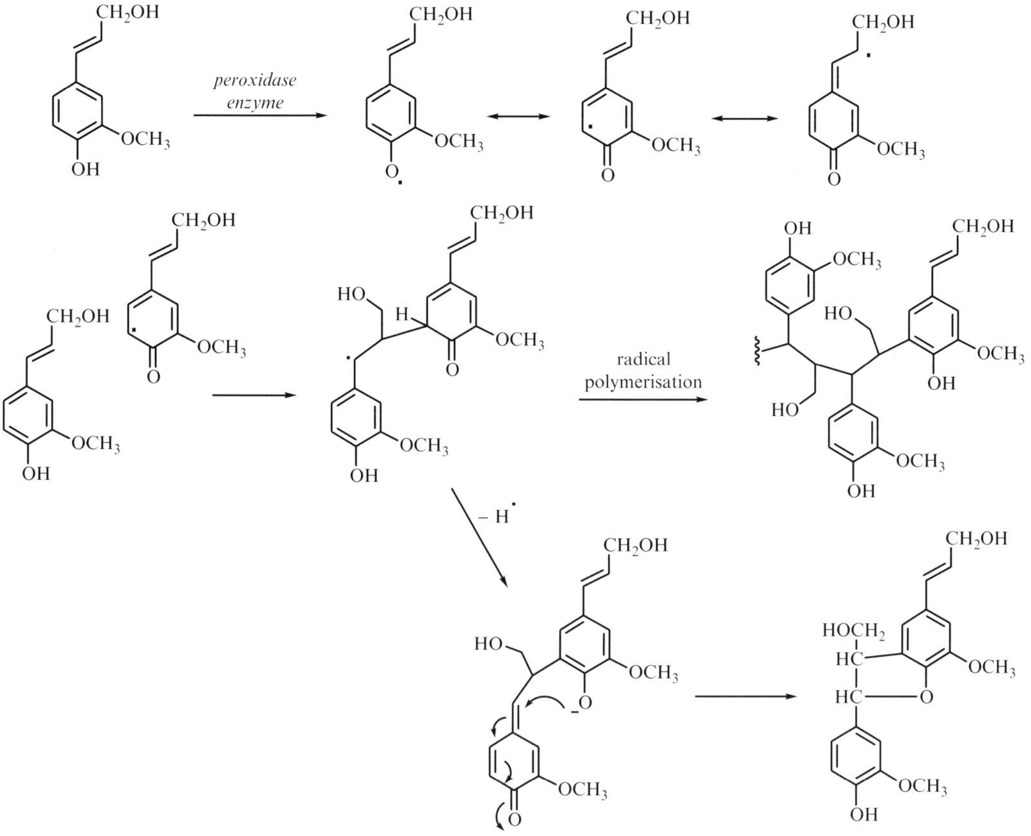

Figure 7.41 Lignin formation via radical coupling.

form a further carbon–carbon bond. Thus the formation of a phenoxy radical initiates a radical polymerisation reaction which forms a highly heterogeneous polymer. Evidence that the polymerisation is a chemical reaction and not an enzyme-catalysed reaction comes from the observation that lignin is not optically active, despite containing many chiral centres.

There are also examples of carbon–carbon forming reactions involving the coenzyme vitamin B_{12} which proceed via radical mechanisms, which will be described in Section 11.1.

Problems

(1) *N*-acetylneuraminic acid aldolase catalyses the reaction shown below. Given that the enzyme requires no cofactors, suggest a mechanism (*Hint*: use the open chain forms of the monosaccharides).

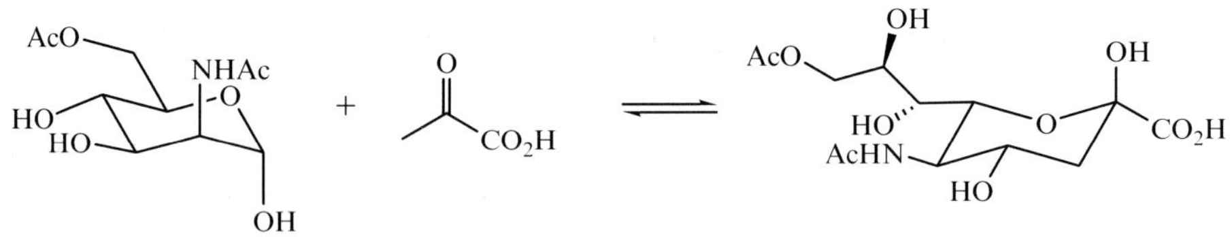

(2) The two plant enzymes chalcone synthase and resveratrol synthase catalyse the reactions shown below. Suggest mechanisms for the two enzymatic reactions. Comment on the observation that when these enzymes were sequenced, they were found to share 70–75% amino acid sequence identity.

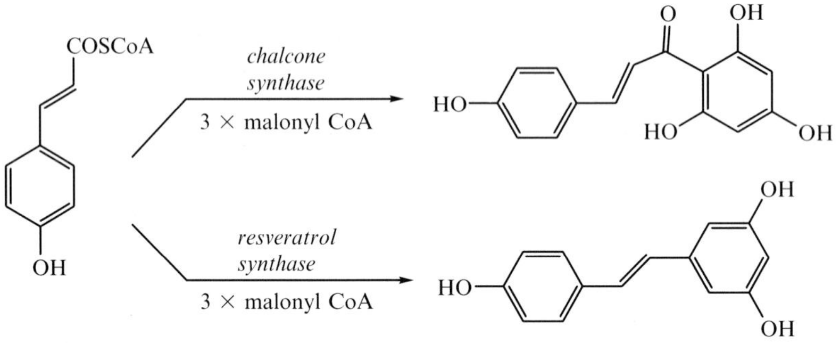

(3) PEP carboxylase catalyses the carboxylation of phosphoenolpyruvate (PEP) using Mn^{2+} as a cofactor. When incubated with $HC^{18}O_3^-$ two atoms of ^{18}O were found in the oxaloacetate product, and one atom of ^{18}O in inorganic phosphate. Suggest a mechanism.

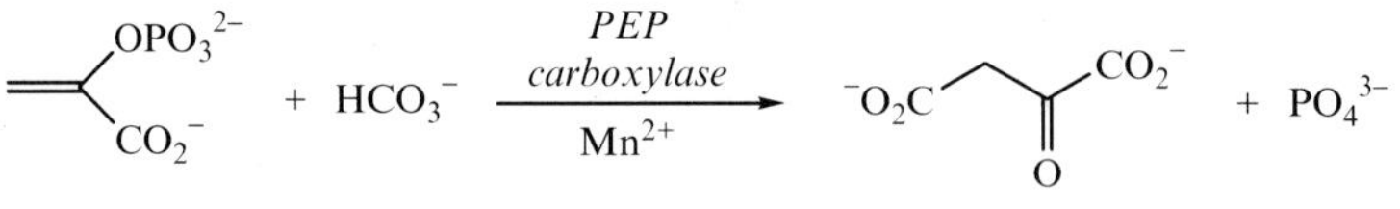

(4) Transcarboxylase catalyses the simultaneous carboxylation of propionyl CoA to methylmalonyl CoA and decarboxylation of oxaloacetate to pyruvate, as shown below. The enzyme requires biotin as a cofactor, but does *not* require ATP. Suggest a mechanism.

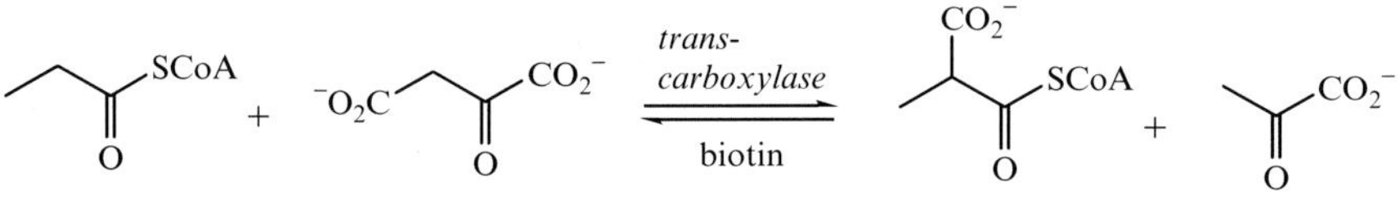

(5) Suggest a mechanism for the TPP-dependent transketolase reaction illustrated in Figure 7.25.

(6) Pyruvate oxidase catalyses the oxidative decarboxylation of pyruvate to acetate (not acetyl CoA). The enzyme requires TPP and oxidised flavin as cofactors, and utilises oxygen as an electron acceptor (i.e. it is converted to hydrogen peroxide). Suggest a mechanism.

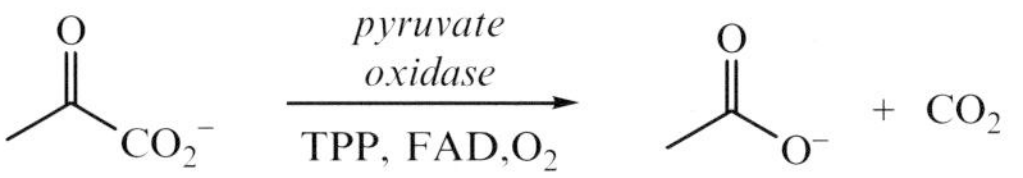

(7) The conversion of geranyl pyrophosphate into bornyl pyrophosphate is catalysed by bornyl pyrophosphate synthetase. Write a mechanism for the enzymatic reaction. When the enzyme was incubated with geranyl pyrophosphate specifically labelled with ^{18}O attached to C-1, the product was found to contain ^{18}O only in the oxygen attached to carbon, and not in any of the other six oxygens. Comment on this remarkable result.

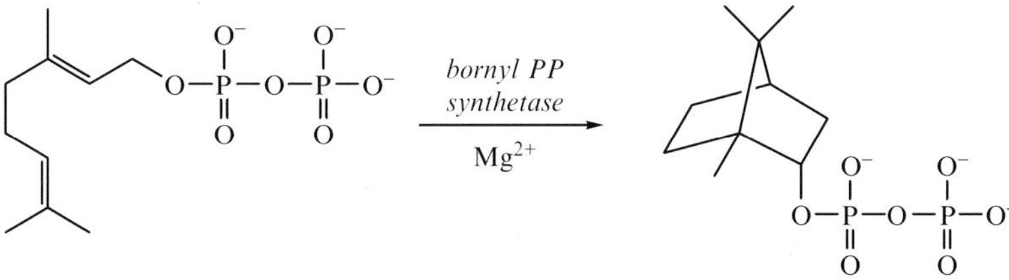

(8) Kaurene synthetase catalyses the cyclisation of geranylgeranyl pyrophosphate (GGPP) to kaurene. An intermediate in the reaction is thought to be copalyl pyrophosphate. Suggest mechanisms for conversion of GGPP into copalyl PP and thence to kaurene. How would you attempt to prove that copalyl pyrophosphate is an intermediate in this reaction?

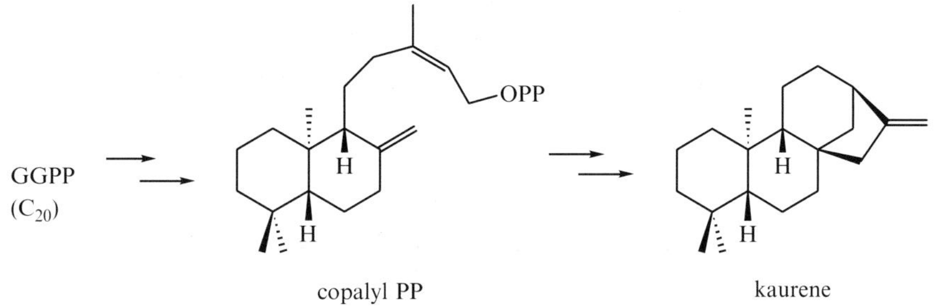

GGPP
(C_{20})

copalyl PP

kaurene

(9) Usnic acid is a natural product thought to be biosynthesised from two molecules of an aromatic precursor. Suggest a biosynthetic pathway to the aromatic precursor from acetyl CoA, and a mechanism for the conversion to usnic acid.

Further reading

General

R.H. Abeles, P.A. Frey & W.P. Jencks (1992) *Biochemistry*. Jones & Bartlett, Boston.
J. Mann (1987) *Secondary Metabolism*, 2nd edn. Clarendon Press, Oxford.
J.S. Staunton (1978) *Primary Metabolism: A Mechanistic Approach*. Clarendon Press, Oxford.
C.T. Walsh (1979) *Enzymatic Reaction Mechanisms*. Freeman, San Francisco.

Aldolases

J.G. Belasco & J.R. Knowles (1983) Polarization of substrate carbonyl groups by yeast aldolase: investigation by Fourier transform infrared spectroscopy. *Biochemistry*, **22**, 122–9.
K.H. Choi, J. Shi, C.E. Hopkins, D.R. Tolan & K.N. Allen (2001) Snapshots of catalysis: the structure of fructose-1,6-bisphosphate aldolase covalently bound to the substrate dihydroxyacetone phosphate. *Biochemistry*, **40**, 13868–75.
D.R. Hall, G.A. Leonard, C.D. Reed, C.I. Watt, A. Berry & W.N. Hunter (1999) The crystal structure of *Escherichia coli* class II fructose-1,6-bisphosphate aldolase in complex with phosphoglycolohydroxamate reveals details of mechanism and specificity. *J. Mol. Biol.*, **287**, 383–94.
J.B. Jones (1986) Enzymes in organic synthesis. *Tetrahedron*, **42**, 3341–403.
A.J. Morris & D.R. Tolan (1994) Lysine-146 of rabbit muscle aldolase is essential for cleavage and condensation of the C3–C4 bond of fructose 1,6-bisphosphate. *Biochemistry*, **33**, 12291–7.
J. Sygusch, D. Beaudry & M. Allaire (1987) Molecular architecture of rabbit skeletal muscle aldolase at 2.7 Å resolution. *Proc. Natl. Acad. Sci. USA*, **84**, 7846–50.
E.J. Toone, E.S. Simon, M.D. Bednarski & G.M. Whitesides (1989) Enzyme-catalysed synthesis of carbohydrates. *Tetrahedron*, **45**, 5365–422.
G.M. Whitesides & C.H. Wong (1985) Enzymes as catalysts in synthetic organic chemistry. *Angew. Chem. Int. Ed. Engl.*, **24**, 617–38.

C.H. Wong & G.M. Whitesides (1994) *Enzymes in Synthetic Organic Chemistry*. Pergamon, Oxford.

Claisen enzymes

S.I. Chang & G.G. Hammes (1990) Structure and mechanism of action of a multifunctional enzyme: fatty acid synthase. *Acc. Chem. Res.*, **23**, 363–9.
J.D. Clark, S.J. O'Keefe & J.R. Knowles (1988) Malate synthase: proof of a stepwise Claisen condensation using the double-isotope fractionation test. *Biochemistry*, **27**, 5961–71.
J.S. Staunton (1991) The extraordinary enzymes involved in erythromycin biosynthesis. *Angew. Chem. Intl. Ed. Engl.*, **30**, 1302–6.

Carboxylases

P.V. Attwood & J.C. Wallace (2002) Chemical and catalytic mechanisms of carboxyl transfer reactions in biotin-dependent enzymes. *Acc. Chem. Res.*, **35**, 113–20.
P. Dowd, R. Hershline, S.W. Ham & S. Naganathan (1994) Mechanism of action of vitamin K. *Nat. Prod. Reports*, **11**, 251–64.
F.C. Hartman & M.R. Harpel (1994) Chemical and genetic probes of the active site of D-ribulose-1,5-bisphosphate carboxylase-oxygenase: a retrospective based on the three-dimensional structure. *Adv. Enzymol.*, **66**, 1–76.
J.R. Knowles (1989) The mechanism of biotin-dependent enzymes. *Annu. Rev. Biochem.*, **58**, 195–222.
N.F.B. Phillips, M.A. Shoswell, A. Chapman-Smith, D.B. Keech & J.C. Wallace (1992) Isolation of a carboxyphosphate intermediate and the locus of acetyl CoA action in the pyruvate carboxylase reaction. *Biochemistry*, **31**, 9445–50.
J.W. Suttie (1985) Vitamin K-dependent carboxylase. *Annu. Rev. Biochem.*, **54**, 459–78.

Thiamine pyrophosphate

F. Jordan (2003) Current mechanistic understanding of thiamin diphosphate-dependent enzymatic reactions. *Nat. Prod. Reports*, **20**, 184–201.
R. Kluger, J. Chin, & T. Smyth (1981) Thiamin-catalyzed decarboxylation of pyruvate. Synthesis and reactivity analysis of the central, elusive intermediate, α-lactylthiamin. *J. Am. Chem. Soc.*, **103**, 884–8.

Terpene cyclases

D.E. Cane (1990) Enzymatic formation of sesquiterpenes. *Chem. Rev.*, **90**, 1089–103.
R.B. Croteau, C.J. Wheeler, D.E. Cane, R. Ebert & H.J. Ha (1987) Isotopically sensitive branching in the formation of cyclic monoterpenes: proof that (−)α-pinene and (−)β-pinene are synthesised by the same monoterpene cyclase via deprotonation of a common intermediate. *Biochemistry*, **26**, 5383–9.
C.A. Lesburg, G. Zhai, D.E. Cane, and D.W. Christianson (1997) Crystal structure of pentalenene synthase: mechanistic insights on terpenoid cyclization reactions in biology. *Science*, **277**, 1820–24.
M. Seemann, G. Zhai, J.W. de Kraker, C.M. Paschall, D.W. Christianson & D.E. Cane (2002) Pentalenene synthase. Analysis of active site residues by site-directed mutagenesis. *J. Am. Chem. Soc.*, **124**, 7681–9.

Radical couplings

W.M. Chou & T.M. Kutchan (1998) Enzymatic oxidations in the biosynthesis of complex alkaloids. *Plant. J.*, **15**, 289–300.

K. Freudenberg (1965) Lignin: its constitution and formation from *p*-hydroxycinnamyl alcohols. *Science*, **148**, 595–600.

T. Higuchi (1971) Formation and biological degradation of lignins. *Adv. Enzymol.*, **34**, 207–83.

H. Nordlöv & S. Gatenbeck (1982) Enzymatic synthesis of (+) and (−)-bisdechlorogeodin with sulochrin oxidase from *Penicillium frequentans* and *Oospora sulphurea-ochracea. Arch. Microbiol.* **131**, 208–11.

8   Enzymatic Addition/Elimination Reactions

8.1   Introduction

The addition and elimination of the elements of water is a common process in biochemical pathways. Particularly common is the dehydration of β-hydroxy-ketones or β-hydroxy-carboxylic acids, which can be constructed by aldolase and Claisen enzymes. In this chapter we shall examine the mechanisms employed by these hydratase/dehydratase enzymes, and the related ammonia lyases.

In particular we shall discuss several enzymes on the shikimate pathway, which is depicted in Figure 8.1. This pathway is responsible for the biosynthesis of the aromatic amino acids L-phenylalanine, L-tyrosine and L-tryptophan in plants and micro-organisms. Nature synthesises the aromatic amino acids starting from D-glucose via a pathway involving a series of interesting elimination reactions. This is an important pathway for plants, since it is also responsible for the biosynthesis of the precursors to the structural polymer lignin encountered in Section 7.10. Animals, which do not utilise this pathway, have to consume the aromatic amino acids as part of their diet.

The elimination of the elements of water involves cleavage of a $C-H$ bond and cleavage of a $C-O$ bond. The timing of $C-H$ versus $C-O$ cleavage

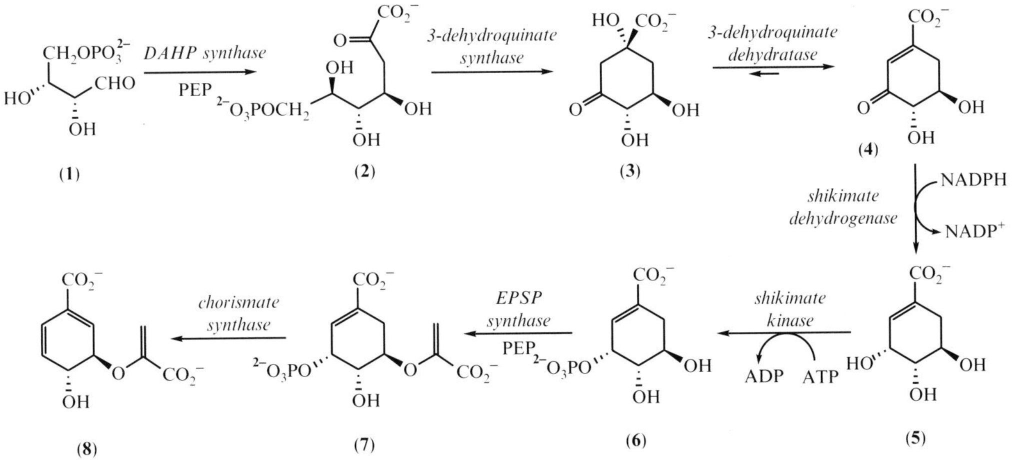

Figure 8.1 The shikimate pathway. (1) erythrose-4-phosphate; (2) 3-deoxy-D-arabinoheptulosonic acid 7-phosphate (DAHP); (3) 3-dehydroquinate; (4) 3-dehydroshikimate; (5) shikimate; (6) shikimate-3-phosphate; (7) 5-enolpyruvyl-shikimate-3-phosphate (EPSP); (8) chorismate.

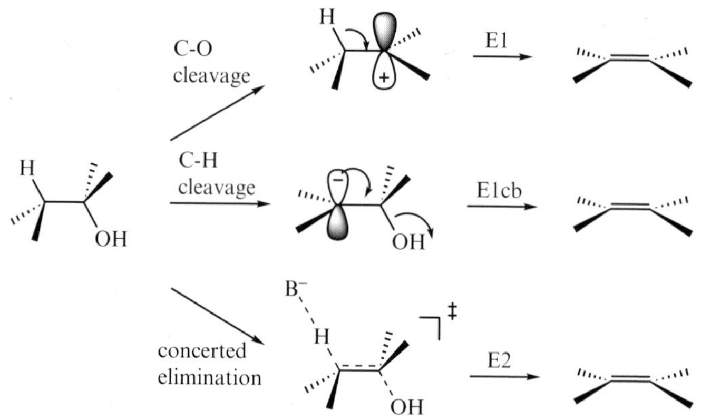

Figure 8.2 Mechanisms for elimination of water.

determines the type of elimination mechanism involved, as illustrated in Figure 8.2. If C−O cleavage takes place first, a carbonium ion is generated which eliminates via cleavage of an adjacent C−H bond. This E1 mechanism is found in organic reactions, but is rare in biological systems. If C−H bond cleavage occurs first a carbanion intermediate is generated, which eliminates via subsequent C−O bond cleavage. This E1cb mechanism is quite common in biological chemistry in cases where the intermediate carbanion is stabilised, as in the elimination of β-hydroxy-ketones or β-hydroxy-thioesters. If C−O and C−H bond cleavage are concerted then a single-step E2 mechanism is followed.

The stereochemical course of enzymatic elimination reactions is strongly dependent upon the mechanism of elimination. If an E2 mechanism is operating then an *anti*-elimination will ensue. This is generally observed in the enzyme-catalysed dehydrations of β-hydroxy-carboxylic acids. However, if an E1cb elimination mechanism is operating then the stereochemistry of the reaction can either be *syn*- or *anti*- depending upon the positioning of the catalytic groups at the enzyme active site. Commonly the *syn*- stereochemistry is observed for enzyme-catalysed elimination of β-hydroxy-ketones, although this is not true in all cases.

8.2   Hydratases and dehydratases

Enzymes that catalyse the elimination of water are usually known as dehydratases. However, since addition/elimination reactions are reversible, sometimes the biologically relevant reaction is the hydration of an alkene, in which case the enzyme would be known as a hydratase. We shall consider in turn enzymes which catalyse the elimination of β-hydroxy-ketones, β-hydroxy-thioesters and β-hydroxy-carboxylic acids.

The dehydration of 3-dehydroquinic acid to 3-dehydroshikimic acid shown in Figure 8.3 is a well-studied example of an enzyme-catalysed elimination of a

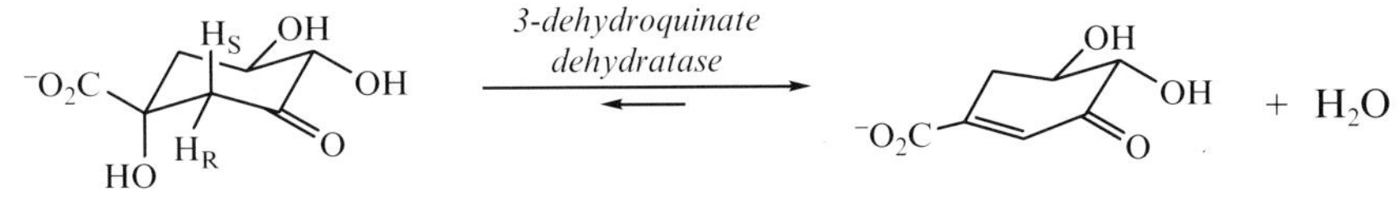

Figure 8.3 Reaction catalysed by 3-dehydroquinate dehydratase.

β-hydroxy-ketone. This reaction is catalysed by 3-dehydroquinate dehydratase, the third enzyme on the shikimate pathway.

The enzymatic reaction is reversible, although the equilibrium constant of 16 lies in favour of the forward reaction. The stereochemical course of the enzymatic reaction in *Escherichia coli* was shown, using stereospecifically labelled substrates, to be a *syn*-elimination of the equatorial C-2 *pro R* hydrogen and the C-1 hydroxyl group. The stereochemistry of the reaction is remarkable, since the axial C-2 *pro S* hydrogen is much more acidic in basic solution than the *pro R* hydrogen, due to favourable overlap with the $C-O$ π bond.

A major clue to the enzyme mechanism is that the enzyme is inactivated irreversibly by treatment with substrate and sodium borohydride. This indicates that an imine linkage is formed between the C-3 ketone and the ε-amino group of an active site lysine residue. Peptide mapping studies have established that this residue is Lys-170.

A mechanism for enzymatic reaction is shown in Figure 8.4. Upon formation of the imine linkage at C-3, a conformational change is thought to take place, giving a twist boat structure. In this conformation the $C-H$ bond of the *pro R* hydrogen lies parallel to the orbital axes of the adjacent $C-N$ π bond, favouring removal of this hydrogen. Proton abstraction by an active site base

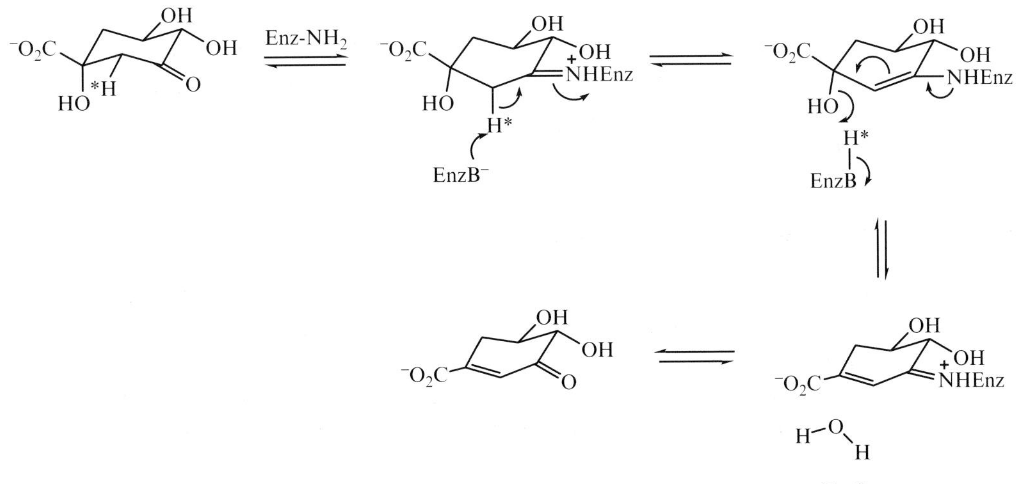

Figure 8.4 Mechanism for reaction catalysed by *Escherichia coli* 3-dehydroquinate dehydratase.

leads to the formation of a planar enamine intermediate. This intermediate acts as a stabilised carbanion intermediate for an E1cb elimination mechanism. Extrusion of water, presumably assisted by general acid catalysis, is followed by hydrolysis of the iminium salt linkage to give 3-dehydroshikimic acid.

The elimination of β-hydroxyacyl coenzyme A (CoA) thioesters is one of the reactions involved in the fatty acid synthase cycle described in Section 7.4. β-Hydroxydecanoyl thioester dehydratase catalyses the reversible dehydration of β-hydroxydecanoyl thioesters to give *trans*-2-decenoyl thioesters, which are subsequently isomerised to give *cis*-3-decenoyl thioesters. This reaction sequence represents a branch point in the biosynthesis of saturated and unsaturated fatty acids, as shown in Figure 8.5.

β-Hydroxydecanoyl thioester dehydratase from *E. coli* is a dimer of sub-unit molecular weight 18 kDa and requires no cofactors for activity. Stereochemical studies have revealed that the *proS* hydrogen is abstracted at C-2, so elimination of the 3*R* alcohol results in an overall *syn*-elimination of water. The subsequent isomerisation involves abstraction of the *proR* hydrogen at C-4. There is considerable evidence for a histidine active site base, which has been identified by peptide mapping studies as His-70. It is thought that His-70 also mediates the isomerisation reaction, as shown in Figure 8.6. However, there is no intramolecular transfer of deuterium from C-4 to C-2, suggesting that the protonated His-70 exchanges readily with solvent water.

This enzyme is specifically inactivated by a substrate analogue 3-decynoyl-*N*-acetyl-cysteamine. The α-proton of the inhibitor is abstracted by His-70 in the normal fashion, but protonation of the alkyne at the γ-position generates a highly reactive allene intermediate. This intermediate is then attacked by the nearby His-70 resulting in irreversible inactivation, as shown in Figure 8.7. Since the inhibitor is activated by deprotonation at C-2 the enzyme brings about its own inactivation by way of the enzyme mechanism. This class of mechanism-based inhibitors are, therefore, often known as 'suicide inhibitors'.

The dehydration of β-hydroxy-carboxylic acids is a reaction which occurs quite frequently in biochemical pathways. Two dehydrations which occur on the citric acid cycle are shown in Figure 8.8: the dehydration of citrate and rehydration of *cis*-aconitate to give isocitrate, catalysed by the single enzyme

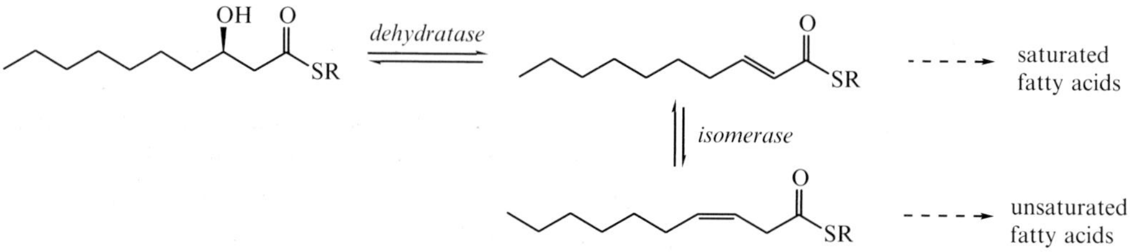

Figure 8.5 Reactions catalysed by β-hydroxydecanoyl thioester dehydratase.

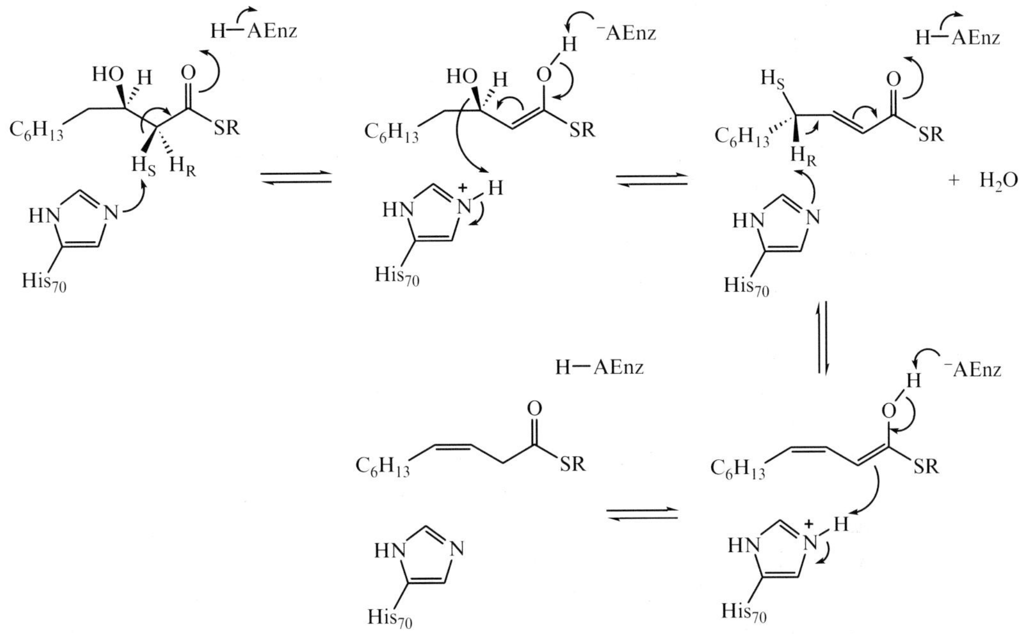

Figure 8.6 Mechanism for β-hydroxydecanoyl thioester dehydratase.

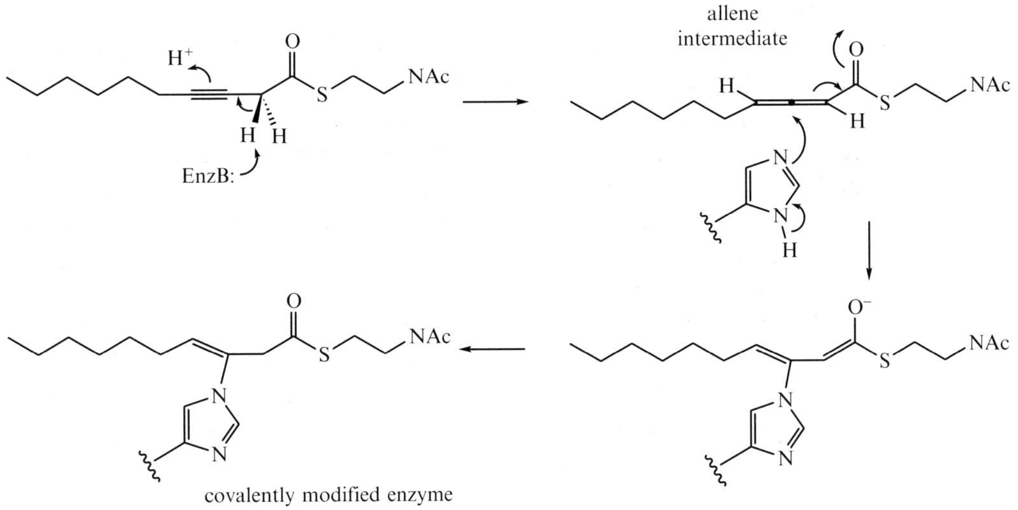

Figure 8.7 Irreversible inhibition of β-hydroxydecanoyl thioester dehydratase.

aconitase; and the hydration of fumarate to malate catalysed by the enzyme fumarase. Both these enzymes were found at an early stage to be dependent upon iron for activity. When the enzymes were purified and characterised, both enzymes were found to contain [4Fe4S] iron–sulphur clusters at their active sites. The usual biological function of iron–sulphur clusters, as explained in

citate   *cis*-aconitate   iso-citrate

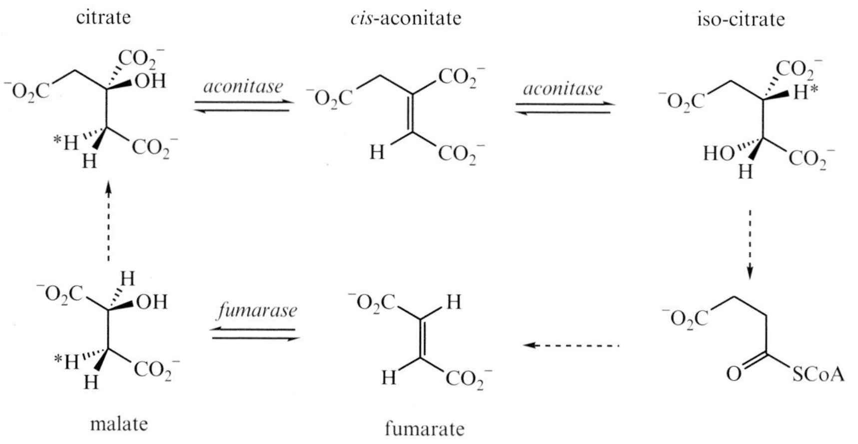

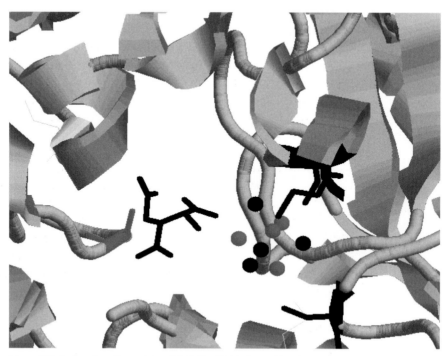

Figure 8.8 Reactions catalysed by aconitase and fumarase.

Section 6.7, is single electron transport. What then is the role of an iron–sulphur cluster in a hydratase enzyme?

Determination of an X-ray crystal structure for mitochondrial aconitase revealed that one of the iron atoms in the cluster co-ordinates the hydroxyl group of isocitrate, as shown in Figure 8.9. This observation suggests that the

Figure 8.9 Active site of aconitase (PDB file 1B0J). Bound isocitrate shown in black. Catalytic residue Ser-642 (on left) shown in red. 4Fe4S cluster: Fe atoms shown in red; S atoms shown in black; cysteine ligands shown in black.

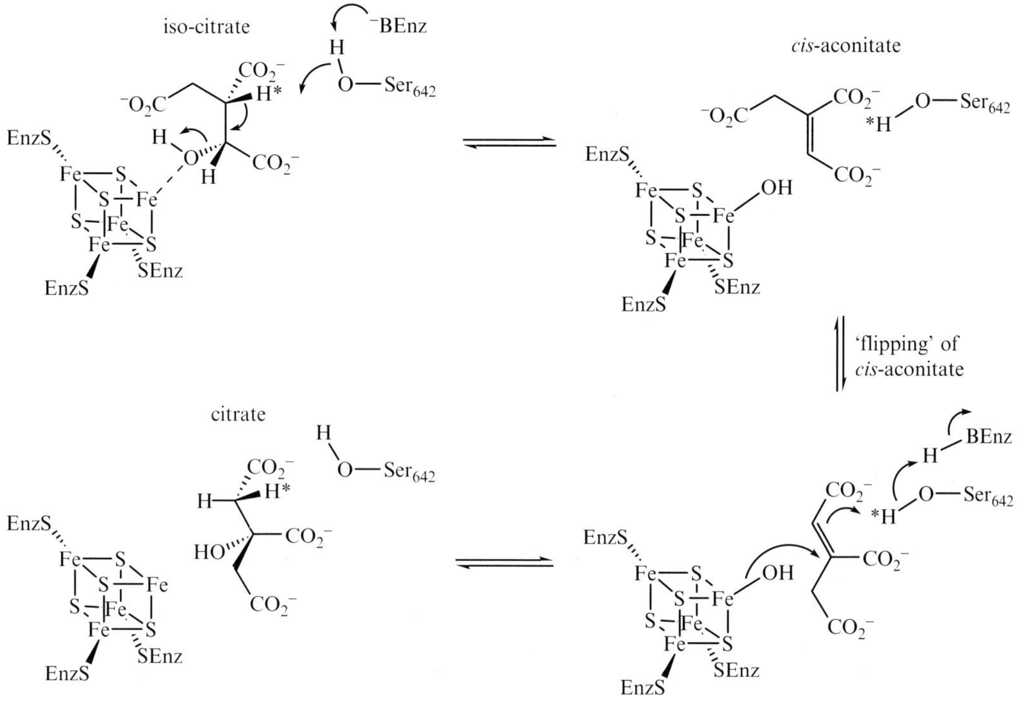

Figure 8.10 Mechanism for aconitase.

role of the iron–sulphur cluster is to function as a Lewis acid group to facilitate C−O cleavage during the catalytic mechanism. Subsequently, the activation of water as iron(II) hydroxide could provide a reactive nucleophile for water addition. Examination of the enzyme crystal structure indicated that Ser-642 is well positioned as the base for proton abstraction of isocitrate. The stereochemical course of both citrate and isocitrate dehydrations has been shown to be *anti-*, thus an E2 mechanism is likely as shown in Figure 8.10. It has also been shown that the proton abstracted by Ser-642 is returned upon hydration of *cis*-aconitate (although the hydroxyl group removed is exchanged). This implies both that Ser-642 is somehow shielded from solvent water and that the intermediate *cis*-aconitate is flipped over in the active site before rehydration.

This brief discussion covers by no means all of the types of hydratase enzymes found in biological systems, but illustrates a few of the most common examples, and the types of mechanistic and stereochemical methods used in other cases.

8.3   Ammonia lyases

The enzymatic elimination of ammonia is carried out on several L-amino acids in biological systems: L-phenylalanine, L-tyrosine, L-histidine and L-aspartic acid, as shown in Figure 8.11. In each of these cases the C_β−H bond being

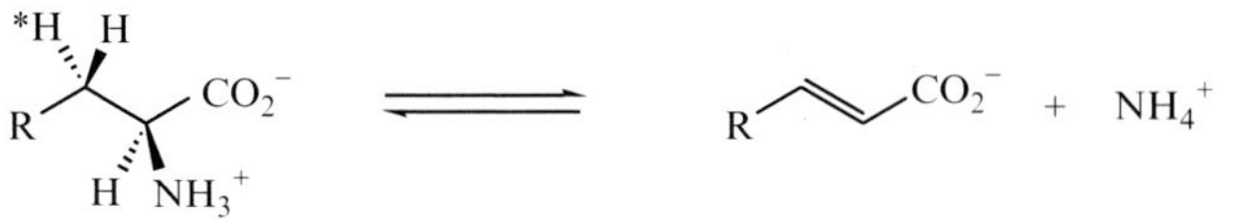

R = Ph   phenylalanine ammonia lyase
R = imidazole histidine ammonia lyase
R = CO$_2$H   aspartase

Figure 8.11 Reactions catalysed by some ammonia lyases.

broken lies adjacent to an activating group in the β-position of the amino acid side chain.

The histidine and phenylalanine ammonia lyases utilise a novel mechanism to assist the leaving group properties of ammonia, which would normally be a poor leaving group. These enzymes are readily inactivated by treatment with nucleophilic reagents, implying that there is an electrophilic group present at the enzyme active site. Examination of purified enzyme revealed that the electrophilic cofactor has a characteristic ultraviolet (UV) absorption at 340 nm.

Inactivation of phenylalanine ammonia lyase with NaB^{3}H$_4$ followed by acidic hydrolysis of the protein gave ^{3}H-alanine, suggesting that the electrophilic group might be a dehydroalanine amino acid residue, formed by dehydration of a serine residue. However, determination of the X-ray crystal structure of *Pseudomonas putida* histidine ammonia lyase in 1999 revealed that the electrophilic prosthetic group is 3,5-dihydro-5-methylene-imidazolone, formed by dehydration of Ser-143, and cyclisation of the amide nitrogen of Gly-144 with the amide carbonyl of Ala-142, as shown in Figure 8.12. Substitution of Ser-143 for threonine using site-directed mutagenesis gives an inactive enzyme; however, a Cys-143 mutant enzyme is able to form the dehydroalanine cofactor by elimination of hydrogen sulphide.

Two possible mechanisms for the phenylalanine ammonia lyase reaction are illustrated in Figure 8.13. Attack of the α-amino group of the substrate upon the methylene group of the cofactor (path A), followed by 1,3-prototropic shift, generates a secondary amine intermediate, in which the amine is activated as an enamide leaving group. *Anti*-elimination of this intermediate leaves the amine covalently attached to the cofactor. Prototropic shift, followed by elimination of ammonia, regenerates the methylene–imidazolone cofactor.

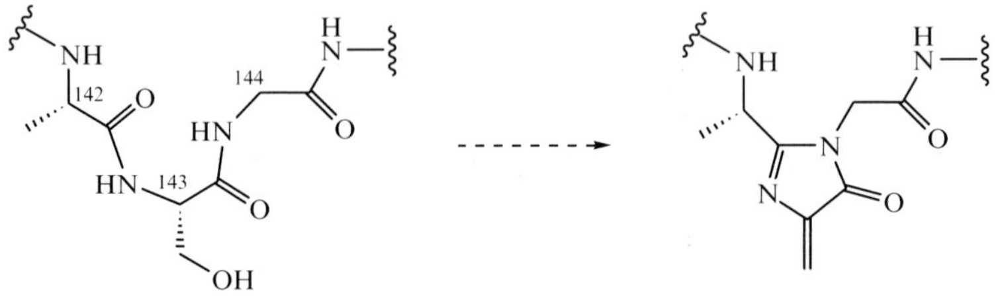

Figure 8.12 Methylene-imidazolone cofactor.

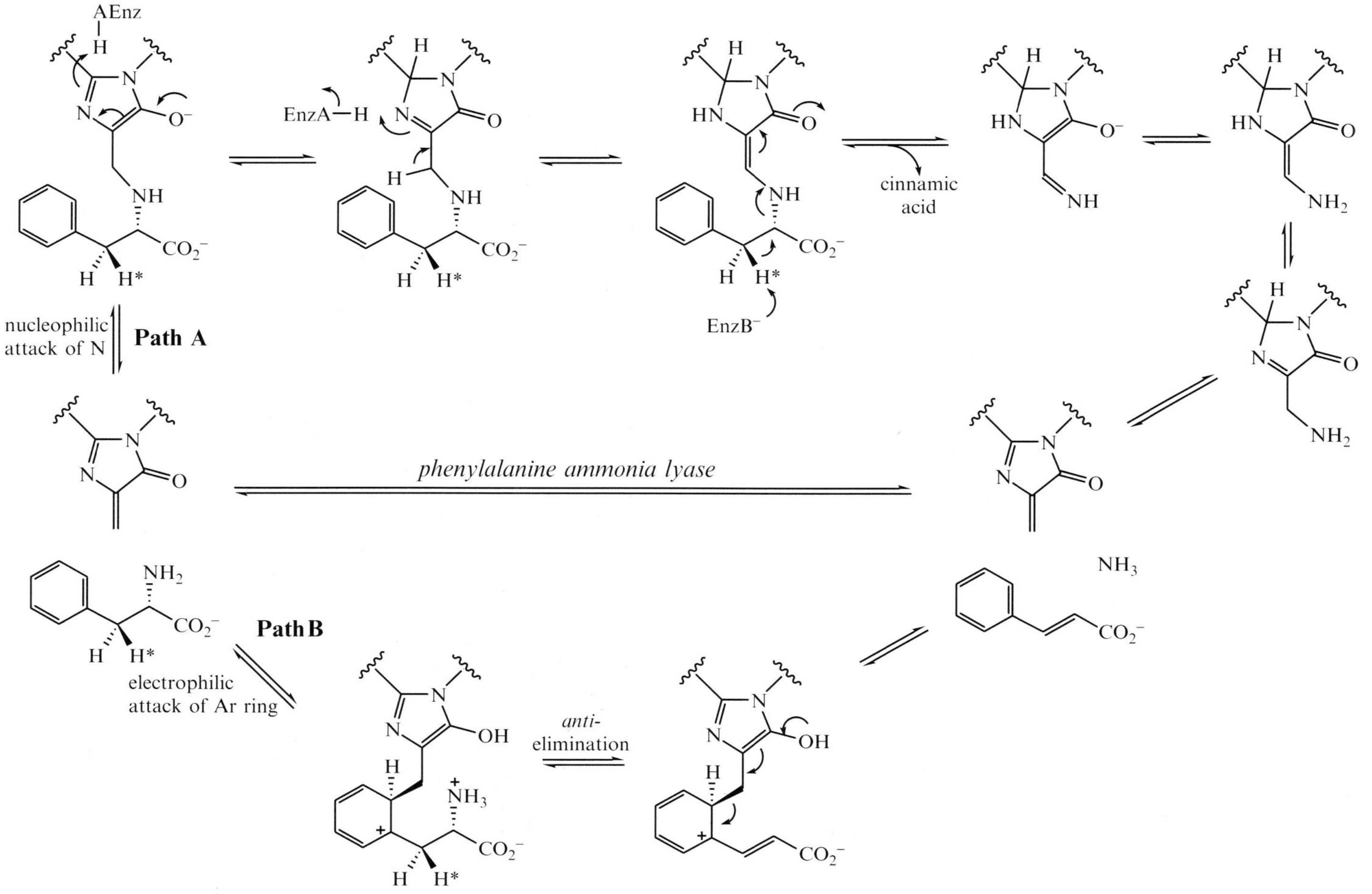

Figure 8.13 Nucleophilic (path A) and Friedl-Crafts type (path B) mechanisms for phenylalanine ammonia lyase.

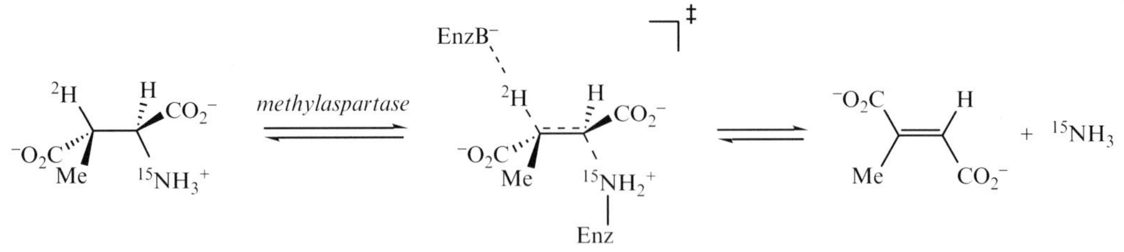

Figure 8.14 Concerted C−H and C−N cleavage in the methylaspartase reaction.

One would expect to see a primary kinetic isotope effect for β-^{2}H-labelled substrates, which is not observed for phenylalanine ammonia lyase, although a kinetic isotope effect is observed for histidine ammonia lyase. An alternative mechanism (path B) involving C−C bond formation with the aryl ring of phenylalanine has recently been proposed for phenylalanine ammonia lyase, analogous to a Friedel–Crafts reaction.

An important mechanistic question is whether the elimination of the substrate is concerted or stepwise? This question has been addressed using kinetic isotope effects in the reaction of methylaspartase, shown in Figure 8.14. This enzyme catalyses the *anti*-elimination of *threo*-β-methyl-aspartic acid, also utilising a dehydroalanine cofactor. Measurement of the rate of the enzymatic reaction using the [3-^{2}H] substrate revealed a primary kinetic isotope effect of 1.7, indicating that C−H bond cleavage is partially rate-determining. However, there was also found to be a ^{15}N isotope effect of 1.025 upon the [2-^{15}N] labelled substrate. If the reaction is concerted, then both of these kinetic isotope effects are operating on the same step, in which case the effects should be additive. So a [3-^{2}H, 2-^{15}N] substrate was prepared and a kinetic isotope effect of 1.042 was observed, indicating that the isotope effects are additive and that the elimination is indeed concerted.

Interestingly, the enzyme aspartase, which catalyses the elimination of aspartic acid, does not contain the methylene-imidazolone cofactor, thus Nature is able to catalyse the elimination of ammonia without the assistance of covalent catalysis.

8.4   Elimination of phosphate and pyrophosphate

In all elimination reactions an important determinant of reaction rate and mechanism is whether a good leaving group is available. Elimination of water is hindered by the fact that the hydroxyl group is a poor leaving group, since the pK$_a$ of the conjugate acid water is 15.7. In the dehydratase enzymes this problem is alleviated by acid or Lewis acid catalysis. However, another strategy found in biochemical pathways for provision of an efficient leaving group is phosphorylation of the leaving group.

The pK$_a$ values for the three dissociation equilibria of phosphoric acid, as shown in Figure 8.15, are 2.1, 7.2 and 12.3. At neutral pH a phosphate mono-

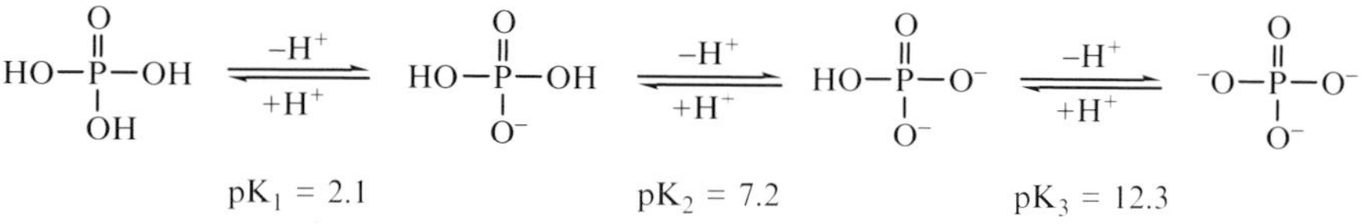

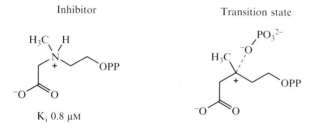

$$pK_1 = 2.1 \qquad\qquad pK_2 = 7.2 \qquad\qquad pK_3 = 12.3$$

Figure 8.15 pK_a values for phosphate anions.

ester would be singly deprotonated, thus for departure of a phosphate leaving group the pK_a of its conjugate acid ($H_2PO_4^-$) would be 7.2 – thus phosphate is a much better leaving group than a hydroxide ion.

Two important examples of the elimination of phosphate are shown in Figure 8.16. The first is the formation of isopentenyl pyrophosphate by an eliminative decarboxylation. The enzyme catalysing this reaction has been purified from baker's yeast, and has been shown to be strongly inhibited by a tertiary amine substrate analogue ($K_i = 0.8\,\mu\text{M}$). At neutral pH the tertiary amine will be protonated, thus the potent inhibition could be explained by this analogue mimicking a tertiary carbonium ion intermediate in the mechanism of this enzyme, as shown in Figure 8.17.

Chorismate synthase catalyses the 1,4-elimination of phosphate from 5-enolpyruvyl-shikimate-3-phosphate (EPSP) to give chorismic acid (see Figure 8.1). The stereochemistry of the enzymatic elimination in *E. coli* has been shown to be *anti-*. The highly unusual feature of this enzymatic reaction is that the enzyme requires reduced flavin as a cofactor, although the overall reaction

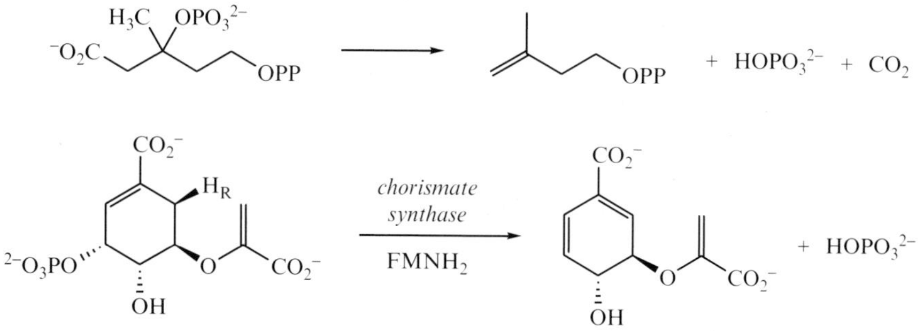

Figure 8.16 Enzymatic eliminations of phosphate. FMNH₂, flavir mononuclectide (reduced).

Figure 8.17 Transition state inhibitor for mevalonate pyrophosphate decarboxylase.

involves no change in redox level. Recent experiments have indicated that the flavin cofactor is involved somehow in the elimination reaction, since incubation of the 6R-6-fluoro analogue (in which the proton removed is substituted with fluorine) with enzyme leads to the observation of a flavin semiquinone species. It, therefore, appears that one-electron transfers may be involved in the mechanism employed by this enzyme.

8.5   CASE STUDY: 5-Enolpyruvyl-shikimate-3-phosphate (EPSP) synthase

There are a number of enzymes that catalyse multi-step addition-elimination reactions, of which the best characterised is the enzyme EPSP synthase. This enzyme catalyses the sixth step on the shikimate pathway, namely the transfer of an enolpyruvyl moiety from phosphoenol pyruvate (PEP) to shikimate-3-phosphate (see Figure 8.1).

Early experiments on the mechanism of this enzyme examined ^{3}H exchange processes. Incubation of [3-^{3}H]-PEP with enzyme and shikimate-3-phosphate led to the release of ^{3}H label into solvent, consistent with the existence of an intermediate containing a rotatable methyl group, as shown in Figure 8.18.

Subsequent overexpression of this enzyme allowed a detailed examination of the reaction using stopped flow kinetics and rapid quench methods. Analysis of a single turnover of the enzymatic reaction by stopped flow methods revealed the existence of an intermediate formed from shikimate-3-phosphate and PEP after 5 ms and eventually consumed after 50 ms. Incubation of large amounts of enzyme with shikimate-3-phosphate and PEP followed by rapid quench of the enzymatic reaction into 100% triethylamine led to the isolation of small amounts of the desired intermediate. This substance was a good substrate for both the forward and reverse reactions of EPSP synthase, satisfying the criteria for establishment of an intermediate in an enzymatic reaction.

Large-scale rapid quench studies using 500 mg of pure enzyme gave 300 μg of intermediate, which was characterised by ^{1}H, ^{13}C and ^{31}P nuclear magnetic

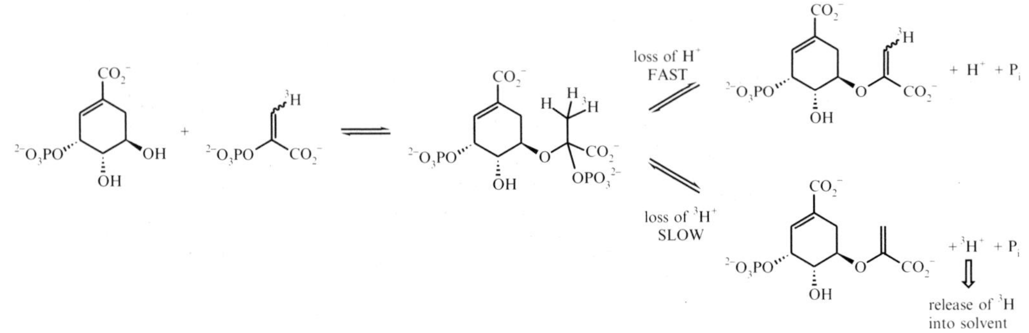

Figure 8.18 ^{3}H exchange in the EPSP synthase reaction.

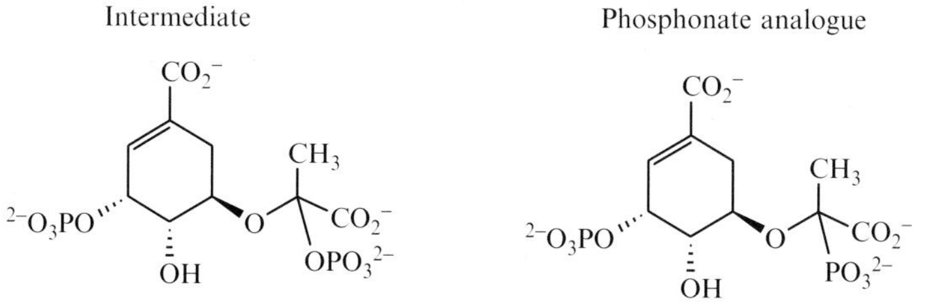

Intermediate   Phosphonate analogue

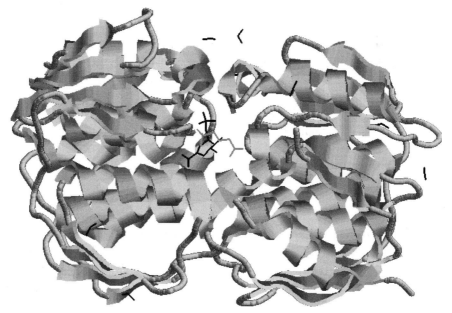

Figure 8.19 Tetrahedral intermediate in the EPSP synthase reaction.

resonance (NMR) spectroscopy, verifying the structure of the tetrahedral intermediate previously suspected. This intermediate was subsequently observed
transiently at the active site of the enzyme by NMR spectroscopy. Synthetic
phosphonate analogues of the tetrahedral intermediate have been synthesised
and were found to be potent inhibitors of the enzyme, as shown in Figure 8.19.

The X-ray crystal structure of EPSP synthase is shown in Figure 8.20.
Examination of the structure reveals that the enzyme contains two domains
connected by a flexible hinge, suggesting that there is a conformational change
of the protein occurring during catalytic turnover.

EPSP synthase is of significant commercial interest, since it is the target for
a herbicide called glyphosate, or phosphonomethyl-glycine. Inhibition of the
shikimate pathway in plants is catastrophic, since the plant can no longer
synthesise the aromatic amino acids required for metabolism and for the con-
struction of the structural polymer lignin. However, glyphosate is non-toxic to

Figure 8.20 Structure of EPSP synthase (PDB file 1G6S), complexed with glyphosate (shown in red)
and shikimate-3-phosphate (shown in black). Situated beneath the active site is a 'hinge' region,
which mediates a protein conformational change.

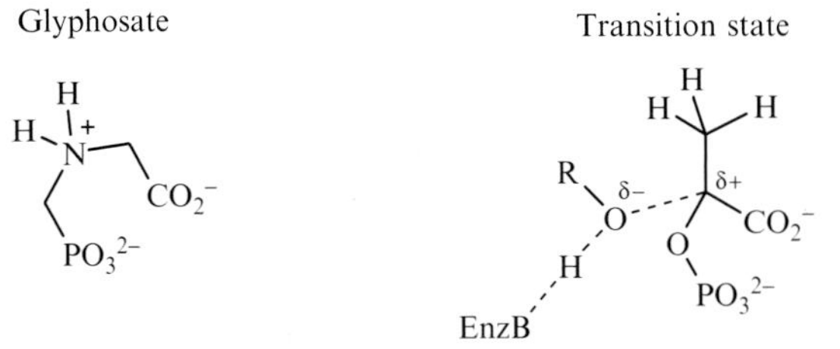

Figure 8.21 Inhibition of EPSP synthase by glyphosate.

animals as they do not utilise the shikimate pathway and hence have to consume the aromatic amino acids as part of their diet. Since glyphosate is non-toxic and remarkably bio-degradable (by enzymes which cleave carbon–phosphorus bonds!), it represents an 'environmentally friendly' herbicide. How does it inhibit EPSP synthase? It is thought that the protonated nitrogen of glyphosate mimics the transition state for attack of shikimate-3-phosphate upon PEP. It is likely that significant positive charge accumulates in this transition state on C-2 of PEP, as shown in Figure 8.21.

There is only one other example of an enolpyruvyl transfer reaction, the conversion of UDP-*N*-acetyl-glucosamine to enolpyruvyl-UDP-*N*-acetyl-glucosamine, involved in bacterial cell wall peptidoglycan biosynthesis. The transferase enzyme responsible for this transformation has been purified, over-expressed, and analysed by similar approaches. A tetrahedral intermediate has been found for this enzyme also, although there is evidence that PEP forms a reversible adduct with an active site cysteine residue. A mechanism for this enzyme is shown in Figure 8.22.

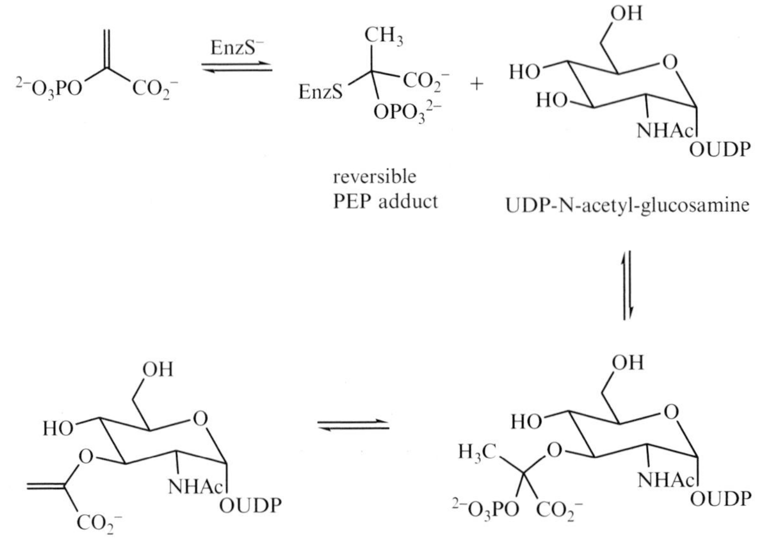

Figure 8.22 Reaction catalysed by UDPGlcNAc enolpyruvyl transferase.

Problems

(1) In the 3-dehydroquinate dehydratase reaction, how would you identify the active site lysine residue which forms an imine linkage with the substrate?

(2) The following intramolecular addition reaction is catalysed by a cyclo-isomerase enzyme from *Pseudomonas putida*. When the cycloisomerase reaction was carried out in ^3H_2O, product was found to contain tritium at C-5 and was found to have the 4S, 5R stereochemistry. Deduce whether the addition reaction occurs with *syn-* or *anti-* stereochemistry, and suggest a mechanism for the reaction.

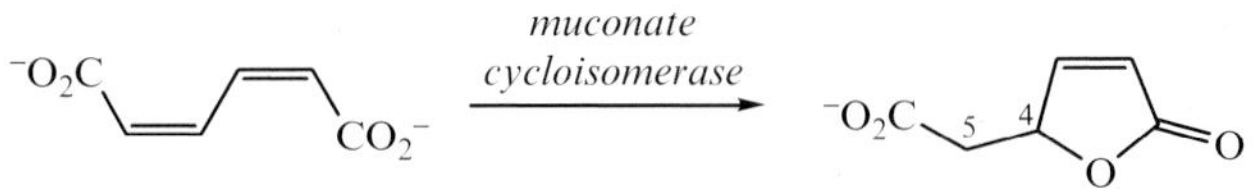

(3) Dihydroxy acid dehydratase is an iron–sulphur cluster-containing hydratase enzyme which catalyses the interconversion of $\alpha\beta$-dihydroxy acids with α-keto acids. Suggest possible mechanisms for this enzyme, and predict what would be observed for each mechanism if an α-^2H dihydroxy acid substrate was incubated with the enzyme.

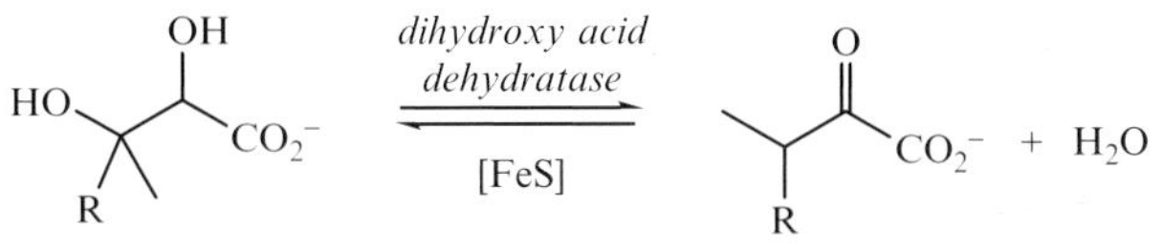

(4) S-Adenosyl homocysteine hydrolase catalyses the conversion of S-adenosyl homocysteine to adenosine and homocysteine. Given the following data, suggest a mechanism for this enzyme: (i) the enzyme contains a catalytic amount (1 mole per mole enzyme sub-unit) of tightly bound NAD^+ (ii) the enzyme catalyses the exchange of the C-4' hydrogen (H*) with ^2H_2O.

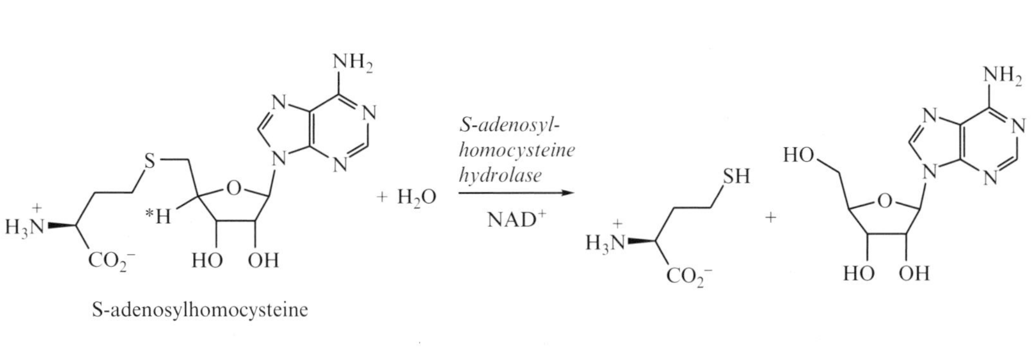

S-adenosylhomocysteine

(5) Chorismic acid (see Figure 8.1) is a substrate for the three enzymes shown below: anthranilate synthase, *para*-aminobenzoate synthase and isochorismate synthase. The amino acid sequences of the three enzymes are similar, suggesting that they may follow a similar mechanistic course. Suggest intermediates and possible mechanisms for these three reactions.

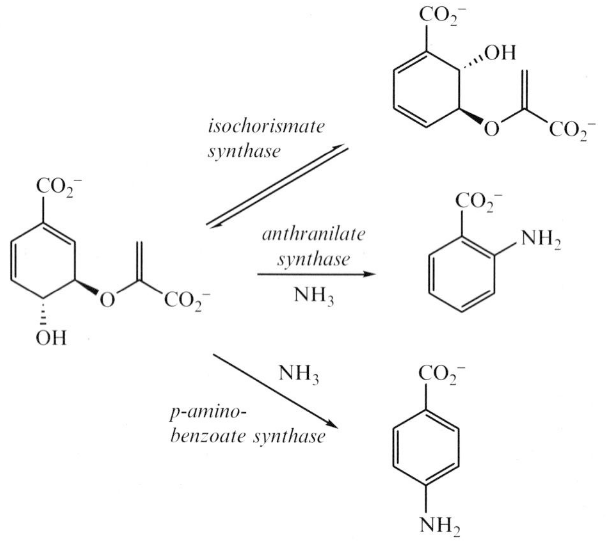

Further reading

General

R.H. Abeles, P.A. Frey & W.P. Jencks (1992) *Biochemistry*. Jones & Bartlett, Boston.
C. T. Walsh (1979) *Enzymatic Reaction Mechanisms*. Freeman, San Francisco.

Dehydratases

J.R. Butler, W.L. Alworth, & M.J. Nugent (1974) Mechanism of dehydroquinase catalyzed dehydration. I. Formation of a Schiff base intermediate. *J. Am. Chem. Soc.*, **96**, 1617–18.
H. Lauble, M.C. Kennedy, H. Beinert & C.D. Stout (1992) Crystal structures of aconitase with isocitrate and nitroisocitrate bound. *Biochemistry*, **31**, 2735–48.
J.M. Schwab & B.S. Henderson (1990) Enzyme-catalysed allylic rearrangements. *Chem. Rev.*, **90**, 1203–45.
M.J. Turner, B.W. Smith & E. Haslam (1975) The shikimate pathway. Part IV. The stereochemistry of the 3-dehydroquinate dehydratase reaction and observations on 3-dehydroquinate synthetase. *J. Chem. Soc. Perkin Trans.*, **1**, 52–5.

Ammonia lyases

N.P. Botting, A.A. Jackson & D. Gani (1989) ^{15}N-isotope and double isotope fractionation studies of the mechanism of 3-methylaspartase: concerted elimination of ammonia from (2*S*,3*S*)-3-methylaspartic acid. *J. Chem. Soc. Chem. Commun.*, 1583–5.

L. Poppe (2001) Methylidene-imidazolone: a novel electrophile for substrate activation. *Curr. Opin. Chem. Biol.*, **5**, 512–24.

T.F. Schwede, J. Retey & G.E. Schulz (1999) Crystal structure of histidine ammonia lyase revealing a novel polypeptide modification as the catalytic electrophile. *Biochemistry*, **38**, 5355–61.

Elimination of phosphates

K.S. Anderson & K.A. Johnson (1990) Kinetic and structural analysis of enzyme intermediates: lessons from EPSP synthase. *Chem. Rev.*, **90**, 1131–49.

S. Dhe-Paganon, J. Magrath & R.H. Abeles (1994) Mechanism of mevalonate pyrophosphate decarboxylase: evidence for a carbocationic transition state. *Biochemistry*, **33**, 13355–62.

M.N. Ramjee, S. Balasubramanian, C. Abell, J.R. Coggins, G.M. Davies, T.R. Hawkes, D.J. Lowe & R.N.F. Thorneley (1992) Reaction of (6*R*)-6-F-EPSP with recombinant *E. coli* chorismate synthase generates a stable flavin mononucleotide semiquinone radical. *J. Am. Chem. Soc.*, **114**, 3151–3.

M.N. Ramjee, J.R. Coggins, T.R. Hawkes, D.J. Lowe & R.N.F. Thorneley (1991) Spectrophotometric detection of a modified flavin mononucleotide (FMN) intermediate formed during the catalytic cycle of chorismate synthase. *J. Am. Chem. Soc.*, **113**, 8566–7.

W.C. Stallings, S.S. Abdel-Meguid, L.W. Lim, H.S. Shieh, H.E. Dayringer, N.K. Leimgruber, R.A. Stegeman, K.S. Anderson, J.A. Sikorski, S.R. Padgette & G.M. Kishore (1991) Structure and topological symmetry of the glyphosate target 5-enolpyruvylshikimate-3-phosphate synthase: a distinctive protein fold. *Proc. Natl. Acad. Sci., USA*, **88**, 5046–50.

D.R. Welch, K.W. Cole & F.H. Gaertner (1994) Chorismate synthase of *Neurospora crassa*: a flavoprotein. *Arch. Biochem. Biophys.*, **165**, 505–18.

9 Enzymatic Transformations of Amino Acids

9.1 Introduction

α-Amino acids are primary cellular metabolites that are required for the assembly of proteins, as discussed in Chapter 2. They are also used for the biosynthesis of alkaloids – a class of nitrogen-containing natural products produced by many organisms. There is, therefore, a sizeable group of enzymatic reactions involved in the biosynthesis, breakdown and transformation of α-amino acids. In this chapter we shall discuss the common enzymatic transformations of α-amino acids, focusing in particular on the role of the coenzyme pyridoxal 5′-phosphate (PLP).

Illustrated in Figure 9.1 are the general types of transformations found for α-amino acids. As discussed in Chapter 2, the amino acids used for protein

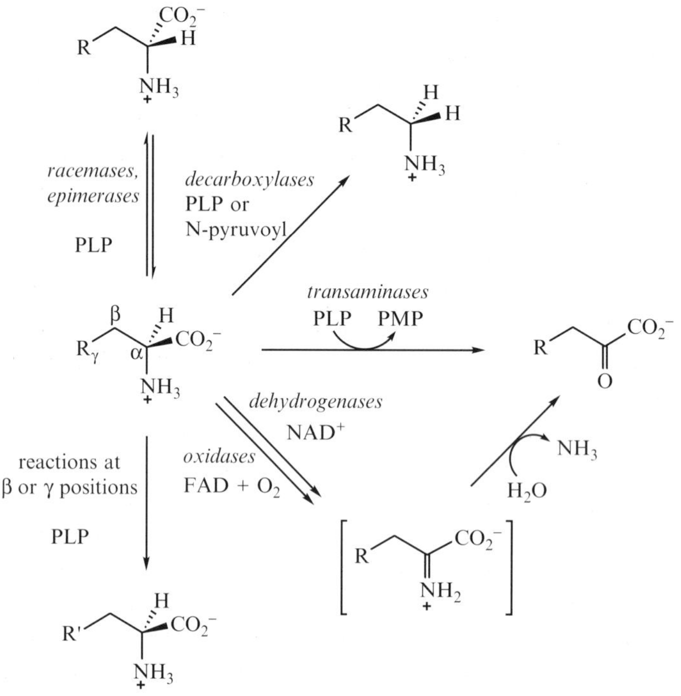

Figure 9.1 Types of enzymatic transformations of α-amino acids. FAD, flavin adenine dinucleotide; NAD, nicotinamide adenine dinucleotide; PLP, pyroxidal 5′-phosphate; PMP, pyridoxamine 5′-phosphate.

biosynthesis have exclusively the L-configuration. However, there are a number of racemase and epimerase enzymes producing D-amino acids that are used for a small number of specific purposes in biological systems. α-Amino acid decarboxylases produce the corresponding primary amines, some of which have important bodily functions in mammals, and others which are used in the biosynthesis of alkaloids in plants. Oxidation of α-amino acids has been mentioned in Chapter 6: there are several NAD^+- and flavin-dependent dehydrogenase and oxidase enzymes that oxidise amino acids via the corresponding iminium salt to the α-keto acid. Imine intermediates then make possible a number of further transformations as we shall see later in the chapter.

9.2   Pyridoxal 5′-phosphate-dependent reactions at the α-position of amino acids

Pyridoxal 5′-phosphate is a coenzyme derived from vitamin B_6 (pyridoxine). Pyridoxine is oxidised and phosphorylated in the body to give the active form of the coenzyme, as shown in Figure 9.2. This vitamin was first isolated from rice bran in 1938, and was found to be active against the deficiency disease pellagra.

A wide range of reaction types are catalysed by PLP-dependent enzymes; however, in general the substrates for these enzymes are α-amino acids. The structural features of the coenzyme which make this chemistry possible are a pyridine ring that acts as an electron sink, and an aldehyde substituent at the C-4 position through which the coenzyme becomes covalently attached to the amino acid substrate. The structure of the coenzyme is shown in Figure 9.2.

Enzymes which utilise PLP bind the cofactor through an imine linkage between the aldehyde group of PLP and the ε-amino group of an active site lysine residue. At neutral pH this imine linkage is protonated to form a more electrophilic iminium ion. Upon binding of the α-amino acid substrate, the α-amino group attacks the iminium ion, displacing the lysine residue and forming an imine linkage itself with the pyridoxal cofactor. This aldimine intermediate, shown in Figure 9.2, is the starting point for each of the mechanisms that we shall meet in the following sections. Although for sake of simplicity I shall write the phenolic hydroxyl group of PLP in protonated form, there is evidence that it is deprotonated when bound to the enzyme, and that the phenolate anion forms a hydrogen bond to the protonated iminium ion, as shown in Figure 9.2.

Formation of the aldimine adduct dramatically increases the acidity of the amino acid α-proton. This activation is utilised by a family of racemase and epimerase enzymes which utilise PLP as a cofactor. In these enzymes formation of the aldimine intermediate is followed by abstraction of the α-proton of the amino acid utilising the pyridine ring as an electron sink and generating a quinonoid species. Delivery of a proton from the opposite face of the molecule

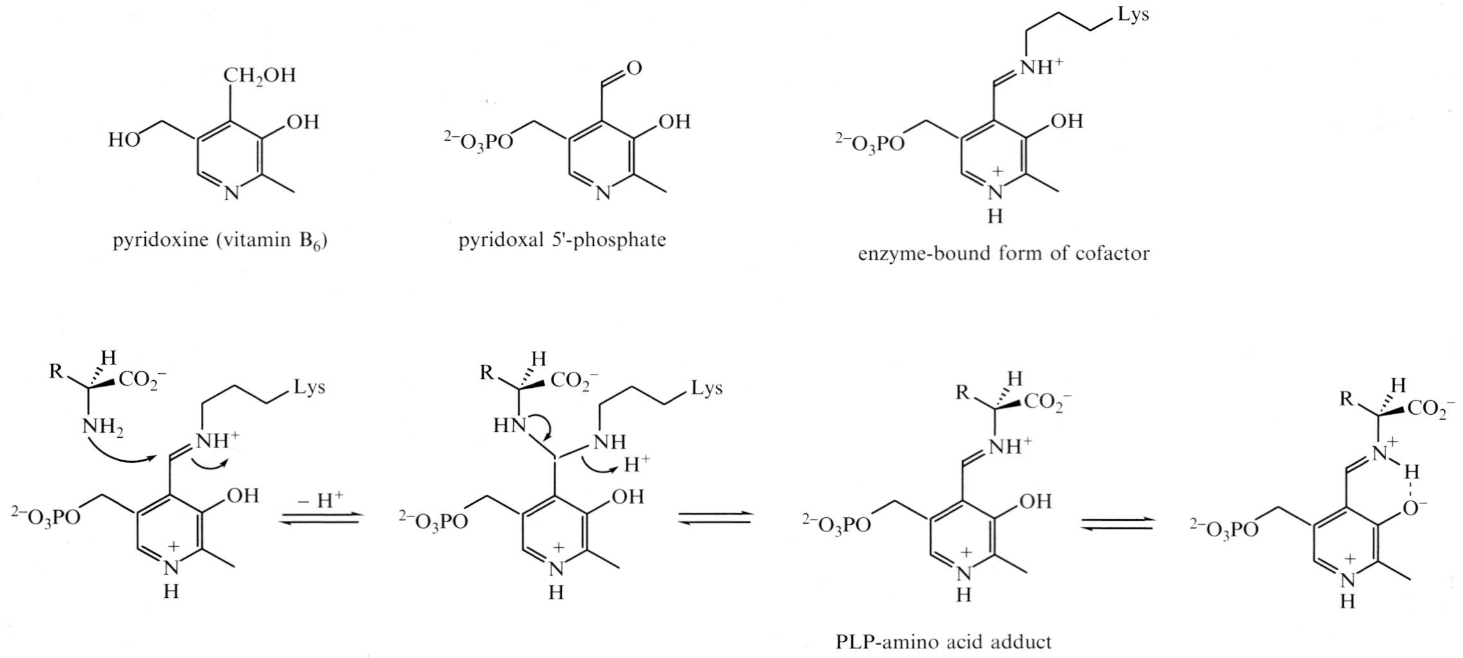

Figure 9.2 Pyridoxal 5'-phosphate and its attachment to PLP-dependent enzymes.

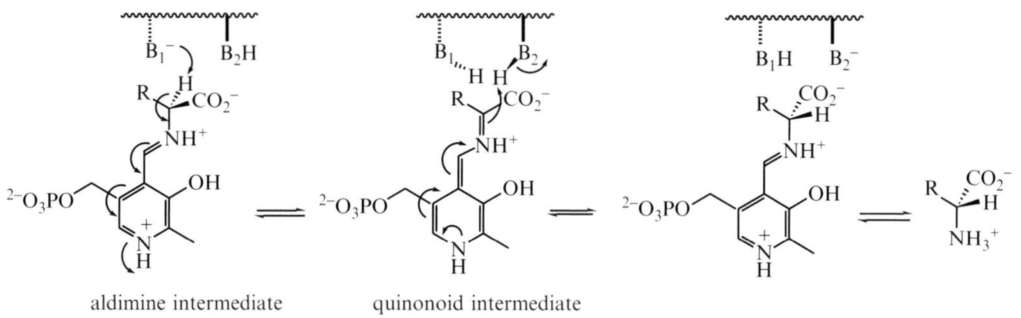

aldimine intermediate   quinonoid intermediate

Figure 9.3 Mechanism for PLP-dependent racemases (two-base mechanism illustrated).

results in inversion of configuration at the α-centre, as shown in Figure 9.3. Detachment of the product from the coenzyme is carried out by attack of the active site lysine residue.

In some racemases reprotonation is carried out by a second active site residue (the 'two-base' mechanism). In other cases deprotonation and reprotonation are carried out by a single active site base that is able to access both faces of the ketimine adduct. The latter 'one-base' mechanism can be demonstrated in a single turnover experiment by incubating a 2-^{2}H-L-amino acid substrate with a stoichiometric amount of enzyme. Isolation of D-amino acid product containing deuterium at the α-position implies intramolecular atom transfer by a single active site base.

One important example of a PLP-dependent racemase is alanine racemase, which is used by bacteria to produce D-alanine. D-alanine is then incorporated into the peptidoglycan layer of bacterial cell walls in the form of a D-Ala-D-Ala dipeptide. Inhibition of alanine racemase is, therefore, lethal to bacteria, since without peptidoglycan the cell walls are too weak to withstand the high osmotic pressure, and the bacteria lyse. One inhibitor of alanine racemase that has antibacterial properties is β-chloro-D-alanine, which inhibits the enzyme via an interesting mechanism shown in Figure 9.4. β-Chloro-D-alanine is accepted as a substrate by the enzyme, which proceeds to bind the inhibitor covalently to its PLP cofactor. However, once in 800 turnovers deprotonation at the α-position is followed by loss of chloride, generating a PLP-bound enamine intermediate. This is detached from the PLP cofactor by attack of the lysine ε-amino group; however, the liberated free enamine reacts with the carbon centre of the PLP-enzyme imine, generating an irreversibly inactivated species. Further examples of such mechanism-based inhibitors will be given in the Problems section. Note also that there is a family of cofactor-independent racemase/epimerase enzymes which will be discussed in Section 10.2.

Amino acid decarboxylases proceed from the PLP-amino acid adduct shown in Figure 9.2, this time using the PLP structure as an electron sink for decarboxylation of this adduct. Reprotonation at (what was) the α-position, followed by detachment of the product from the PLP cofactor, generates the

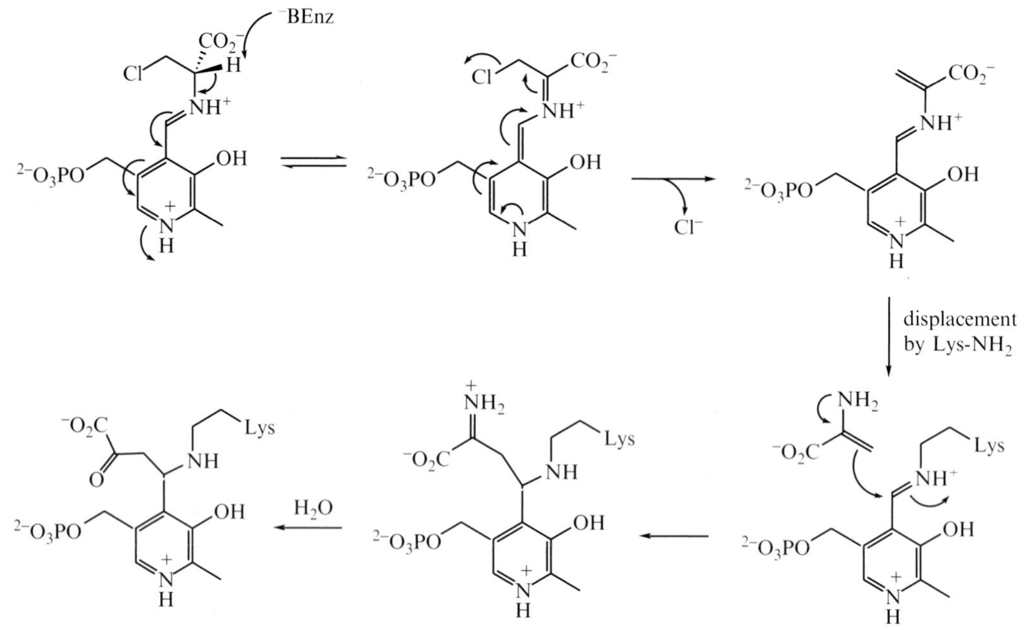

Figure 9.4 Mechanism for inhibition of alanine racemase by β-chloro-D-alanine.

corresponding primary amine. Reprotonation usually takes place with retention of configuration at the α-position, as shown in Figure 9.5. One important example of a PLP-dependent decarboxylase is the mammalian 3,4-dihydroxy-phenylalanine (dopa) decarboxylase, which catalyses the decarboxylation of 3,4-dihydroxyphenylalanine to dopamine. Dopamine is a precursor to the neurotransmitters epinephrine and norepinephrine. This enzyme also catalyses the decarboxylation of 5-hydroxytryptophan to give another neurotransmitter serotonin, as shown in Figure 9.6.

The third transformation carried out at the α-position of amino acids by PLP-dependent enzymes is transamination: conversion of the α-amino acid to an α-keto acid. This class of enzymes will be illustrated by the case study of aspartate aminotransferase.

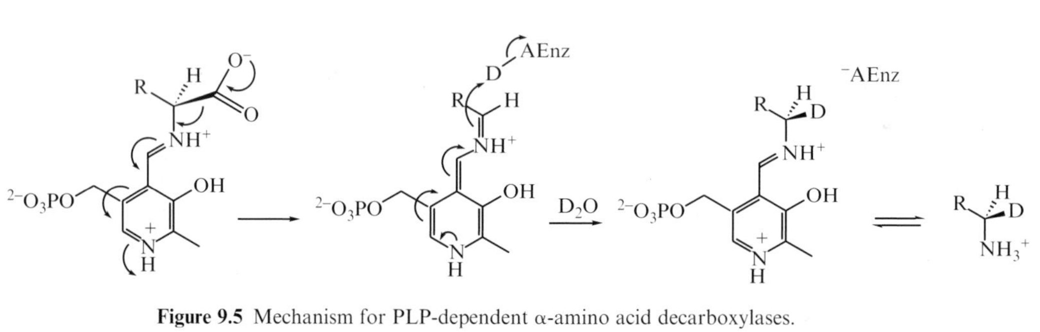

Figure 9.5 Mechanism for PLP-dependent α-amino acid decarboxylases.

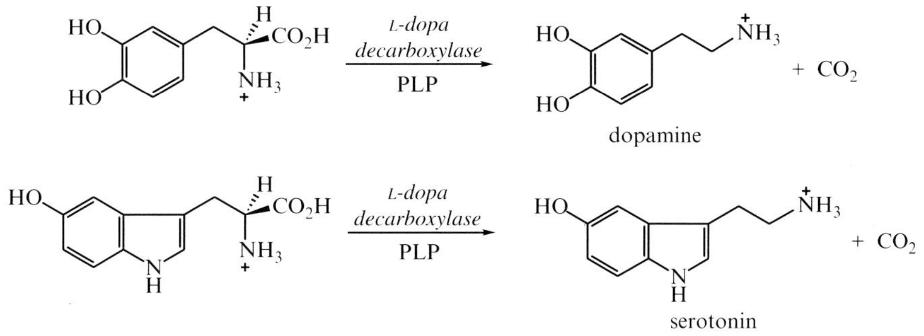

Figure 9.6 Reactions catalysed by L-dopa decarboxylase.

9.3   CASE STUDY: Aspartate aminotransferase

Aspartate aminotransferase catalyses the transamination of L-aspartic acid into oxaloacetate, at the same time converting α-ketoglutarate into L-glutamic acid. The enzymatic reaction, therefore, consists of two half-reactions, shown in Figure 9.7.

Mammals contain two forms of aspartate aminotransferase, a cytosolic form and a mitochondrial form, both of which have been purified and studied extensively. The bacterial enzyme from *Escherichia coli* has also been purified, overexpressed and crystallised, allowing a detailed study of its mechanism of action, which is depicted in Figure 9.8. The resting state of the enzyme in the absence of substrate contains the Lys-258–PLP–aldimine adduct, which absorbs at 430 nm. Upon binding of L-aspartate, the PLP–aldimine intermediate is formed, which also absorbs at 430 nm. The ε-amino group of Lys-258, released from binding the PLP cofactor, acts as a base for deprotonation of the α-hydrogen. This forms the quinonoid intermediate (visible at 490 nm by stopped flow kinetics) also found in the PLP-dependent racemases. However, in this case the quinonoid intermediate is reprotonated adjacent to the heterocyclic ring, generating a ketimine intermediate which can be observed at 340 nm by stopped flow kinetics. Hydrolysis of the ketimine intermediate releases the product oxaloacetate, and generates a modified form of the cofactor known as pyridoxamine 5'-phosphate (PMP), visible at 330 nm.

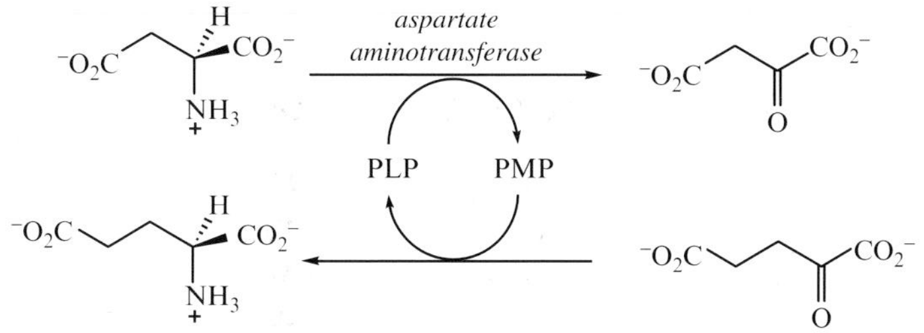

Figure 9.7 Two half-reactions catalysed by aspartate aminotransferase.

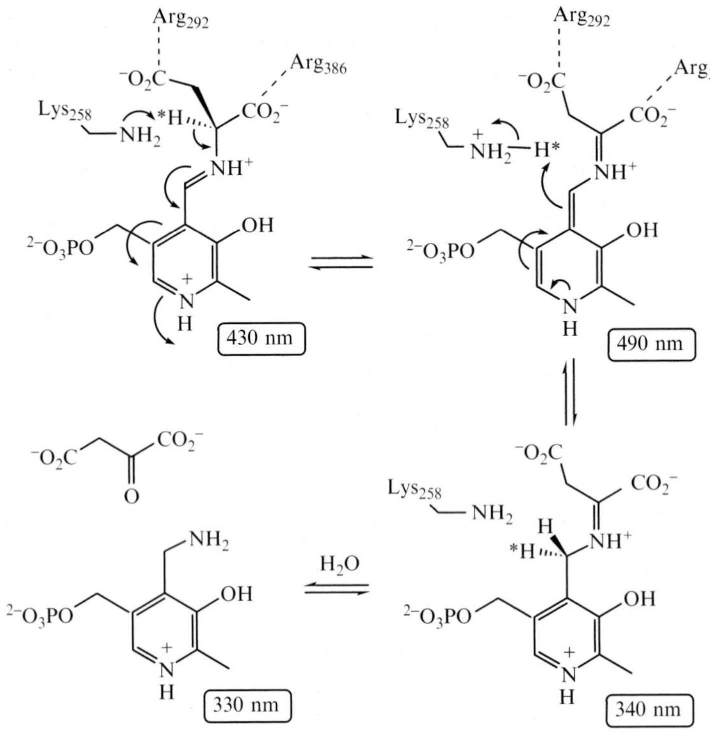

Figure 9.8 Mechanism for aspartate aminotransferase half-reaction.

The reaction is completed by carrying out the reverse transamination on the other α-keto acid substrate for this enzyme. α-Ketoglutarate is bound via a ketimine linkage, which is isomerised as before to the aldimine intermediate. Displacement of the aldimine linkage by Lys-258 releases L-glutamic acid and regenerates the PLP form of the cofactor.

Examination of the X-ray crystal structure of aspartate aminotransferase, illustrated in Figure 9.9, reveals that Lys-258 is suitably positioned to act as an intramolecular base for proton transfer. Replacement of Lys-258 for alanine by site-directed mutagenesis gives a completely inactive mutant enzyme, as expected, since there is no point of attachment or active site base. A Cys-258 mutant enzyme is similarly inactive. However, if this mutant is alkylated with 2-bromoethylamine an active enzyme is obtained which contains a thioether analogue of lysine at its active site, as shown in Figure 9.10. This enzyme has 7% of the activity of the wild-type enzyme with a slightly shifted pH/rate profile of enzymatic activity, since the thioether-containing lysine analogue is slightly less basic than lysine.

Examination of the active site of aspartate aminotransferase, shown in Figure 9.9, reveals that both carboxylate groups of L-aspartate are bound by electrostatic interactions to active site arginine residues: the α-carboxylate

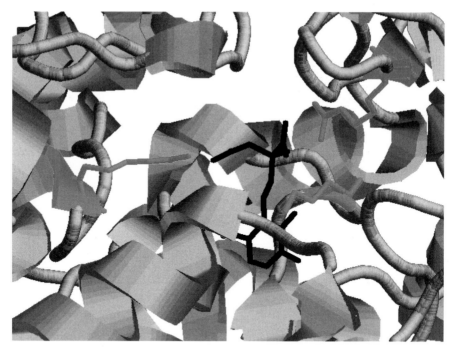

Figure 9.9 Active site of asparate aminotransferase (PDB file 1AJS). L-Aspartic acid bound to pyridoxal 5′-phosphate coenzyme shown in black. Arg-292 (left), Arg-386 (top right) and catalytic base Lys-258 (right) shown in red.

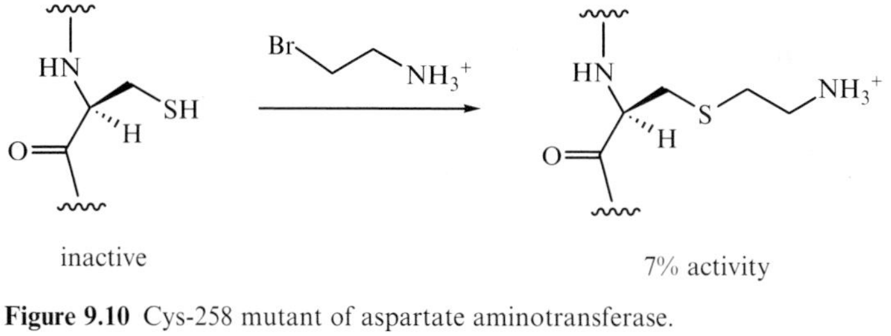

inactive   7% activity

Figure 9.10 Cys-258 mutant of aspartate aminotransferase.

by Arg-386 and the β-carboxylate by Arg-292. In principle, the substrate specificity of this enzyme could be changed by replacing Arg-292 by other amino acids. Mutation of Arg-292 to an aspartate residue gave an enzyme whose catalytic efficiency for L-aspartate had dropped from $34\,500\,\text{M}^{-1}\text{s}^{-1}$ to 0.07 $\text{M}^{-1}\text{s}^{-1}$. However, the mutant enzyme was found to be capable of processing L-amino acid substrates containing positively charged side chains which could interact favourably with Asp-292, illustrated in Figure 9.11. So L-arginine, L-lysine and L-ornithine (one carbon shorter side chain than lysine) were all processed by the mutant enzyme, the best substrate being L-arginine with a k_{cat}/K_M of $0.43\,\text{M}^{-1}\text{s}^{-1}$. This is an example of how the method of site-directed

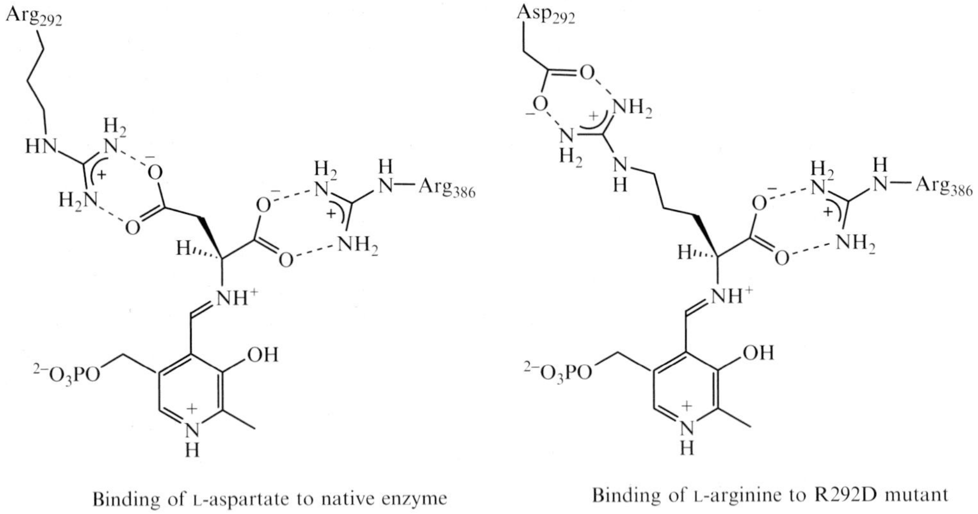

Binding of L-aspartate to native enzyme   Binding of L-arginine to R292D mutant

Figure 9.11 Interaction of L-arginine substrate with Asp-292 of R292D mutant enzyme.

mutagenesis can be used to 'engineer' the substrate specificity of enzyme active sites.

9.4   Reactions at the β- and γ-positions of amino acids

There is a smaller group of enzymatic reactions that take place at the β- and γ-positions of α-amino acids which are also dependent upon PLP. These reactions also make use of the PLP cofactor as an electron sink, but we shall see that there are examples in this class in which PLP acts as a four-electron sink rather than a two-electron sink. I shall illustrate one example of a reaction at the β-position, threonine dehydratase, and one at the γ-position, methionine γ-lyase (see Figure 9.12).

Threonine dehydratase catalyses the conversion of L-threonine into α-ketobutyrate and ammonia. The enzymatic reaction starts from the aldimine

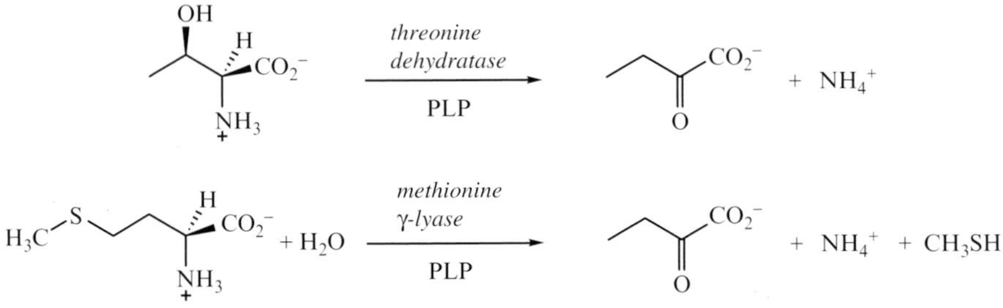

Figure 9.12 Reactions at the β- and γ-positions of α-amino acids.

adduct of PLP with L-threonine, which is deprotonated as above to generate the familiar quinonoid species. In this case the hydroxyl substituent at the β-position acts as a leaving group, presumably with acid catalysis, and an α,β-elimination reaction ensues. Displacement of the imine linkage by the active site lysine residue releases the enamine equivalent of α-ketobutyrate. Hydrolysis of the enamine generates α-ketobutyrate and ammonia, and regenerates the PLP cofactor. The mechanism is depicted in Figure 9.13.

Methionine γ-lyase catalyses the conversion of L-methionine into α-ketobutyrate, ammonia and methanethiol (a particularly smelly enzyme to work with!). The mechanism of this enzyme starts from the aldimine adduct of PLP with L-methionine, which is deprotonated to generate the quinonoid intermediate. However, at this point a *second* deprotonation takes place at the β-position, utilising the adjacent iminium species as a second electron sink to stabilise the β-carbanion. Elimination of the γ-substituent can then take place, followed by reprotonation at the γ-position, to generate the enamine intermediate seen above. The enamine equivalent of α-ketobutyrate is released, which after hydrolysis generates α-ketobutyrate and ammonia. The mechanism is depicted in Figure 9.14.

There are a number of other enzymes in this class, most of which also employ a second deprotonation step, effectively utilising the PLP-amino acid iminium salt as a four-electron sink. The cellular role of these enzymes is often for the degradation of the respective amino acids and recycling of their nitrogen content.

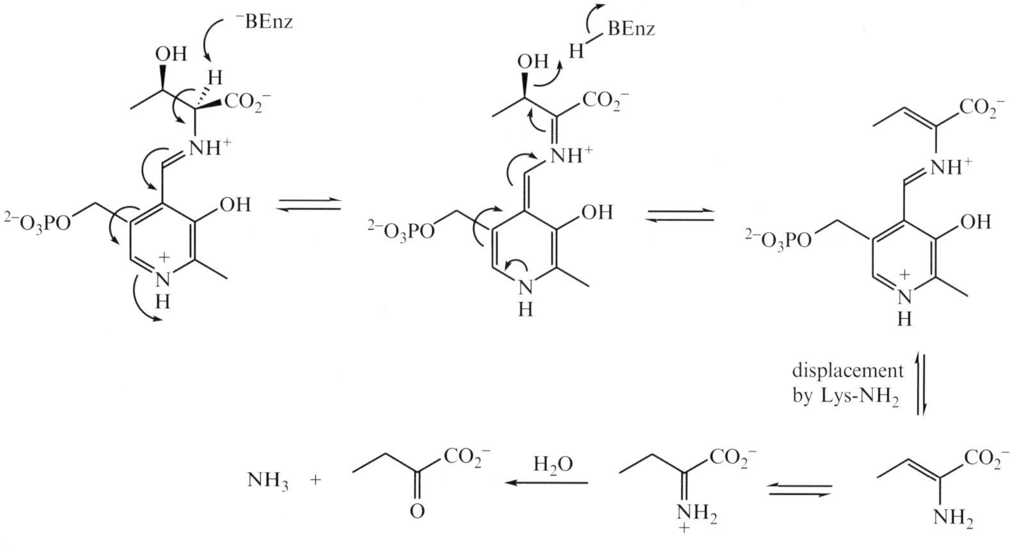

Figure 9.13 Mechanism for threonine dehydratase.

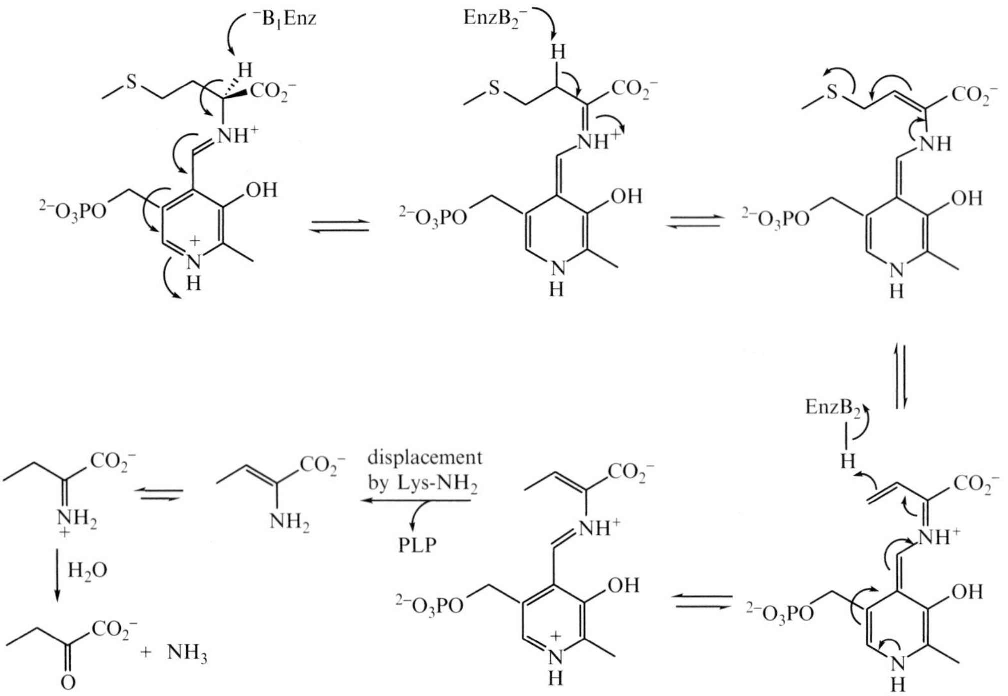

Figure 9.14 Mechanism for methionine γ-lyase.

9.5   Serine hydroxymethyltransferase

Serine hydroxymethyltransferase catalyses the interconversion of glycine and L-serine, using PLP and tetrahydrofolate as cofactors. This enzyme is unusual in that it utilises the PLP coenzyme for carbon–carbon bond formation.

The mechanism for this reaction is depicted in Figure 9.15, in the glycine-to-serine direction (the reaction is freely reversible). Following attachment of glycine to the PLP cofactor, deprotonation generates a quinonoid intermediate as seen above. However, this intermediate now reacts with N_5-methylene tetrahydrofolate, forming the carbon–carbon bond. A second deprotonation at the α-position allows the elimination of the tetrahydrofolate cofactor, generating a PLP–enamine adduct. This intermediate is attacked by water to generate the hydroxymethyl side chain of L-serine.

This is an important cellular reaction, since in the reverse direction it can be used to generate N_5-methylene tetrahydrofolate from L-serine, and this enzyme is largely responsible for the provision of cellular one-carbon methylene equivalents.

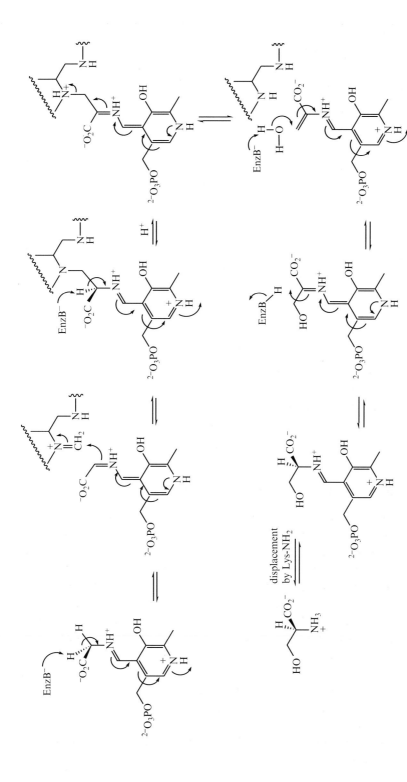

Figure 9.15 Mechanism for serine hydroxymethyltransferase.

9.6   *N*-Pyruvoyl-dependent amino acid decarboxylases

A small number of amino acid decarboxylases have been found which show no requirement for PLP. Historically, the first of these enzymes to be discovered was histidine decarboxylase from *Lactobacillus*. The enzyme can, however, be inactivated by treatment with either sodium borohydride or phenylhydrazine, suggesting the presence of an electrophilic carbonyl cofactor. This cofactor has been shown to be an N-terminal pyruvoyl group, which forms an imine linkage with the α-amino group of L-histidine. The amide carbonyl group then acts as an electron sink for decarboxylation, as shown in Figure 9.16.

Other examples of *N*-pyruvoyl-dependent decarboxylases include an S-adenosyl methionine decarboxylase and an aspartate decarboxylase from *E. coli*.

9.7   Imines and enamines in alkaloid biosynthesis

Certain α-amino acids are used as biosynthetic precursors for the assembly of alkaloid natural products. Examples of alkaloid natural products include nicotine, morphine and strychnine, shown in Figure 9.17.

The complex carbocyclic skeletons of these alkaloid natural products are assembled by intricate biosynthetic pathways. Many of the carbon–carbon bond-forming steps involve imines and enamines derived from α-amino acids. For example, lupinine is a quinolizidine alkaloid found in *Lupinus* plants which is assembled from two molecules of L-lysine as shown in Figure 9.18. The amino acid is first decarboxylated by a PLP-dependent decarboxylase enzyme to give a

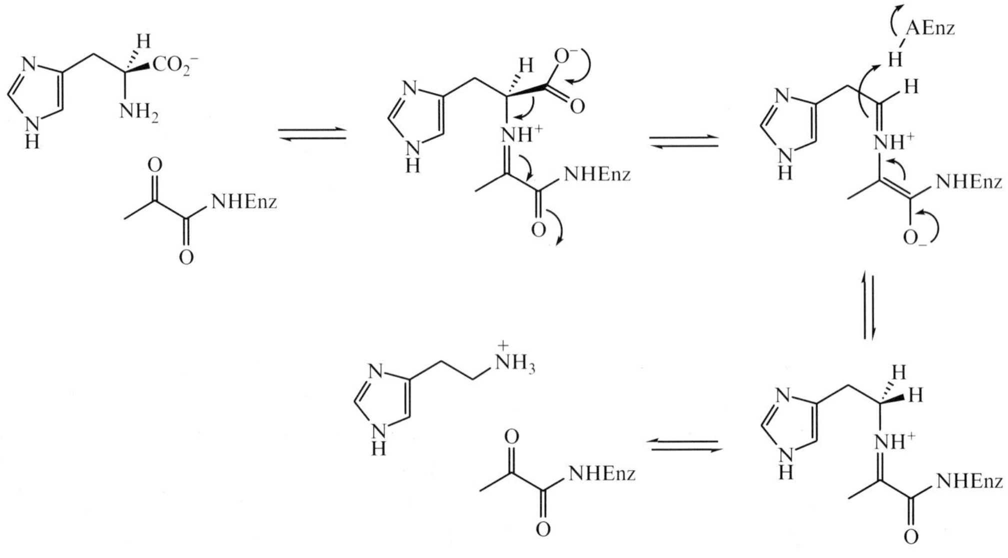

Figure 9.16 Mechanism for N-pyruvoyl-dependent histidine decarboxylase.

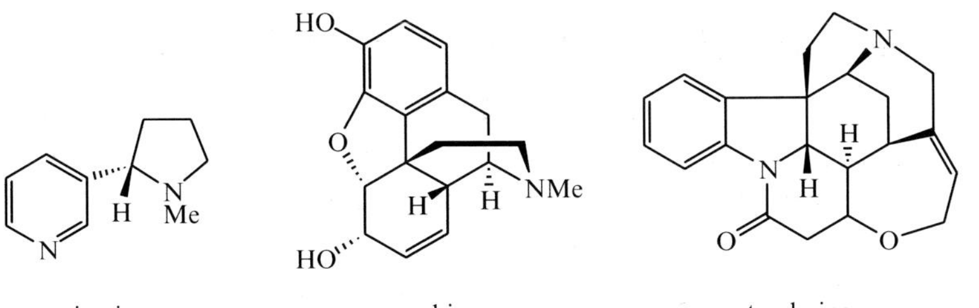

nicotine   morphine   strychnine

Figure 9.17 Alkaloid natural products.

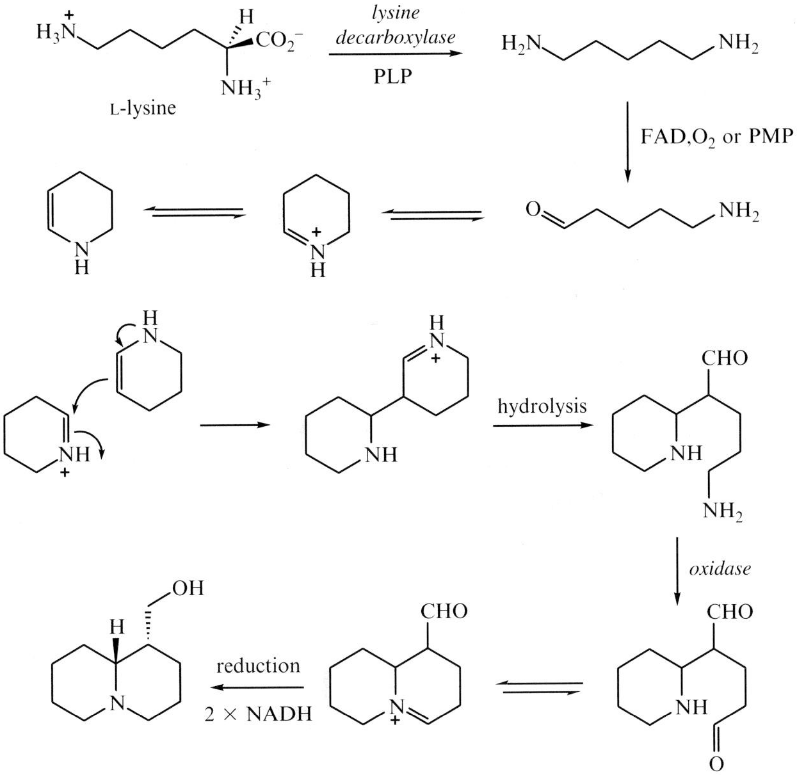

Figure 9.18 Biosynthesis of lupinine.

symmetrical diamine. Oxidative deamination of one amino group, probably by a flavin-dependent oxidase, generates an amino-aldehyde. This condenses to give a cyclic imine, which in turn isomerises to give a cyclic enamine. Reaction of one equivalent of the enamine with another equivalent of the imine forms a carbon–carbon bond between the two lysine-derived species. Hydrolysis of the resulting imine, followed by formation of a second imine linkage, leads to the natural product lupinine as shown in Figure 9.18.

The biosyntheses of nicotine (from L-lysine and L-ornithine), morphine (from two molecules of L-tyrosine) and strychnine (from L-tryptophan) also involve intermediate imine and enamine species, as well as examples of phenolic radical couplings encountered in Section 7.10. Readers interested further in alkaloid biosynthesis should consult the Further reading below.

Problems

(1) Suggest a mechanism for the reaction catalysed by threonine β-epimerase, a PLP-dependent enzyme.

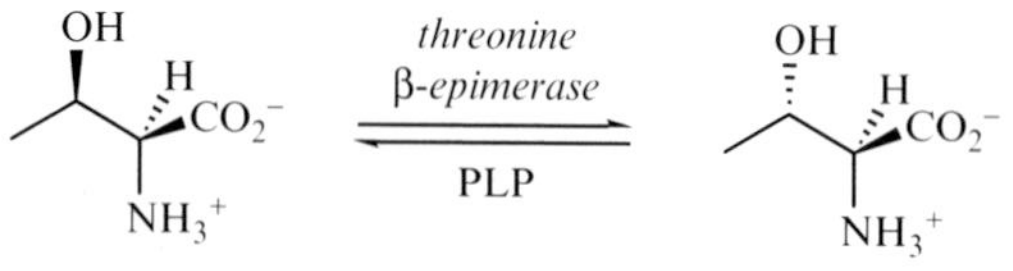

(2) Aspartate aminotransferase is inactivated in time-dependent fashion by L-vinyl-glycine. Suggest a mechanism of inactivation.

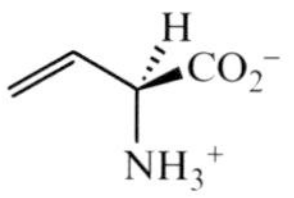

L-vinyl-glycine

(3) γ-Aminobutyrate transaminase has an important function in the human brain, where it catalyses the conversion of γ-aminobutyrate (a neurotransmitter) to succinate semialdehyde. This enzyme is inactivated (K_i 0.6 μM) by gabaculine. Suggest a mechanism of inactivation.

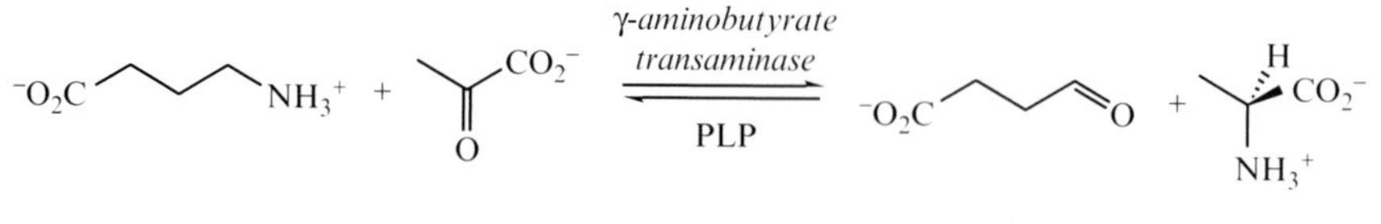

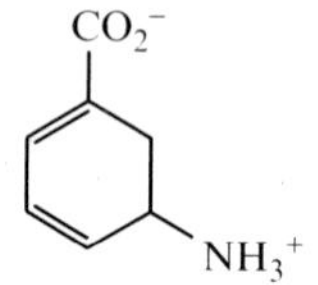

gabaculine

(4) The enzyme threonine synthase catalyses the conversion of L-homoserine phosphate to L-threonine using **PLP** as a cofactor. Write a mechanism for the enzymatic reaction that is consistent with the following observations: (i) carrying out the enzymatic reaction in $H_2^{18}O$ leads to the incorporation of ^{18}O into the β-hydroxyl group of L-threonine; (ii) incubation of 3S-[3-^{3}H]-homoserine phosphate with the enzyme leads to loss of the ^{3}H label in the product, whilst incubation of the 3R-[3-^{3}H] isomer leads to retention of the ^{3}H label.

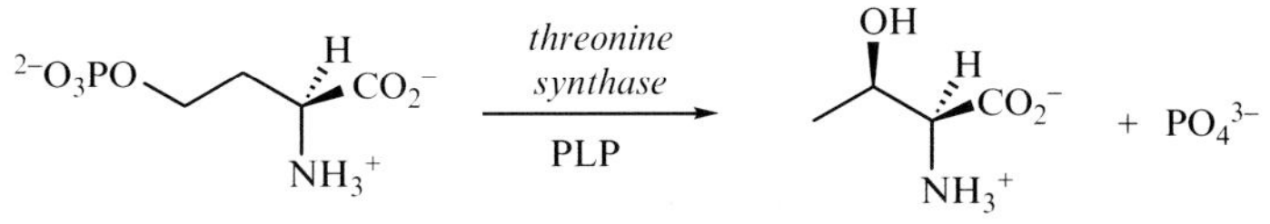

(5) Kynureninase is a PLP-dependent enzyme that catalyses the hydrolytic cleavage of L-kynurenine to give anthranilic acid and L-alanine. Given that the enzyme catalyses the exchange of the C_α hydrogen with solvent, suggest a mechanism for this enzyme. A sample of stereospecifically labelled L-kynurenine (see below) was incubated with the enzyme in ^2H_2O, and the labelled L-alanine product was converted to labelled acetic acid. The resulting chiral methyl group was found to have the S configuration. Deduce whether the enzymatic reaction proceeds with retention or inversion of configuration at the labelled centre.

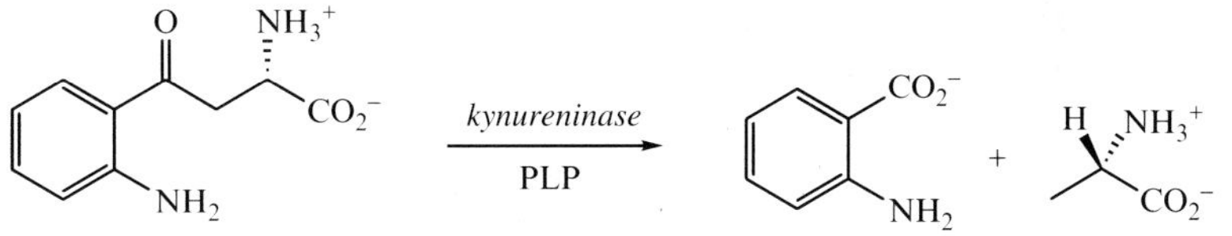

L-kynurenine

kynureninase
PLP

anthranilic acid

enzyme
D_2O

$[^2H, ^3H]$-L-alanine - - - - ▸ S-$[^2H, ^3H]$-acetate

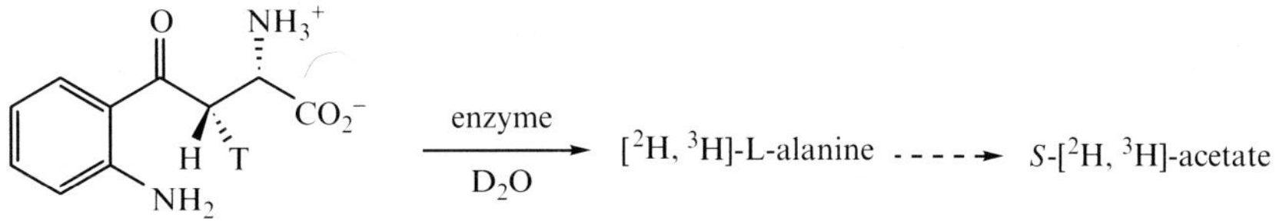

Further reading

General

C.T. Walsh (1979) *Enzymatic Reaction Mechanisms.* Freeman, San Francisco.

PLP-dependent enzymes

A.E. Braunstein & E.V. Goryachenkova (1984) The β-replacement-specific pyridoxal-P-dependent lyases. *Adv. Enzymol.*, **56**, 1–90.
C.N. Cronin & J.F. Kirsch (1988) Role of arginine-292 in the substrate specificity of aspartate aminotransferase as examined by site-directed mutagenesis. *Biochemistry*, **27**, 4572–9.
H. Hayashi, H. Wada, T. Yoshimura, N. Esaki & K. Soda (1990) Recent topics in pyridoxal 5′-phosphate enzyme studies. *Annu. Rev. Biochem.*, **59**, 87–110.
J. Jäger, M. Moser, U. Sauder & J.N. Jansonius (1994) Crystal structures of *Escherichia coli* aspartate aminotransferase in two conformations. *J. Mol. Biol.*, **239**, 285–305.
A.E. Martell (1982) Reaction pathways and mechanisms of pyridoxal catalysis. *Adv. Enzymol.*, **53**, 163–200.
A. Planas & J.F. Kirsch (1991) Re-engineering the catalytic lysine of aspartate amino-transferase by chemical elaboration of a genetically introduced cysteine. *Biochemistry*, **30**, 8268–76.
L. Schirch (1982) Serine hydroxymethyltransferase. *Adv. Enzymol.*, **53**, 83–112.
J.C. Vederas & H.G. Floss (1980) Stereochemistry of pyridoxal phosphate-catalysed enzyme reactions. *Acc. Chem. Res.*, **13**, 455–63.
C.T. Walsh (1982) Suicide substrates: mechanism-based enzyme inactivators. *Tetrahedron*, **38**, 871–909.

Pyruvoyl-dependent enzymes

P.A. Recsei & E.E. Snell (1984) Pyruvoyl enzymes *Annu. Rev. Biochem.*, **53**, 357–88.
P.D. van Poelje & E.E. Snell (1990) Pyruvoyl-dependent enzymes. *Annu. Rev. Biochem.*, **59**, 29–60.

Alkaloid biosynthesis

J. Mann (1987) *Secondary Metabolism*, 2nd edn., Clarendon Press, Oxford.
T. Hashimoto & Y. Yamada (1994) Alkaloid biosynthesis: molecular aspects. *Annu. Rev. Plant Physiol. Plant Mol. Biol.*, **45**, 257–85.
R.B. Herbert (1989) *The Biosynthesis of Secondary Metabolites*, 2nd edn., Chapman & Hall, London.

10 Isomerases

10.1 Introduction

The final group of enzymatic transformations that we shall meet are the isomerisation reactions: the interconversion of two isomeric forms of a molecule. The interconversion of enantiomers is catalysed by racemase enzymes, as discussed in Section 9.2. In this chapter we shall meet a second family of racemases that require no cofactors. We shall also meet enzymatic proton transfer reactions which involve the interconversion of tautomeric forms of ketones, and the interconversion of positional isomers of allylic compounds. These isomerisation reactions are summarised in Figure 10.1.

10.2 Cofactor-independent racemases and epimerases

In Section 9.2 we met the family of α-amino acid racemase enzymes which utilise the coenzyme pyridoxal 5′-phosphate (PLP). In addition there is a family of racemase and epimerase enzymes which require no cofactors at all. A separate section is being devoted to these enzymes because of a single important mechanistic issue: how do these enzymes achieve the deprotonation of the α-proton of an α-amino acid? In the PLP-dependent racemases the formation of an imine linkage with the α-amino group dramatically increases the acidity of the α-proton. If these enzymes contain no PLP then no such assistance is possible, so they must have some alternative way of carrying out this deprotonation. This question is of wider significance, since there is evidence for enzymatic deprotonation adjacent to carboxylic acids in other enzymes such as the flavin-dependent amino acid oxidase enzymes (see Section 6.3).

Examples of cofactor-independent α-amino acid racemases are glutamate racemase from *Lactobacillus fermenti* and aspartate racemase from *Streptococ-*

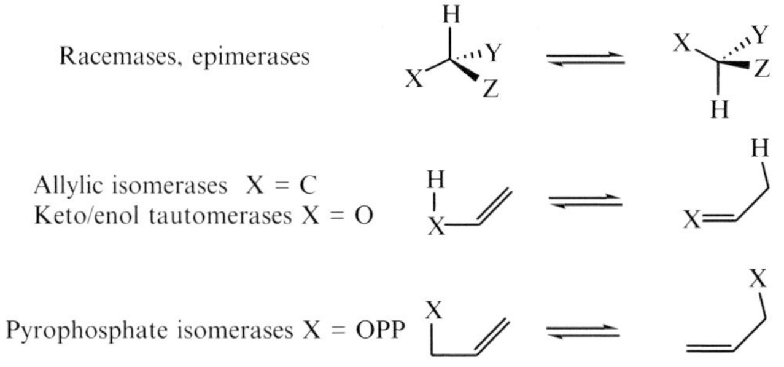

Racemases, epimerases

Allylic isomerases X = C
Keto/enol tautomerases X = O

Pyrophosphate isomerases X = OPP

Figure 10.1 Summary of enzyme-catalysed isomerisation reactions.

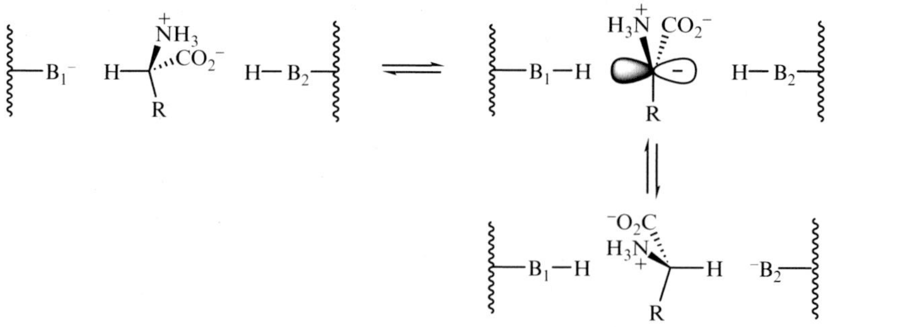

Figure 10.2 Two-base mechanism for cofactor-independent racemases.

cus thermophilus. There is evidence in both these enzyme-catalysed reactions for a two-base mechanism of racemisation, as shown in Figure 10.2. In this mechanism an α-proton is removed from one face of the amino acid by an active site base and a proton delivered onto the other face by a protonated base on the opposite side of the enzyme active site. In this mechanism the unstable α-carbanion would exist only fleetingly.

How is experimental evidence obtained for such a two-base mechanism? One such method is illustrated in Figure 10.3. In this experiment a stoichiometric amount of enzyme is incubated with substrate for a short period of time in tritiated water, and the distribution of ^{3}H label examined in the products. In a one-base active site the α-hydrogen is removed from the L-enantiomer and returned to the opposite face by the same base, giving the D-enantiomer. Hence any ^{3}H label incorporated by exchange of the active site base with ^{3}H$_2$O would be delivered equally to both L- and D-enantiomers. However, in a single catalytic cycle of a two-base active site one would expect the α-proton of the L-enantiomer to be abstracted by one base and a ^{3}H label to be delivered from the opposite face by the other base, resulting in increased incorporation of ^{3}H label into the D-enantiomer. Thus, starting with L-glutamate, glutamate racemase catalyses the preferential incorporation of ^{3}H from ^{3}H$_2$O into D-glutamic acid.

This approach has established the likelihood of a two-base mechanism for a number of cofactor-independent racemases and epimerases. However, this does not explain how the intermediate carbanion, however fleeting, is stabilised.

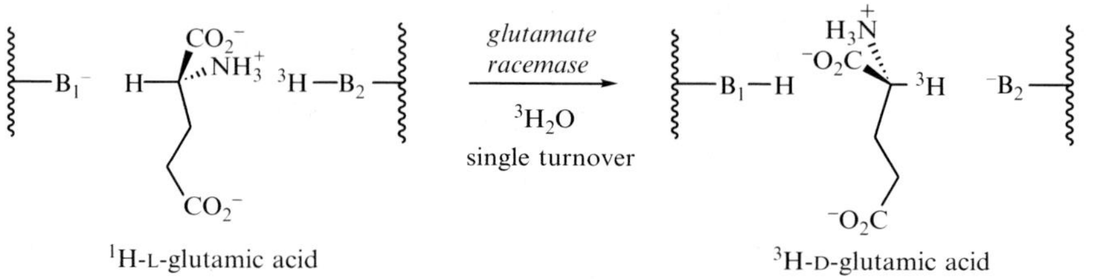

Figure 10.3 Experimental evidence in favour of a two-base mechanism.

Recent studies on mandelate racemase have provided an insight into this problem. Mandelate racemase is a cofactor-independent enzyme that catalyses the interconversion of R- and S-mandelic acid. A two-base mechanism has been implicated for this enzyme, and is supported by an X-ray crystal structure of the enzyme in which active site residues Lys-166 and His-297 are suitably positioned to act as the two bases (see Figure 10.4). Replacement of His-297 by glutamine using site-directed mutagenesis gives a mutant enzyme that is unable to catalyse the racemisation reaction. This mutant enzyme (still containing Lys-166) is able to catalyse the exchange of the α-proton of S-mandelic acid with ^2H_2O to give ^2H-S-mandelic acid, but does not catalyse exchange with R-mandelic acid. This result implies that Lys-166 deprotonates the α-proton of S-mandelic acid, forming a carbanion intermediate. This intermediate is sufficiently stable to exchange with the ε-NHD_2^+ of Lys-166.

How is the carbanion intermediate stabilised? It was evident from the X-ray crystal structure that the carboxylate group of the substrate forms hydrogen bonds with a protonated Lys-164 residue and a protonated Glu-317 residue. Replacement of Glu-317 by a glutamine residue using site-directed mutagenesis gives a mutant enzyme with 10^4-fold reduced catalytic efficiency. It has been

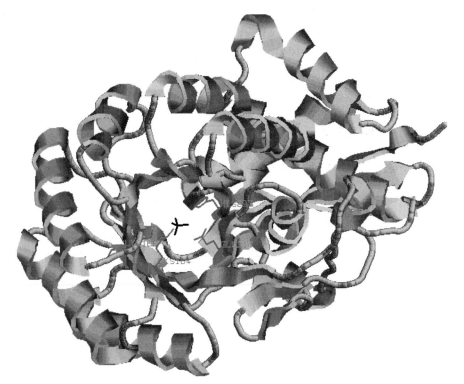

Figure 10.4 Structure of mandelate racemase (PDB file 2MNR), showing the α,β-barrel structure of the protein. Catalytic bases Lys-166 and His-297, and active site residues Lys-164 and Glu-317, are shown in red. Bound SO_4^{2-} ion shown in black.

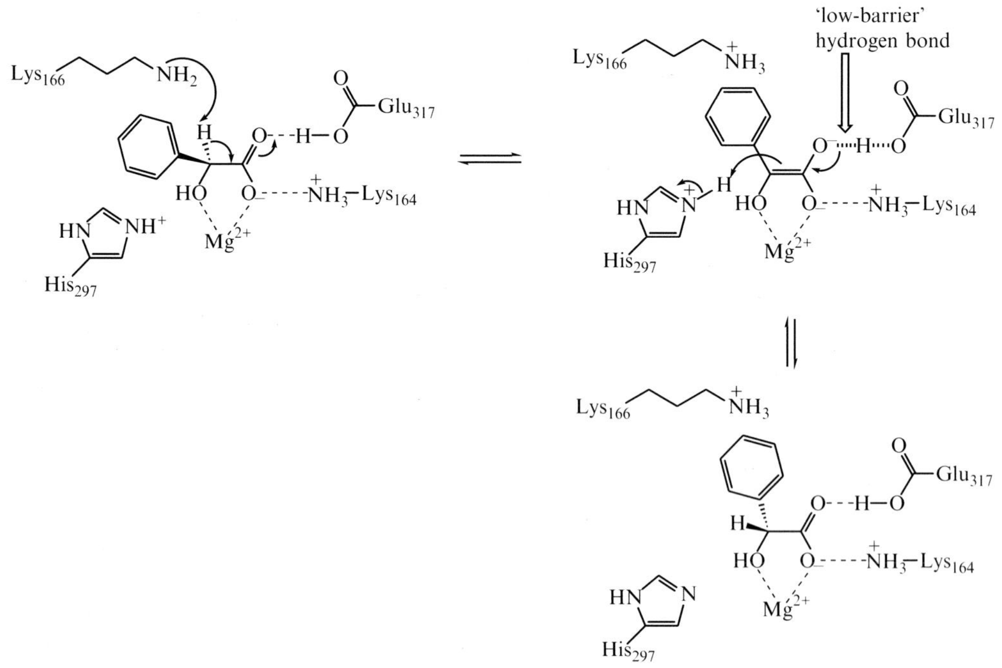

Figure 10.5 Mechanism for mandelate racemase.

proposed that Glu-317 forms a strong 'low barrier' hydrogen bond with the substrate carboxylate which stabilises the formation of the α-carbanion, as shown in Figure 10.5. It may be that similar methods of stabilisation are employed in other cases of enzymatic deprotonation adjacent to carboxylate groups.

10.3   Keto–enol tautomerases

The enolisation of ketones is a well-known reaction in organic chemistry that is utilised as an intermediate process in many enzyme-catalysed reactions, notably the aldolases encountered in Section 7.2. In most cases enols and enolate anions are thermodynamically unstable species that are not isolable. However, in a few cases enol tautomers of ketones are sufficiently stabilised to be isolable, and there are several enzymes which catalyse the interconversion of keto and enol tautomers.

One simple example is that of phenylpyruvate tautomerase, which catalyses the interconversion of phenylpyruvic acid with its stabilised enol form, as shown in Figure 10.6. The enzyme catalyses the stereospecific exchange of the *proR* hydrogen with ^2H_2O via the *E* enol isomer, using acid/base active site chemistry.

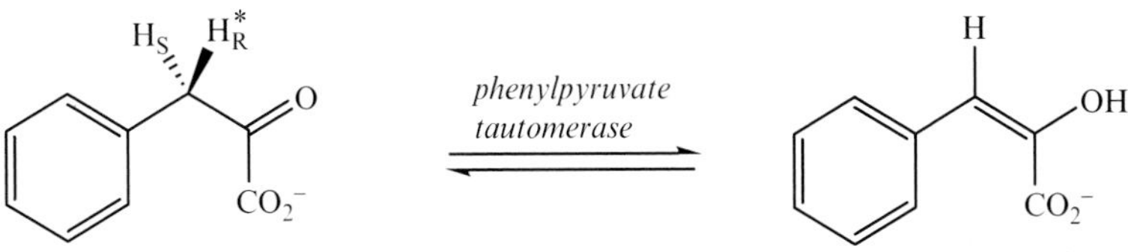

Figure 10.6 Phenylpyruvate tautomerase.

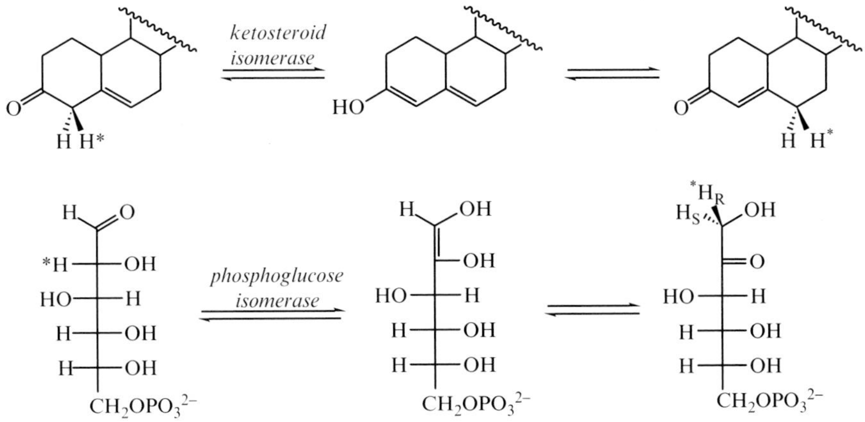

glucose-6-phosphate fructose-6-phosphate

Figure 10.7 Ketosteroid isomerase and phosphoglucose isomerase.

Two-step keto–enol isomerisations are involved in the mechanisms of a number of isomerase enzymes. Ketosteroid isomerase (see Section 3.5) catalyses the isomerisation of Δ-2-ketosteroid to Δ-3-ketosteroid through a dienol intermediate, as shown in Figure 10.7. This reaction involves deprotonation by an active site base Asp-38 and concerted protonation by Tyr-14 to generate the dienol intermediate. Asp-38 then returns the abstracted proton to the γ-position to complete the isomerisation reaction.

There is also a family of aldose–ketose isomerases that catalyse the interconversion of aldose sugars containing a C-1 aldehyde with ketose sugars containing a C-2 ketone. Phosphoglucose isomerase catalyses the interconversion of glucose-6-phosphate with fructose-6-phosphate, as shown in Figure 10.7. This reaction proceeds by abstraction of the C-2 proton to generate an enediol intermediate, followed by return of the proton specifically at the *proR* position of C-1. Intramolecular transfer of H^* is observed in both directions, indicating that a single active site base is responsible.

10.4 Allylic isomerases

We have already met a few examples of allylic (or 1,3-) migrations: the ketosteroid isomerase reaction illustrated in the previous section is effectively

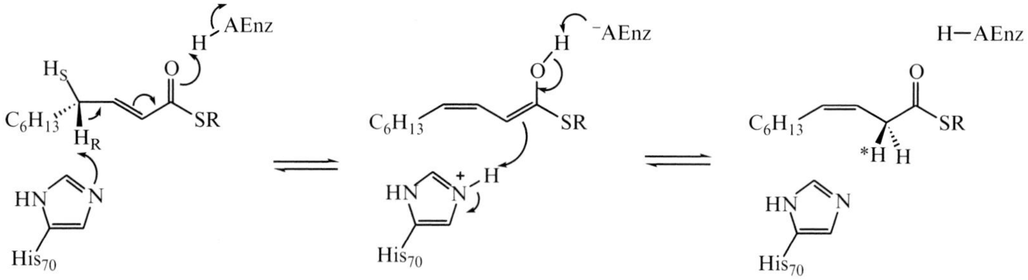

Figure 10.8 Mechanism for β-hydroxydecanoyl thioester isomerase.

a 1,3-hydrogen migration; and we have encountered 1,3-migrations of allylic pyrophosphates in the terpene cyclase reactions of Section 7.9. There are many allylic isomerases which operate in many different areas of biological chemistry: I will briefly mention two examples that are important for the cellular assembly of fatty acids and terpenoid natural products, respectively.

In Section 8.2 we met the enzyme β-hydroxydecanoyl thioester dehydratase, which catalyses the elimination of β-hydroxy-thioesters in fatty acid assembly. This enzyme also catalyses the isomerisation of the *trans*-2,3-alkenyl thioester to the *cis*-3,4-alkenyl thioester, as shown in Figure 10.8. The dehydration reaction catalysed by this enzyme utilises an active site histidine base, His-70 (see Section 8.2). This base is also thought to be involved in the isomerisation reaction. The reaction involves the transfer of the C-4 *proR* hydrogen to the C-2 *proS* position. This is defined as a 1,3-suprafacial hydrogen shift, since the hydrogen is transferred across the same face of the molecule. It is thought that His-70 mediates this transfer via an enediol intermediate as shown in Figure 10.8. However, it is worthy of note that no intramolecular hydrogen transfer is observed in this case, unlike examples such as ketosteroid isomerase, suggesting that the protonated histidine exchanges readily with solvent water.

Isopentenyl pyrophosphate isomerase catalyses the interconversion of isopentenyl pyrophosphate (IPP) and dimethylallyl pyrophosphate (DMAPP), the two five-carbon building blocks for the assembly of terpene natural products (also see Section 7.9). The enzymatic reaction involves stereospecific removal of the *proR* hydrogen at C-2, and delivery of a hydrogen onto the *re*-face of the 4,5-double bond. The stereochemistry was established by conversion of a stereospecifically labelled substrate in ^2H_2O (as shown in Figure 10.9), generating a chiral methyl group of *R* configuration at C-4. This reaction is defined as a

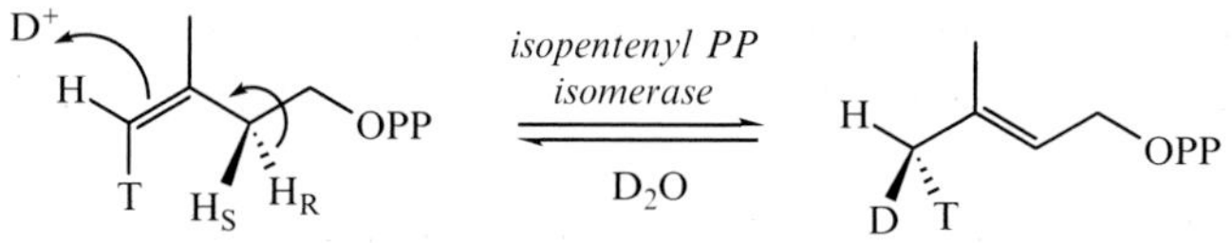

Figure 10.9 Stereochemistry of isopentenyl pyrophosphate isomerase.

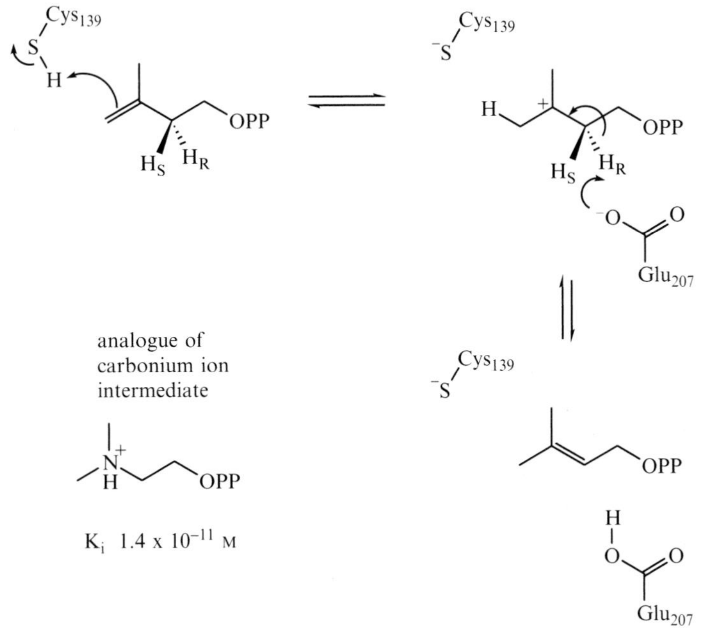

Cys139

analogue of
carbonium ion
intermediate

K_i 1.4×10^{-11} M

Cys139

Glu207

Cys139

OPP

H

Glu207

Figure 10.10 Mechanism for isopentenyl pyrophosphate isomerase.

1,3-antarafacial hydrogen shift, since the hydrogen is re-inserted on the opposite face of the molecule.

The antarafacial stereochemistry of hydrogen transfer suggests the involvement of two active site groups in the mechanism. This was confirmed by subsequent chemical modification studies. Site-directed mutagenesis experiments have implicated Cys-139 as an acidic group which protonates the 4,5-double bond, generating a tertiary carbonium ion intermediate. Stereospecific deprotonation is then carried out by Glu-207, as shown in Figure 10.10. The existence of the carbocation intermediate is supported by the potent inhibition of this enzyme (K_i 1.4×10^{-11} M) by a tertiary amine substrate analogue. At neutral pH this amine is protonated, and the ammonium cation mimics the carbocation intermediate which is bound tightly by the enzyme.

10.5   CASE STUDY: Chorismate mutase

Pericyclic reactions are commonly used in organic synthesis; the Diels–Alder reaction being an important synthetic reaction for the stereoselective synthesis of cyclohexane rings. However, only one enzyme-catalysed pericyclic reaction has been well characterised, which is the reaction catalysed by chorismate mutase.

Chorismate mutase catalyses the Claisen rearrangement of chorismic acid into prephenic acid. This reaction is important for plants and micro-organisms,

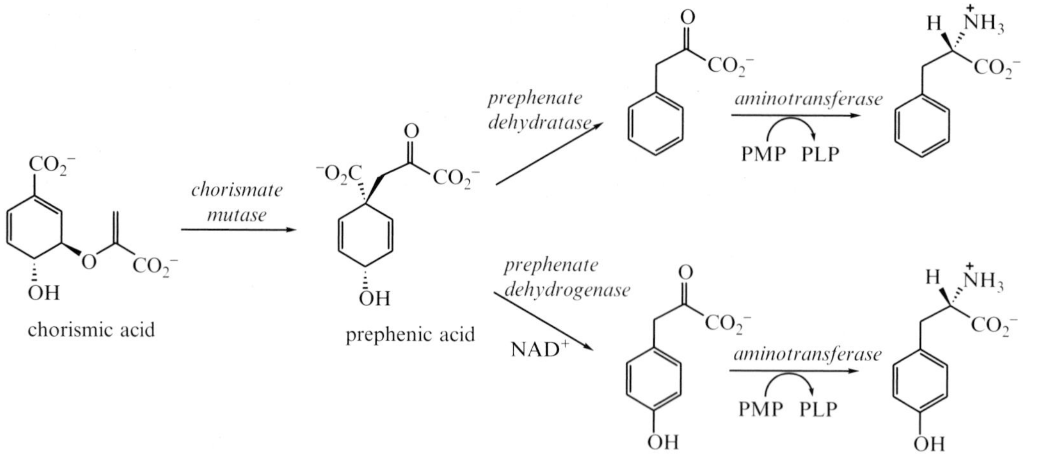

Figure 10.11 Reaction catalysed by chorismate mutase.

which utilise prephenic acid as a precursor to the amino acids L-phenylalanine and L-tyrosine (see Figure 10.11). This reaction occurs non-enzymatically at a significant rate, contributing to the singular instability of the important metabolite chorismic acid. However, the reaction is accelerated 10^6-fold by the enzyme chorismate mutase.

The stereochemistry of the enzyme-catalysed and uncatalysed reactions has been investigated, revealing that both reactions proceed through a chair-like transition state involving the *trans*-diaxial conformer of the substrate, as shown in Figure 10.12. Tritium substitution at C-5 or C-9 gives no secondary kinetic isotope effect on the enzyme-catalysed reaction, whereas such isotope effects are observed on the uncatalysed reaction. This suggests that the rate-determining step of the enzyme-catalysed reaction is substrate binding, hence isotope effects on subsequent steps would not be observed. This situation, in which there is a high preference for catalytic processing of a bound substrate, rather than dissociation, is termed a 'high commitment to catalysis'.

How does the enzyme achieve the 10^6-fold rate acceleration? If the mechanisms of the uncatalysed and enzyme-catalysed reactions are similar, does the

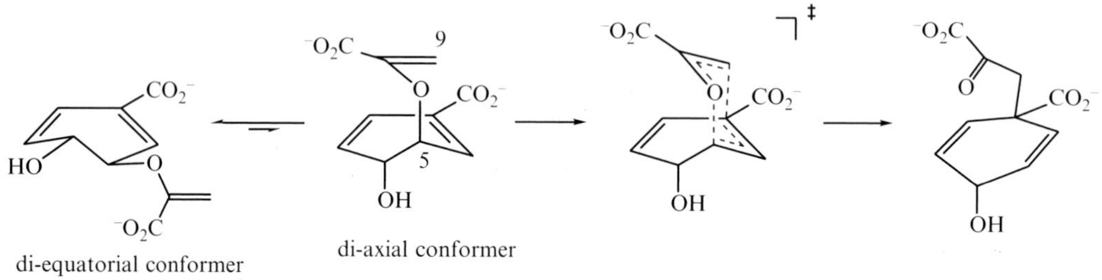

Figure 10.12 Mechanism for chorismate mutase.

rate acceleration come from transition state stabilisation? A bicyclic transition state analogue has been synthesised for the chorismate mutase reaction, as shown in Figure 10.13. This analogue inhibits the enzymatic reaction strongly (K_i 3 μM), suggesting that the enzyme does selectively bind the transition state of the reaction. This analogue has been used to prepare catalytic antibodies (see Section 12.3) capable of catalysing the chorismate mutase reaction to the extent of 10^4-fold over the uncatalysed reaction.

The chorismate mutase enzyme from *Bacillus subtilis* has been crystallised in the presence of the transition state analogue, and the X-ray crystal structure solved (see Figure 10.14). This enzyme is a trimer of identical 127-amino acid

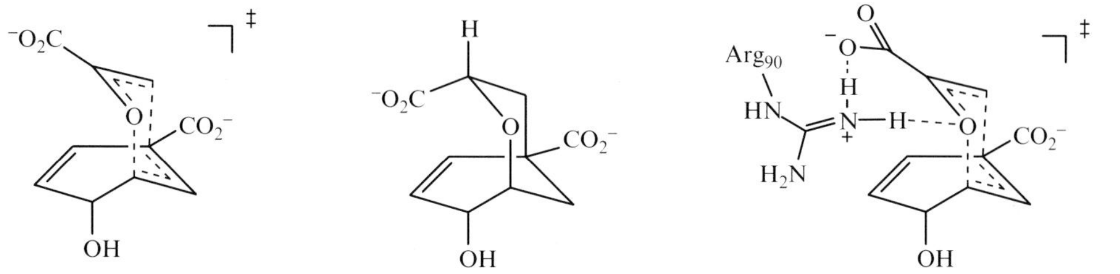

chair-like transition state   transition state analogue   transition state stabilisation by Arg-90

Figure 10.13 Transition state analogue for chorismate mutase.

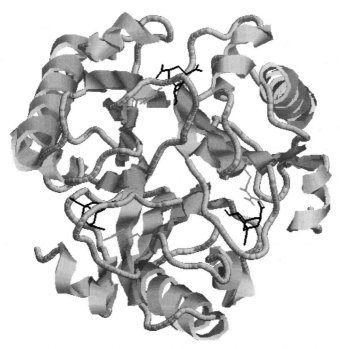

Figure 10.14 Structure of *Bacillus subtilis* chorismate mutase (PDB file 2CHT) trimer, with bound transition state analogue shown in black. Arg-90 is shown in red. Each active site is at the interface of the protein sub-units.

sub-units. Each sub-unit contains five strands of β-sheet and two α-helices. The β-strands of each sub-unit pack together to form the trimer structure, with active sites at the interface of each pair of sub-units.

Examination of the active site structure has revealed that the ether oxygen of the analogue lies within hydrogen-bonding distance of the guanidinium side chain of Arg-90, as shown in Figure 10.13. Molecular modelling studies on the enzyme–substrate and enzyme–transition state complexes have revealed that Arg-90 can form a favourable electrostatic/hydrogen-bonding interaction with the transition state. This interaction and other transition-state binding appear to provide sufficient stabilisation to account for the 10^6-fold rate acceleration. This type of electrophilic catalysis is precedented in synthetic organic chemistry in the form of Lewis acid catalysis of pericyclic reactions such as the Claisen rearrangement.

There is one additional class of isomerisation reactions, the vitamin B_{12}-dependent skeletal rearrangements, which will be described in Section 11.2, since they involve radical chemistry.

Problems

(1) 2,6-Diaminopimelic acid is a naturally occurring amino acid involved in the biosynthesis of L-lysine. *SS*-Diaminopimelate is converted to *RS*-diaminopimelate by a cofactor-independent epimerase enzyme, and *RS*-diaminopimelate is then decarboxylated to L-lysine by a PLP-dependent decarboxylase. What product would you expect if these two reactions were carried out consecutively in ^2H_2O?

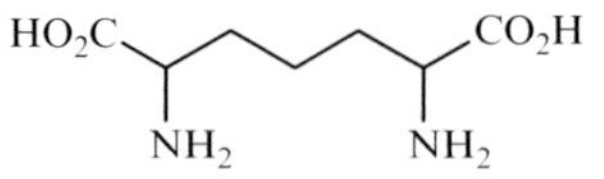

2,6-diaminopimelic acid

(2) The following reactions are part of a pathway for degradation of phenol in *Pseudomonas putida*. Suggest a mechanism for the isomerase reaction. This organism is found to degrade 2-chlorophenol quite readily. What do you think the fate of the additional chlorine atom is?

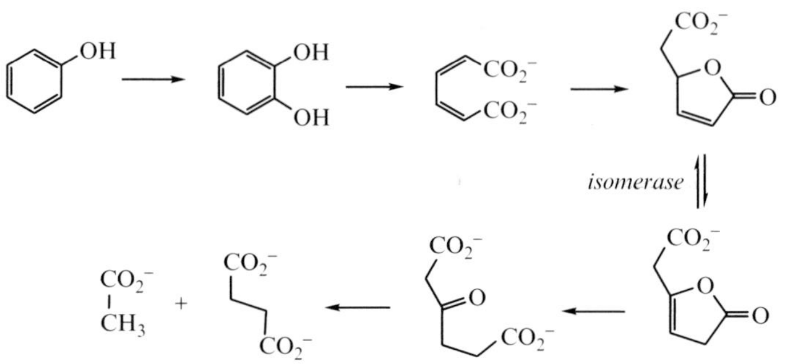

(3) Suggest a mechanism for the hydrolase enzyme below, which is involved in an aromatic degradation pathway in *Escherichia coli*.

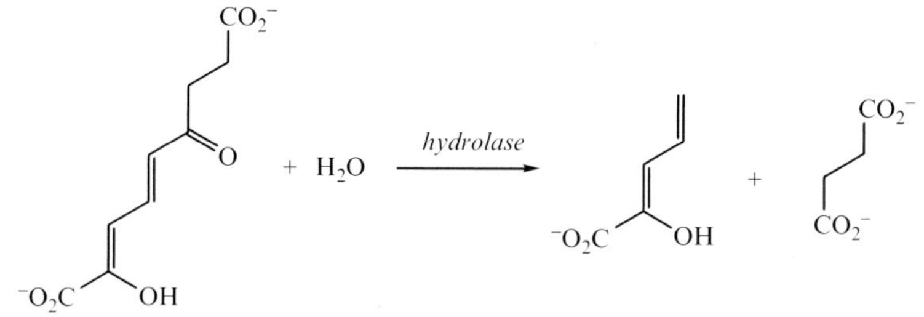

(4) Peptidyl–proline amide bonds are sometimes found in a kinetically stable *cis*-conformation rather than the thermodynamically more favourable *trans*-conformation. An enzyme activity has been found that is capable of catalysing the *cis–trans* isomerisation of such peptidyl–proline amide bonds, as shown below (*Note*: This enzyme is strongly inhibited by the immunosuppressant cyclosporin A shown in Figure 1.3.) Suggest possible mechanisms for this isomerase reaction.

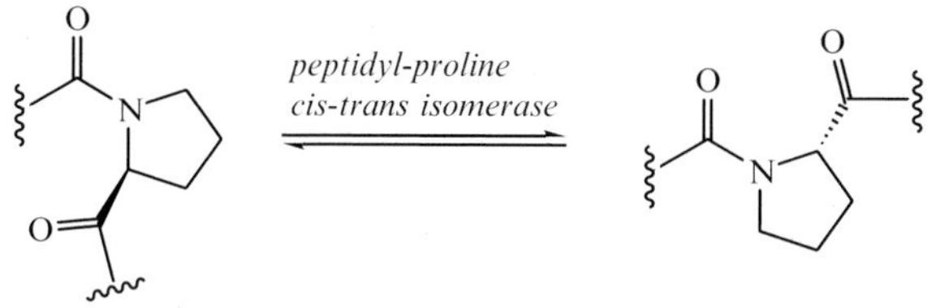

(5) Suggest step-wise and concerted mechanisms for the phosphoenol pyruvate (PEP) mutase reaction shown below. How might you distinguish between these mechanisms?

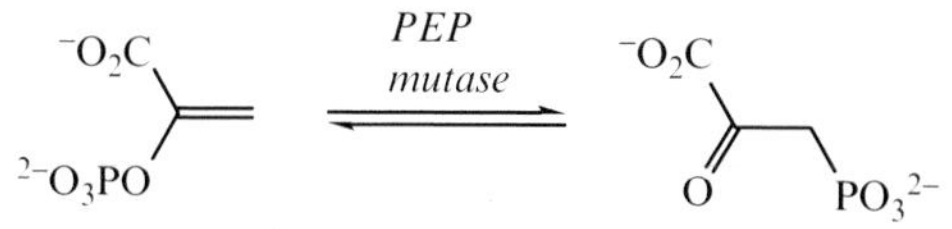

Further reading

General

C.T. Walsh (1979) Enzymatic Reaction Mechanisms. Freeman, San Francisco.

Racemases

E. Adams (1976) Enzymatic racemization. *Adv. Enzymol.*, **44**, 69–138.

K.A. Gallo, M.E. Tanner & J.R. Knowles (1993) Mechanism of the reaction catalysed by glutamate racemase. *Biochemistry*, **32**, 3991–7.

G.L. Kenyon, J.A. Gerlt, G.A. Petsko & J.W. Kozarich (1995) Mandelate racemase: structure-function studies of a pseudosymmetric enzyme. *Acc. Chem. Res.*, **28**, 178–86.

J.S. Wiseman & J.S. Nichols (1984) Purification and properties of diaminopimelic acid epimerase from *Escherichia coli*. *J. Biol. Chem.*, **259**, 8907–14.

M.E. Tanner (2002) Understanding Nature's strategies for enzyme-catalysed racemization and epimerization. *Acc. Chem. Res.*, **35**, 237–46.

Allylic isomerases

A. Kuliopulos, A.S. Mildvan, D. Shortle & P. Talalay (1989) Kinetic and ultraviolet spectroscopic studies of active-site mutants of Δ^5-3-ketosteroid isomerase. *Biochemistry*, **28**, 149–59.

J.E. Reardon & R.H. Abeles (1985) Time-dependent inhibition of isopentenyl pyrophosphate isomerase by 2-(dimethylamino)ethyl pyrophosphate. *J. Am. Chem. Soc.*, **105**, 4078–9.

J.E. Reardon & R.H. Abeles (1986) Mechanism of action of isopentenyl pyrophosphate isomerase: evidence for a carbonium ion intermediate. *Biochemistry*, **25**, 5609–16.

I.A. Rose (1975) Mechanism of the aldose-ketose isomerase reactions. *Adv. Enzymol.*, **43**, 491–518.

J.M. Schwab & B.S. Henderson (1990) Enzyme-catalysed allylic rearrangements. *Chem. Rev.*, **90**, 1203–45.

I.P. Street, H.R. Coffman, J.A. Baker & C.D. Poulter (1994) Identification of cysteine-139 and gluatamate-207 as catalytically important groups in the active site of isopentenyl pyrophosphate/dimethylallyl pyrophosphate isomerase. *Biochemistry*, **33**, 4212–17.

L. Xue, P. Talalay & A.S. Mildvan (1990) Studies of the mechanism of the Δ^5-3-ketosteroid isomerase reaction by substrate, solvent, and combined kinetic deuterium isotope effects on wild-type and mutant enzymes. *Biochemistry*, **29**, 7491–500.

Chorismate mutase

Y.M. Chook, H. Ke & W.N. Lipscomb (1993) Crystal structures of the monofunctional chorismate mutase from *Bacillus subtilis* and its complex with a transition state analog. *Proc. Natl. Acad. Sci. USA*, **90**, 8600–603.

S.D. Copley & J.R. Knowles (1985) The uncatalysed Claisen rearrangement of chorismate to prephenate prefers a transition state of chairlike geometry. *J. Am. Chem. Soc.*, **107**, 5306–8.

M.N. Davidson, I.R. Gould & I.H. Hillier (1995) Contribution of transition-state binding to the catalytic activity of *Bacillus subtilis* chorismate mutase. *J. Chem. Soc. Chem. Commun.*, 63–4.

S.G. Sogo, T.S. Widlanski, J.H. Hoare, C.E. Grimshaw, G.A. Berchtold & J.R. Knowles (1984) Stereochemistry of the rearrangement of chorismate to prephenate: chorismate mutase involves a chair transition state. *J. Am. Chem. Soc.*, **106**, 2701–3.

11 Radicals in Enzyme Catalysis

11.1 Introduction

A special chapter is being devoted to the topic of radical chemistry in enzyme catalysis because of a series of remarkable discoveries in this area that have occurred since 1990. Before then, it was known that certain enzymes *could* generate free radical intermediates, such as cytochrome P_{450} mono-oxygenases (Section 6.8), certain metallo-enzymes (Sections 6.9, 6.10, 7.10), and certain flavo-enzymes (Section 6.3), but that such intermediates were generally short-lived reaction intermediates, generated fleetingly in special circumstances. It has now emerged that enzymes can generate a variety of radical species, some of which are long-lived, using several different strategies.

11.2 Vitamin B₁₂-dependent rearrangements

Vitamin B_{12} has the most complex structure of all of the vitamins. The X-ray crystal structure of vitamin B_{12} was solved in 1961 by Hodgkin. The structure, shown in Figure 11.1, consists of an extensively modified porphyrin ring system, containing a central Co^{3+} ion. The two axial ligands are a benzimidazole nucleotide and an adenosyl group. The cobalt–carbon bond formed with the adenosyl ligand is weak and susceptible to homolysis, and this is the initiation step for the radical-mediated vitamin B_{12}-dependent reactions.

We shall consider three vitamin B_{12}-dependent rearrangements: propanediol dehydrase, methylmalonyl coenzyme A (CoA) mutase, and glutamate mutase. Both reactions involve the 1,2-migration of a hydrogen atom, and the corresponding 1,2-migration of another substituent, either $-OH$ or $-CO_2H$, as shown in Figure 11.2.

Propanediol dehydrase catalyses the rearrangement of propane-1,2-diol to propionaldehyde. There is no incorporation of solvent hydrogens during the reaction, implying that there is an intramolecular hydrogen transfer. Stereo-specific labelling studies have shown that the reaction involves the removal of the *proS* hydrogen at C-1. This hydrogen atom is transferred specifically to the *proS* position at C-2, giving an inversion of configuration at C-2.

Tritium labelling of the C-1 *proS* hydrogen gives rise to exchange of ^3H into the adenosyl 5′-position. This implies that there is an adenosyl 5′-CH_3 intermediate in the enzyme mechanism formed by homolysis of the adenosyl–cobalt bond and hydrogen atom transfer from the substrate. Homolysis of the adenosyl–cobalt bond is further supported by the detection of Co^{2+} intermedi-

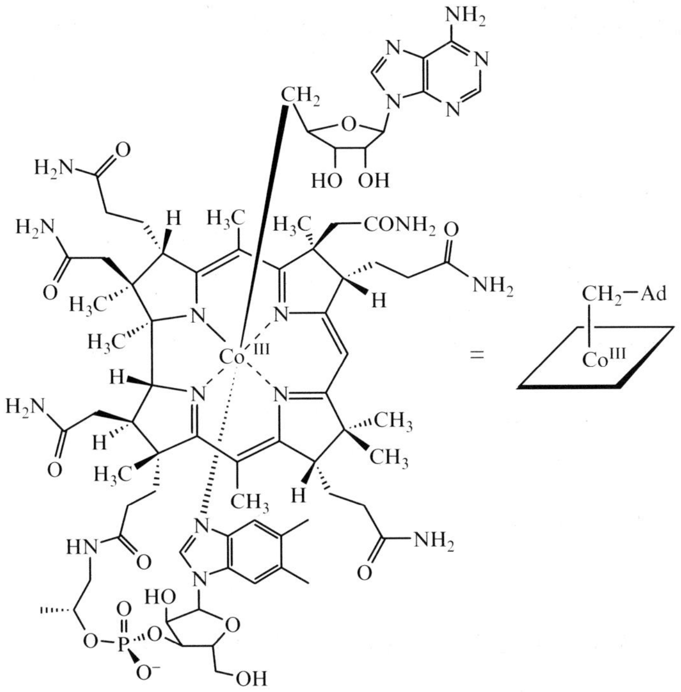

Figure 11.1 Structure of vitamin B_{12} cofactor.

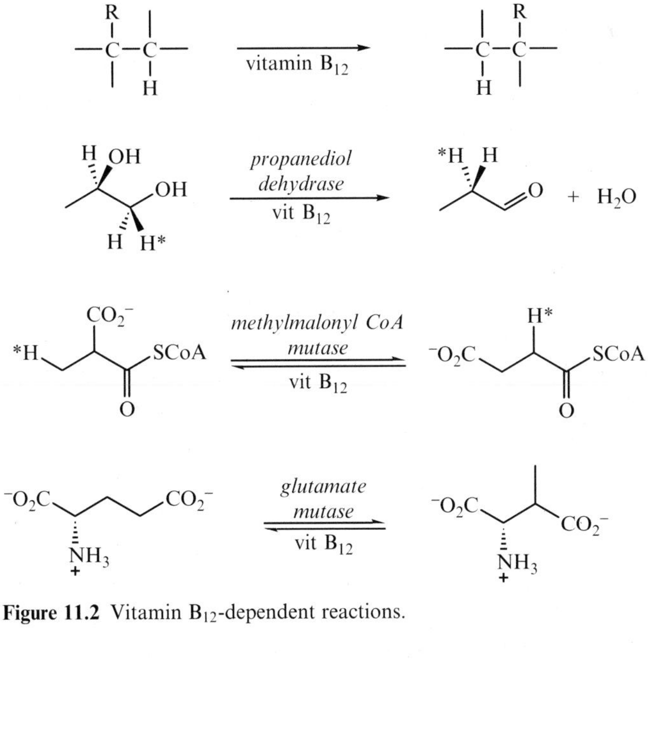

Figure 11.2 Vitamin B_{12}-dependent reactions.

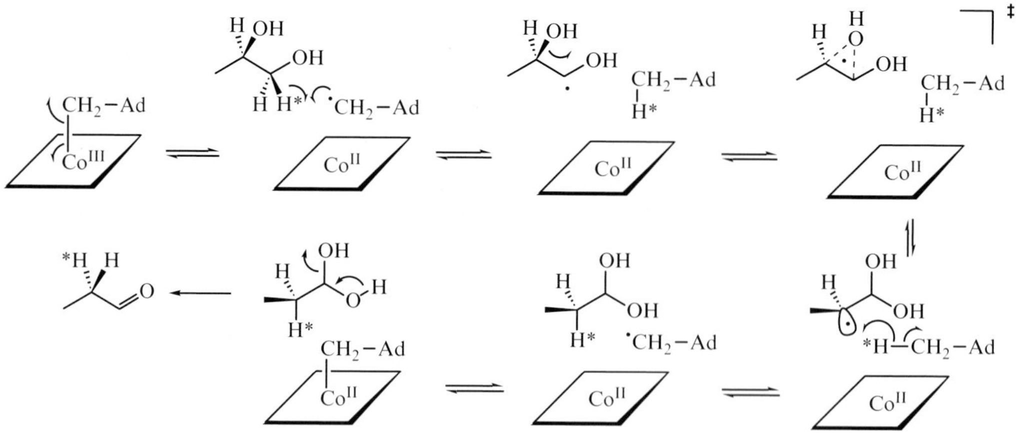

Figure 11.3 Mechanism for propanediol dehydrase.

ates in the reaction by stopped flow electron spin resonance (ESR) spectroscopy studies.

The proposed mechanism for propanediol dehydrase is shown in Figure 11.3. Initiation of the reaction by homolysis of the carbon–cobalt bond generates an adenosyl radical, which abstracts the C-1 *proS* hydrogen, to give a substrate-derived radical. Migration of the OH group then occurs, from C-2 to C-1, probably through a cyclic transition state, to give a product radical. The reaction is completed by abstraction of H• from the 5′-deoxy-adenosyl group, and regeneration of the B_{12} cofactor. The hydrate product is then dehydrated stereospecifically, losing the migrating hydroxyl group to give the product aldehyde.

The methylmalonyl CoA mutase reaction is also initiated by formation of the 5′-deoxy-adenosyl radical. This abstracts a hydrogen atom from the substrate methyl group as shown in Figure 11.4. Intramolecular ring closure of this primary radical onto the thioester carbonyl forms a cyclopropyl-oxy radical intermediate. Fragmentation of this strained intermediate by cleavage of a different C–C bond generates a more stable secondary radical adjacent to the carboxylate group. Attachment of the abstracted hydrogen atom gives the succinyl CoA product, and regenerates the vitamin B_{12} cofactor. There is convincing precedent for this mechanism from organic reactions that perform this type of rearrangement via free radical intermediates. The X-ray crystal structure of methylmalonyl CoA mutase has been determined, and is shown in Figure 11.5.

There are a small number of other vitamin B_{12}-dependent rearrangements that can be also rationalised by the cyclisation of a radical intermediate onto an sp^2 carbon centre. The reaction catalysed by glutamate mutase, however, cannot be rationalised in this way, since the migrating group is an sp^3 centre, the α-carbon of glutamic acid. An alternative radical mechanism has been

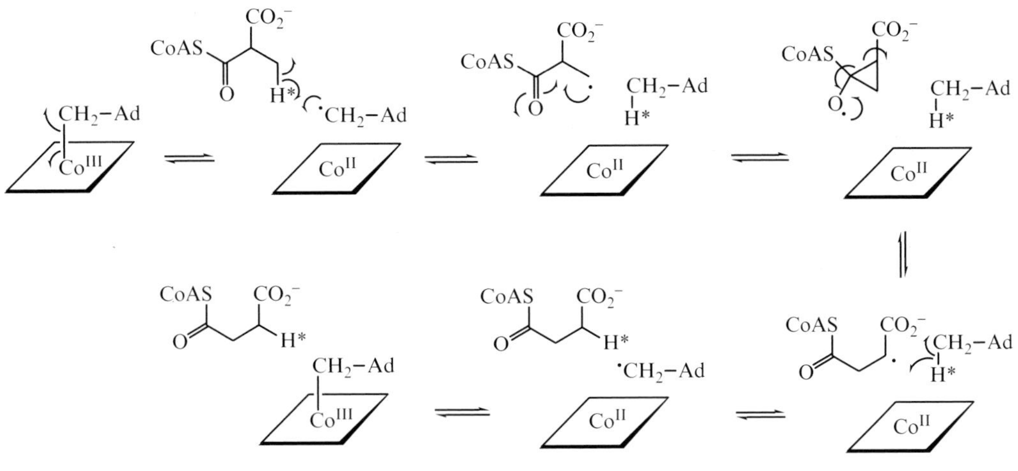

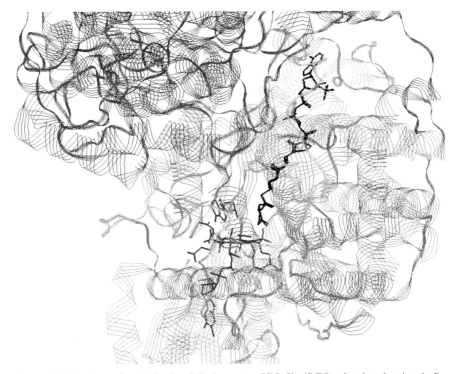

Figure 11.4 Mechanism for methylmalonyl CoA mutase.

Figure 11.5 Structure of methylmalonyl CoA mutase (PDB file 4REQ), showing the vitamin B_{12} cofactor in red, and bound methylmalonyl CoA in black. The adenosyl ligand can be seen above the porphyrin ring of the cofactor. The protein structure is shown in strands, since the active site is buried in the centre of the structure.

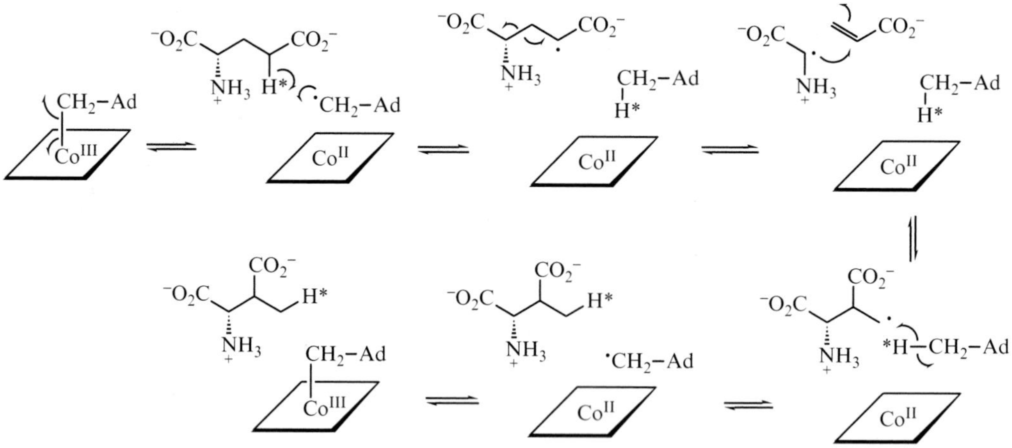

Figure 11.6 Mechanism for glutamate mutase.

proposed for this enzyme, involving the C−C fragmentation of the substrate radical to give acrylic acid and a glycyl radical intermediate, which reattaches to C-2 of acrylic acid to give the product radical. The mechanism is shown in Figure 11.6.

The elucidation of X-ray crystal structures for vitamin B_{12}-dependent rearrangements is now permitting detailed studies of many unanswered questions: how is the Co−C homolysis initiated? How does the enzyme accelerate radical-based rearrangements?

11.3   The involvement of protein radicals in enzyme catalysis

The enzyme ribonucleotide reductase catalyses the conversion of ribonucleotides (used for RNA biosynthesis) to 2′-deoxyribonucleotides (used for DNA biosynthesis), shown in Figure 11.7. In 1973 it was discovered that incubation of the R2 sub-unit of *Escherichia coli* ribonucleotide reductase with iron(II) salts, oxygen and ascorbate led to the formation of a stable, long-lived radical species. Detailed studies using electron paramagnetic resonance (EPR) spectroscopy established that this species is a tyrosyl radical, formed in close proximity to a binuclear iron cluster. This was the first example of a stable protein radical.

Reduction of the tyrosyl radical was shown to lead to the loss of enzyme activity, thus it is required for catalysis; however, structural studies showed that Tyr-122 bearing the protein radical was located approximately 35 Å away from the active site of the enzyme, which is located on the R1 sub-unit of the enzyme. It is now known that single electron transfers within the protein lead to the formation of a cysteine radical on Cys-439 in the active site, which then abstracts the C-3′ hydrogen to initiate a rather complicated radical reaction,

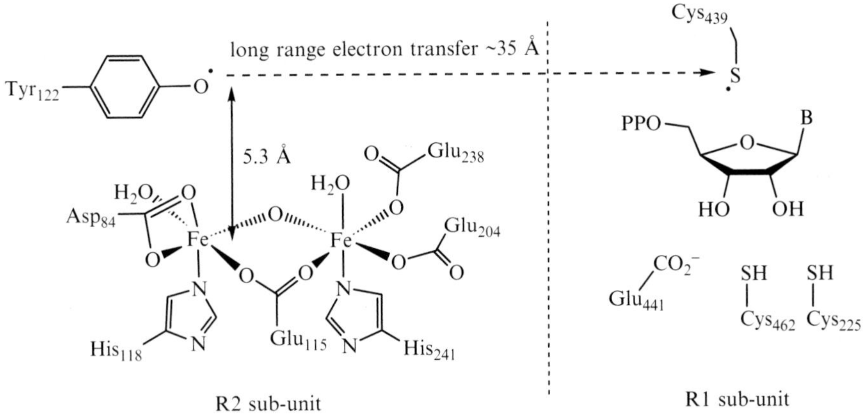

Figure 11.7 Tyrosyl and cysteinyl radicals in ribonucleotide reductase.

shown in Figure 11.8. Protonation of the C-2' hydroxyl group by Cys-462, and one-electron transfer from C-3' to C-2', generates a radical at C-2'. Abstraction of a hydrogen atom from a second cysteine, Cys-225, gives the reduced centre at C-2', and a Cys–Cys disulphide radical anion, which then transfers one electron to C-3' to generate a further radical at C-3'. Abstraction of a hydrogen atom from Cys-439 regenerates the cysteinyl radical and completes the catalytic cycle.

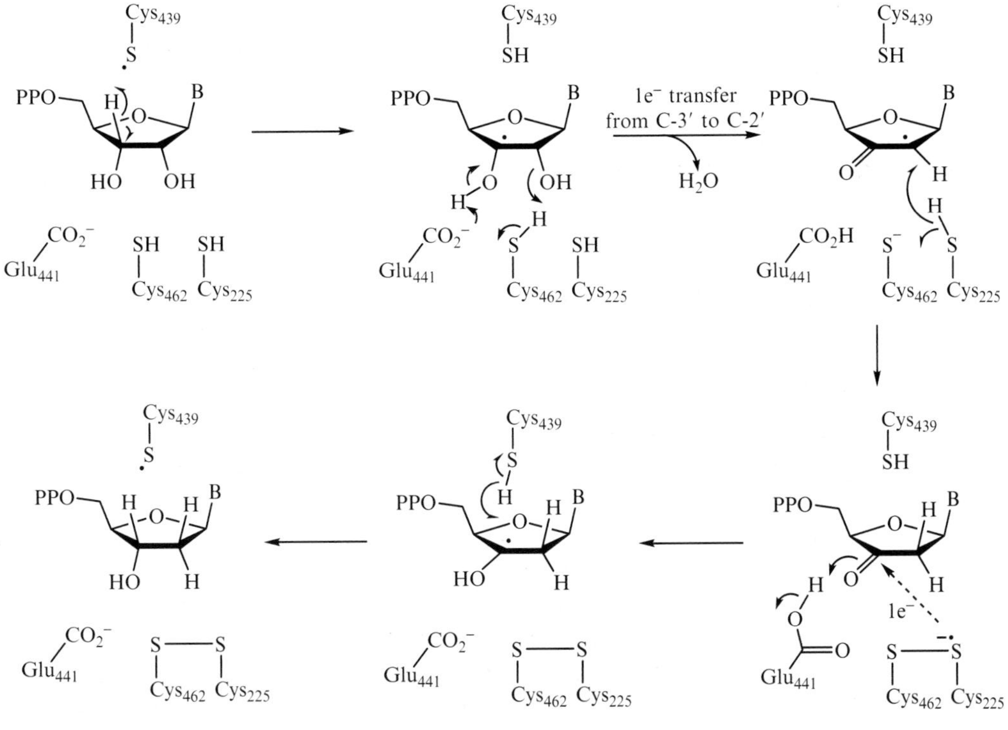

Figure 11.8 Mechanism for ribonucleotide reductase.

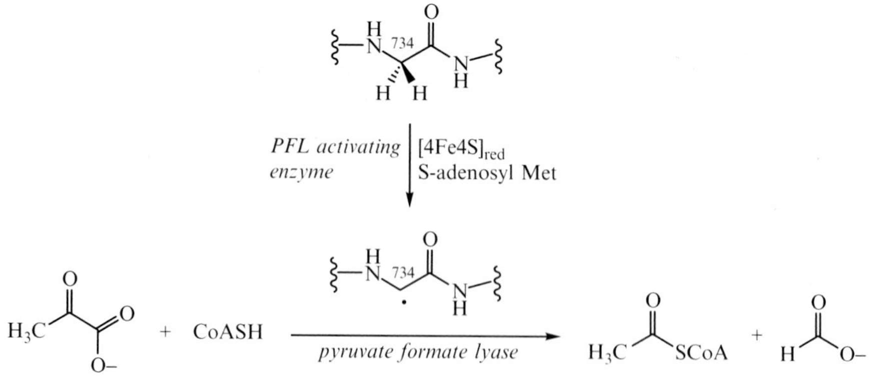

Figure 11.9 Pyruvate formate lyase.

The second protein radical to be discovered was in the enzyme pyruvate formate lyase, which catalyses the reversible interconversion of pyruvate and CoA with formate and acetyl CoA, as shown in Figure 11.9. This is a key step in the anaerobic fermentation of glucose, converting pyruvate into acetyl CoA. In 1989 a stable radical species was detected in *E. coli* pyruvate formate lyase, which was subsequently shown to be located on the α-carbon of a glycine residue, at Gly-734. The glycine radical is generated by a separate activating enzyme.

The glycine radical is present at the active site of the enzyme, which also contains two cysteine residues, Cys-418 and Cys-419. The catalytic mechanism is thought to involve the formation of a thiyl radical at Cys-419, followed by attack on the keto group of pyruvate. C−C fragmentation then generates a formate radical, which abstracts a hydrogen atom from Gly-734 to complete the catalytic cycle, as shown in Figure 11.10.

Several other examples of protein radicals in enzyme-catalysed reactions have since been discovered. A stable tyrosyl radical is involved in the reaction catalysed by prostaglandin H synthase, and a glycine radical is implicated in the anaerobic ribonucleotide reductase. A stable tryptophan radical has been implicated in cytochrome C peroxidase, in proximity to a haem cofactor. Thus, it has been firmly established that protein radicals can be involved directly in enzyme catalysis.

11.4   S-adenosyl methionine-dependent radical reactions

Shortly after the discovery of stable protein radicals described above, several observations emerged that implicated the coenzyme S-adenosyl methionine (SAM) in radical-mediated enzyme reactions. The activating enzyme for pyruvate formate lyase was found to require SAM for activity (see Figure 11.9). Similarly, the activating enzyme for anaerobic ribonucleotide reductase

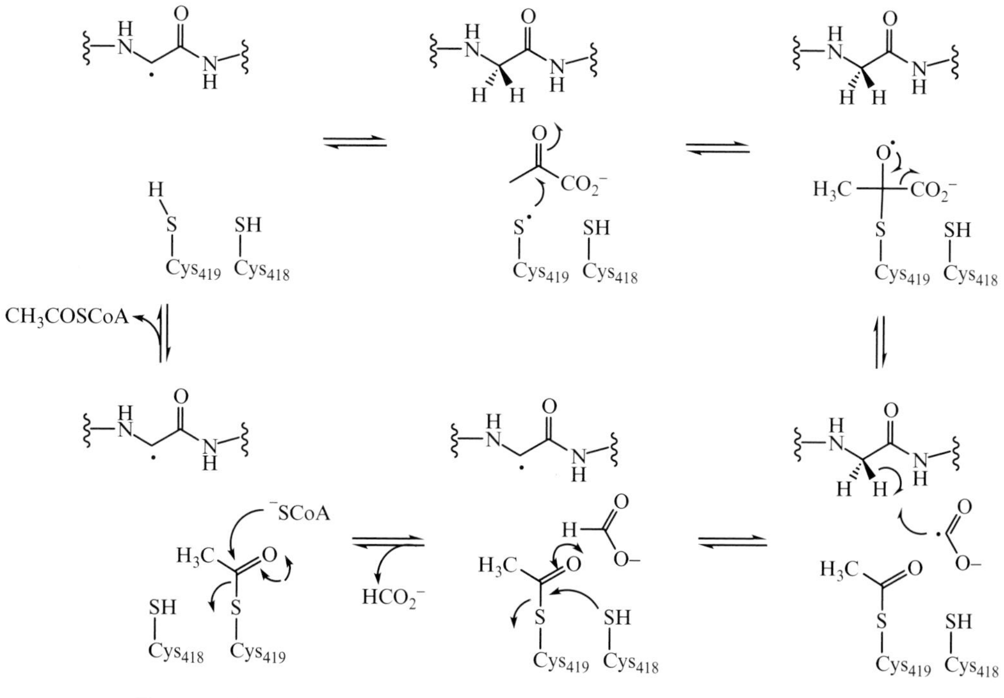

Figure 11.10 Mechanism for pyruvate formate lyase.

required SAM for activity. Until this point, the only known role for SAM was for methyl group transfer (see Section 5.8), thus it was surprising to find it involved in redox chemistry.

The molecular details of the involvement of SAM in radical chemistry were first established for the enzyme lysine 2,3-aminomutase, from *Clostridium subterminale*. This enzyme catalyses the interconversion of L-lysine with β-lysine (see Figure 11.11), a 1,2-rearrangement similar in kind to the vitamin B_{12}-dependent reactions described in Section 11.2. The enzyme did not, however, utilise vitamin B_{12} as a coenzyme, but instead required pyridoxal 5′-phosphate (PLP) and SAM for activity, and a reduced [4Fe4S] cluster.

Detailed studies of this enzyme by rapid-quench EPR spectroscopy have established the existence of radical intermediates in the catalytic mechanism. Following the attachment of L-lysine to the PLP cofactor (see Section 9.2), hydrogen atom abstraction at the β-position generates a C-3 radical, which attacks the imine linkage to give a three-membered aziridine intermediate.

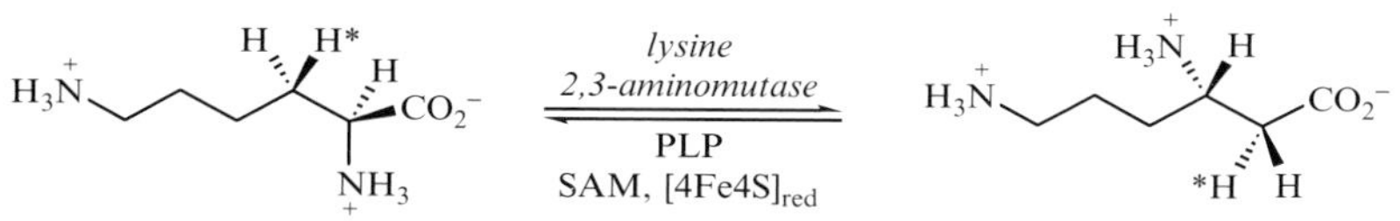

Figure 11.11 Reaction catalysed by lysine 2,3-aminomutase.

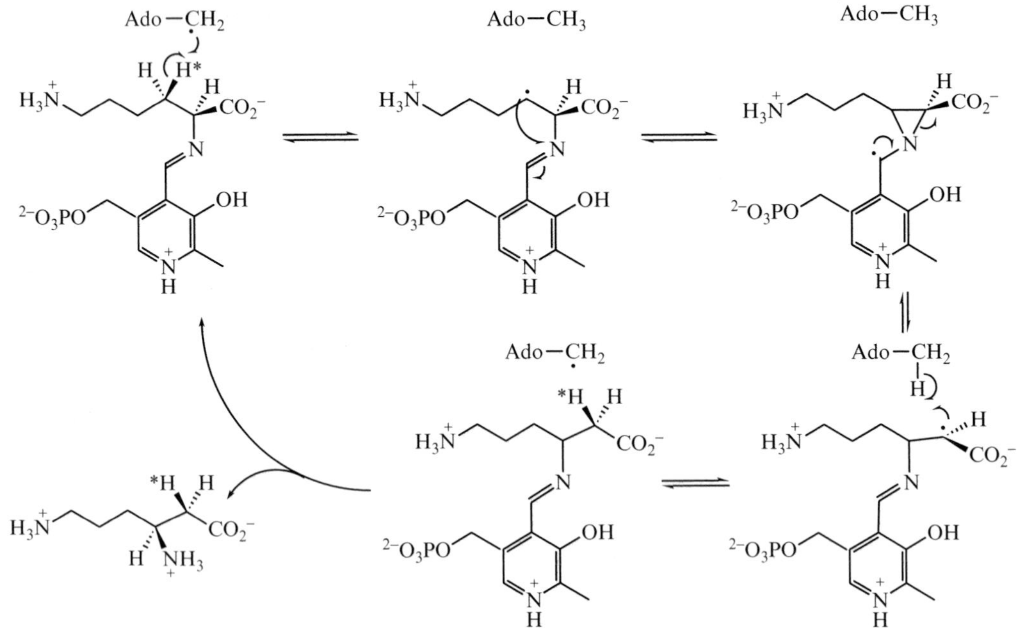

Figure 11.12 Catalytic mechanism for lysine 2,3-aminomutase.

Fragmentation of the other C—N bond then gives a C-2 radical, which re-acquires a hydrogen atom to give the β-lysine-PLP adduct, as shown in Figure 11.12. What species is responsible for hydrogen atom abstraction? By analogy with the vitamin B_{12}-dependent reactions, the reactive species is thought to be a 5′-deoxyadenosyl radical, formed by reductive cleavage C—S bond cleavage of SAM. Generation of this radical species must be carried out by the [4Fe4S] cluster, but the precise mechanism is of interest, since the C—S bond is much stronger than the C—Co bond of vitamin B_{12}. It is known that the α-amino and carboxyl substituents of SAM are co-ordinated to the [4Fe4S] cluster, thus a mechanism for activation can be drawn, as shown in Figure 11.13.

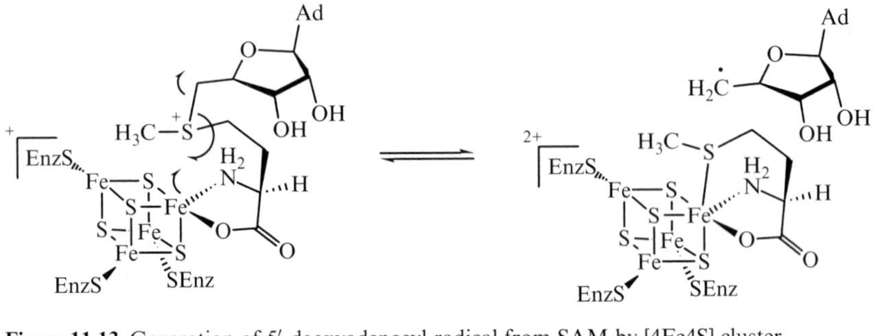

Figure 11.13 Generation of 5′-deoxyadenosyl radical from SAM by [4Fe4S] cluster.

The same combination of SAM and an [4Fe4S] cluster is found in the activating enzymes for pyruvate formate lyase and anaerobic ribonucleotide reductase, and, as we shall see in the next section, in the enzyme biotin synthase. Thus, although at first it was termed 'the poor man's vitamin B_{12}', the involvement of SAM in radical chemistry is proving to be at least as great as its more elaborate cousin.

11.5   Biotin synthase and sulphur insertion reactions

The coenzyme biotin is an essential cofactor in enzymes such as acetyl CoA carboxylase responsible for α-carboxylation reactions, described in Section 7.5. Mammals must consume biotin in their diet, but bacteria are able to biosynthesise biotin using a remarkable pathway. The final step in the pathway, catalysed by biotin synthase, is a sulphur insertion reaction shown in Figure 11.14. This type of reaction has little chemical precedent, but is also used for the biosynthesis of the cofactor lipoic acid, as shown in Figure 11.14.

The mechanism of biotin synthase is still a topic of active research, but several parallels have emerged with the reactions described in Section 11.4. The enzyme requires SAM for activity, and a reduced [4Fe4S] cluster, which is assembled from two [2Fe2S] clusters. Enzyme activity also requires several other redox proteins and redox cofactors for activation, whose precise role has yet to be established; however, it seems likely that, as described in Section 11.4, the [4Fe4S] is somehow able to activate SAM to generate a 5′-deoxyadenosine radical. It is thought that this radical abstracts a hydrogen atom at C-9 to generate a radical intermediate, as shown in Figure 11.15, which abstracts a sulphur atom from the [4Fe4S] cluster. Abstraction of a second hydrogen atom at C-6 generates a second radical intermediate, which cyclises to form the five-membered ring. Abstraction of a sulphur atom from the [4Fe4S] cluster means

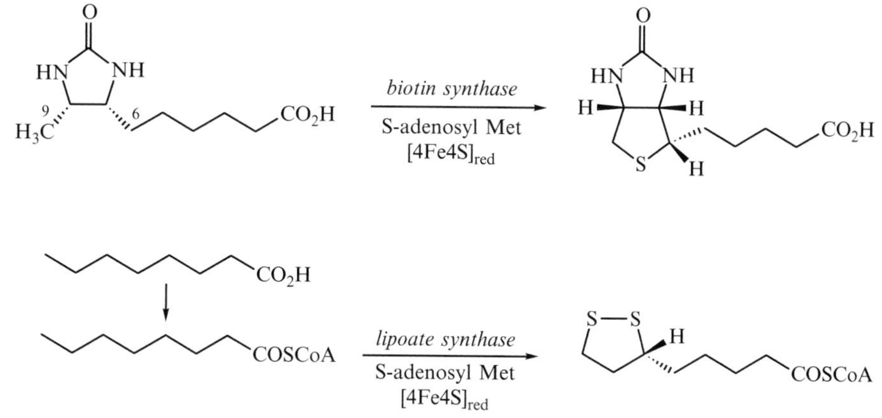

Figure 11.14 Enzymatic sulphur insertion reactions.

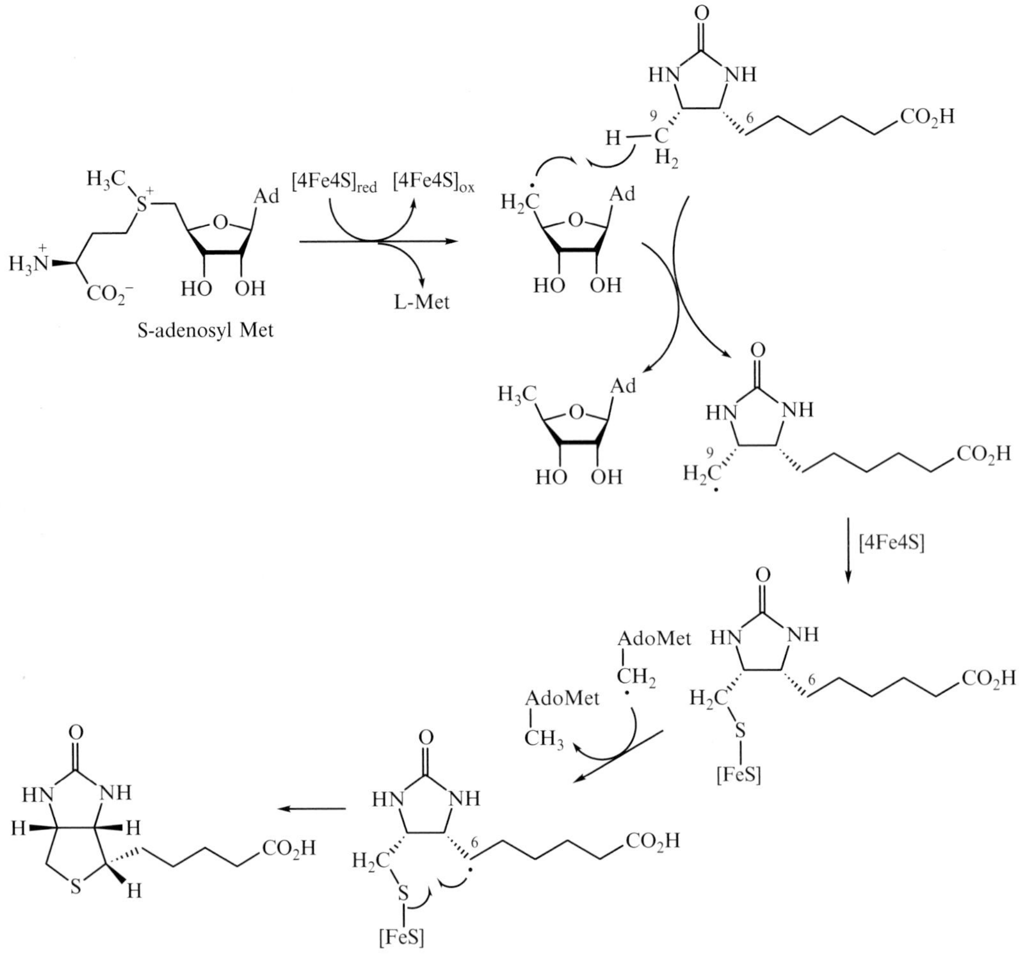

Figure 11.15 Mechanism for biotin synthase.

that the enzyme effectively self-destructs in order to complete one catalytic cycle, which is one of several remarkable features of this enzyme.

Sulphur insertion reactions have also appeared in the biosynthesis of coenzymes thiamine and molybdopterin, and in the modified RNA base thio-uridine. This area of enzymology is therefore expanding rapidly at present, and the emergence of X-ray crystal structures of these enzymes is eagerly awaited.

11.6   Oxidised amino acid cofactors and quinoproteins

At the same time that protein radicals were being discovered in enzyme active sites, several unusual amino acid cofactors were being studied. In the active site of galactose oxidase from *Fusarium* spp., a copper metallo-enzyme, a modified tyrosine residue, was observed in the X-ray crystal structure of the enzyme, in which Tyr-272 is cross-linked via a thioether linkage to the side chain of Cys-228 (see Figure 11.16). Mechanistic studies have indicated that this modified tyrosine cofactor is involved in the enzyme's catalytic mechanism, which appears to proceed via a radical mechanism.

A number of oxidases and dehydrogenases have been discovered that contain novel cofactors containing ortho-quinone oxidised aromatic rings. Collectively this family of proteins are known as quinoproteins, and the cofactors are illustrated in Figure 11.16. The cofactor pyrroloquinone (PQQ) is found in methanol dehydrogenase from *Methylobacterium extorquens*, which catalyses the oxidation of methanol to formaldehyde. Methylamine dehydrogenase from *M. extorquens*, contains the cofactor tryptophylquinone (TTQ), formed by oxidative cyclisation of Trp-55 and Trp-106, also in the vicinity of a copper centre. Finally, the copper-containing amine oxidases, found in bacteria and yeast, contain 2,4,5-trihydroxyphenylalanine quinone (TPQ), formed by oxidation of an active site tyrosine residue by the copper centre. Thus, there is a small family of cofactors in which tyrosine and tryptophan amino acids have been oxidised to form ortho-quinones by a nearby metal centre. In each case, the

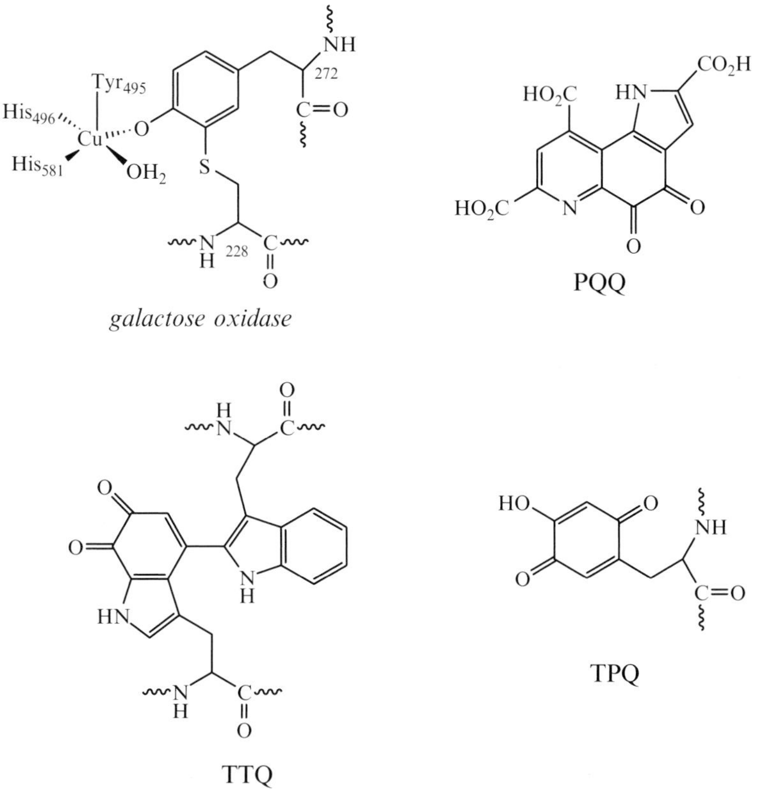

galactose oxidase

PQQ

TTQ

TPQ

Figure 11.16 Structures of ortho-quinone and oxidised amino acid cofactors.

cofactor is involved in the catalytic mechanism, and in each case the formation of the novel cofactor is likely to involve oxidative radical chemistry.

Problems

(1) The enzyme methyleneglutarate mutase requires vitamin B_{12} as a cofactor. (a) Write a mechanism for this reaction involving cyclisation of a substrate radical intermediate. (b) Also suggest an alternative mechanism involving C–C fragmentation of a radical intermediate.

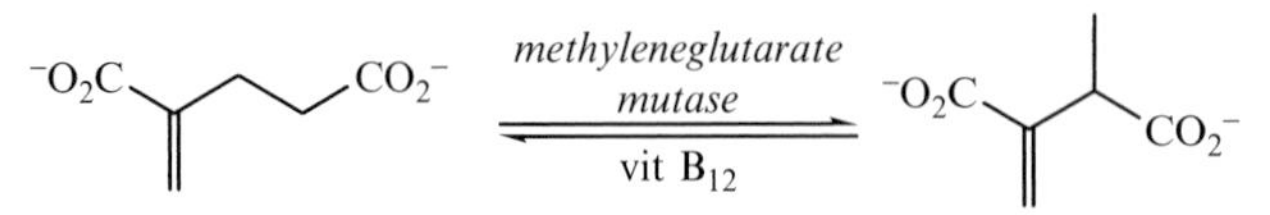

(2) The plant alkaloid hyoscyamine is generated by an enzyme-catalysed intra-molecular rearrangement of littorine, as shown below. The stereochemistry of the conversion is also shown. It was previously thought that this reaction involved vitamin B_{12} as a cofactor; however, plants are thought not to contain vitamin B_{12}. (a) Suggest a radical mediated mechanism for this transformation. (b) If vitamin B_{12} is not involved, how else might a radical intermediate be generated?

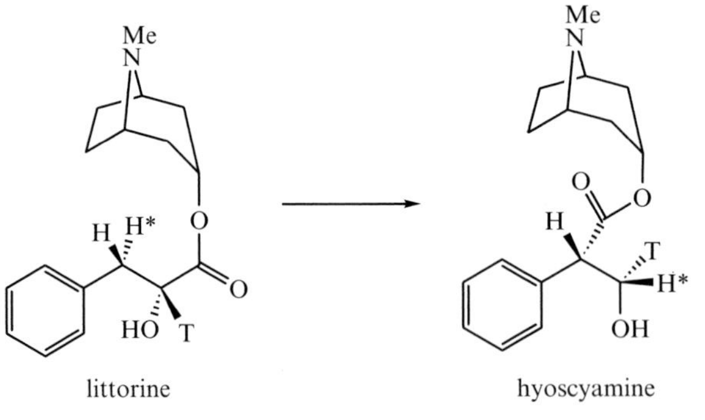

littorine   hyoscyamine

(3) The enzyme benzylsuccinate synthase catalyses the addition of toluene to succinate, as shown below. The enzyme is thought to contain a glycyl radical at its active site. Suggest a catalytic mechanism involving the glycyl radical.

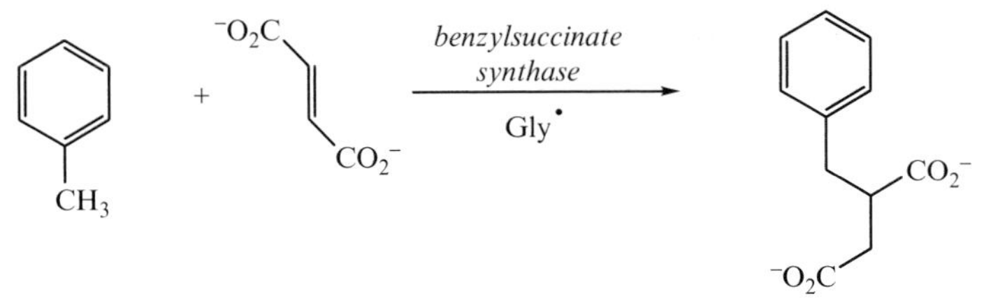

(4) Suggest a radical mechanism for the formation of the modified tyrosine cofactor of galactose oxidase, shown in Figure 11.16, from the native structures of Tyr-122 and Cys-228, employing the nearby copper centre. You can assume that the copper centre can access Cu(I) and Cu(II) oxidation states, and that air oxidation is involved.

(5) Suggest a catalytic mechanism for the quinoprotein methanol dehydrogenase from *Methylobacterium extorquens*, which catalyses the oxidation of methanol to formaldehyde, involving formation of a covalent linkage with the pyrroloquinone (PQQ) cofactor.

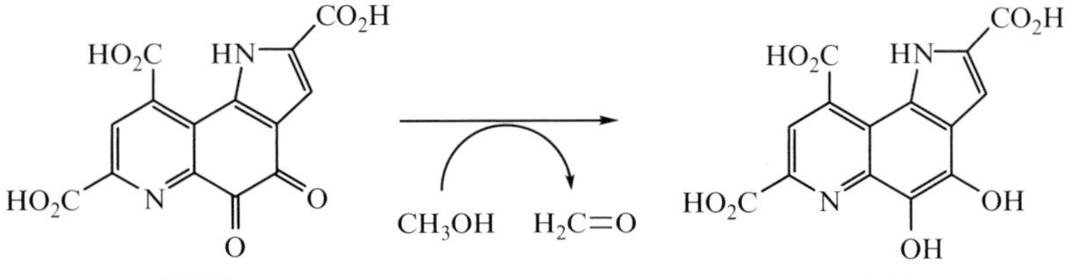

Further reading

Vitamin B$_{12}$-dependent enzymes

R.H. Abeles & D. Dolphin (1976) The vitamin B$_{12}$ cofactor. *Acc. Chem. Res.*, **9**, 114–20.
R. Banerjee (2003) Radical carbon skeletal rearrangements: catalysis by coenzyme B$_{12}$-dependent mutases. *Chem. Rev.*, **103**, 2083–94.
W. Buckel & B.T. Golding (1996) Glutamate and 2-methyleneglutarate mutase: from microbial curiosities to paradigms for coenzyme B$_{12}$-dependent enzymes. *Chem. Soc. Rev.*, **26**, 329–37.
E.N.G. Marsh & C.L. Drennan (2001) Adenosylcobalamin-dependent isomerases: new insights into structure and mechanism. *Curr. Opin. Chem. Biol.*, **5**, 499–505.
T. Toraya (2003) Radical catalysis in coenzyme B$_{12}$-dependent isomerization (eliminating) reactions. *Chem. Rev.*, **103**, 2095–127.

Protein Radicals

J. Stubbe & W.A. van der Donk (1998) Protein radicals in enzyme catalysis. *Chem. Rev.*, **98**, 705–62.

S-Adenosylmethionine-dependent radical processes

M. Fontecave, E. Mulliez & S. Ollagnier de Choudens (2001) Adenosylmethionine as a source of 5′-deoxyadenosyl radicals. *Curr. Opin. Chem. Biol.*, **5**, 506–11.
P.A. Frey & O.T. Magnusson (2003) S-adenosylmethionine: a wolf in sheep's clothing, or a rich man's adenosylcobalamin? *Chem. Rev.*, **103**, 2129–48.
A. Marquet (2001) Enzymology of carbon–sulfur bond formation. *Curr. Opin. Chem. Biol.*, **5**, 541–9.

Quinoproteins

C. Anthony (1996) Quinoprotein-catalysed reactions. *Biochem. J.*, **320**, 697–711.

12 Non-Enzymatic Biological Catalysis

12.1 Introduction

In the final chapter I wish to address the question: are enzymes unique in their ability to catalyse biochemical reactions? The discovery of enzymes and the elucidation of their function provides a basis for understanding how cellular biological chemistry can be catalysed at rates sufficient to sustain life. Many of the reactions catalysed by enzymes are highly complex, yet they can be rationalised in terms of selective enzyme–substrate interactions and well-precedented chemical reactions. Could such selective catalysis be carried out by other biological macromolecules? The answer is yes: we shall meet examples of naturally occurring ribonucleic acid (RNA) molecules capable of catalysing selective self-splicing reactions, and we shall see how the immune system can be used to produce catalytic antibodies.

Finally, a challenging problem for the biological chemist is: if you think you understand how enzymes work, can you design synthetic molecules capable of enzyme-like catalytic properties? This is an exciting area of current research, and we shall see a few ingenious examples of how this problem has been addressed.

12.2 Catalytic RNA

Ribonucleic acid is an important cellular material involved in various aspects of protein biosynthesis. There are three types of RNA which are found in all cells: messenger RNA (mRNA) into which the deoxyribonucleic acid (DNA) sequence of a gene is transcribed before being translated into protein; transfer RNA (tRNA) which is used to activate amino acids for protein biosynthesis; and ribosomal RNA (rRNA) which comprises the ribosome protein assembly apparatus. Thus, information transfer and cellular structures were thought in the 1970s to be the only functions of RNA.

In the early 1980s the groups of Cech and Altman reported the startling discovery that certain RNA molecules were capable of catalysing chemical reactions without the assistance of proteins. The first catalytic RNA (or 'ribozyme') to be identified was found in the large sub-unit ribosomal RNA of a ciliated protozoan *Tetrahymena thermophila*. This RNA molecule has the remarkable ability to cut itself out of a larger piece of RNA, leaving a shortened RNA template which is subsequently used for protein biosynthesis. This self-splicing reaction requires no protein-based molecules, but requires two cofactors: a divalent metal ion such as Mg^{2+}; and a guanosine monomer, which can either be phosphorylated or bear a free 5'-hydroxyl group.

The mechanism of this self-splicing reaction has been elucidated, and is shown in Figure 12.1. The RNA precursor is able to bind the guanosine cofactor in the presence of Mg^{2+} ions. The 3'-hydroxyl group of the guanosine co-factor then attacks the phosphodiester linkage at the 5' cleavage site, displacing a free 3'-OH and becoming covalently attached to the RNA molecule. It is thought that the Mg^{2+} acts as a Lewis acid to assist the leaving group properties of the departing 3'-OH. The cleaved RNA strand remains non-covalently bound to the ribozyme through a six-base complementary 'guide sequence', whilst a conformational change brings G_{414} into the guanosine binding site. The free uridine 3'-OH then attacks the G_{414}–U phosphodiester bond, forming the new phosphodiester bond of the shortened RNA, and releasing the ribozyme which terminates in G_{414}.

How effective is this 'ribozyme' as a catalyst? The site of cleavage and sequence selectivity of the *Tetrahymena* ribozyme is very high, and the rate constant calculated for the transesterification step is an impressive 350 min^{-1}, some 10^{11}-fold greater than the rate of the corresponding uncatalysed reaction.

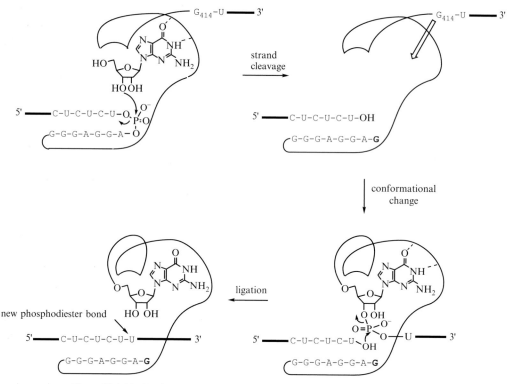

Figure 12.1 Mechanism of self-splicing reaction of *Tetrahymena* ribozyme.

The rate of reaction is in fact limited by formation of the starting complex rather than the phosphotransfer step, so the *Tetrahymena* ribozyme is highly efficient at catalysing its own self-splicing. However, it is debatable whether the *Tetrahymena* ribozyme can be classified as a true catalyst, since it is modified in structure by the reaction.

There are now a number of examples of such catalytic RNA species, whose mechanisms of catalysis can be studied. The hairpin ribozyme is a 92-nucleotide catalytic RNA which catalyses the reversible, site-specific cleavage of the phosphodiester backbone of RNA through transesterification. Although this catalytic RNA requires Mg^{2+} for activity (a common feature of catalytic RNA), replacement with $[Co(NH_3)_6]^{3+}$ retains activity, implying that Mg^{2+} ions are not directly involved in catalysis. Determination of the X-ray crystal structure of the hairpin ribozyme in complex with a pentavalent vanadate transition state analogue (shown in Figure 12.2) has shown that a rigid active site makes additional hydrogen bonds to the transition state. These hydrogen bonds are formed by nitrogen atoms of active site RNA bases, as shown in Figure 12.3. These observations suggest that RNA catalysis can employ transition state stabilisation by RNA bases, in a similar fashion to protein-based enzymes.

It has long been established that the ribosome, the cellular machinery responsible for protein biosynthesis, contains RNA. Biochemical studies of the peptidyl transferase activity of the ribosome have indicated that RNA is responsible for catalysis, which was proved in dramatic fashion in 2000 by the determination of the X-ray crystal structure of the large ribosomal sub-unit of the *Haloarcula marismortui* ribosome. Co-crystallisation with substrate analogues has shown that the catalytic site is composed entirely of RNA, with N-3 of A2486 apparently involved in acid–base catalysis during peptide bond formation, as shown in Figure 12.4.

The fact that RNA can catalyse chemical reactions *and* carry genetic information offers the possibility that it might have been the information storage system of primitive pre-cellular life. This would solve a paradox, that whilst DNA is a superb carrier of genetic information, it requires protein in order for the information to be expressed; whereas protein is highly proficient in catalysis, but it requires DNA to encode its sequence. There is, therefore, considerable interest in the study of catalytic RNA as an evolutionary forerunner of DNA-based information storage and protein catalysis. In this respect it is interesting to note that many of the coenzymes that are used today in enzymatic reactions contain ribonucleotides: adenosine triphosphate (ATP), nicotinamide adenine dinucleotide (NAD), flavin adenine dinucleotide (FAD), S-adenosyl methionine (SAM) and vitamin B_{12}. Maybe these molecules (or their ancestors) were key players in pre-biotic chemistry.

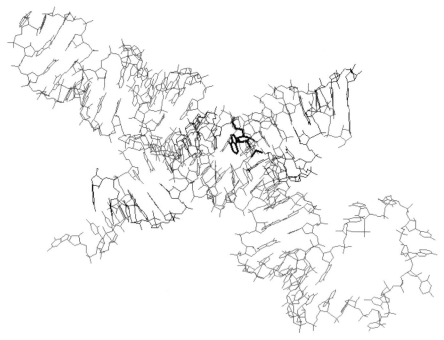

Figure 12.2 Structure of hairpin ribozyme, complexed with vanadate transition state analogue (PDB file 1M5O). The structure consists of two ribonucleotides: chain A (in black), which is the substrate for phosphodiester cleavage; and RNA chain B (in red), which is the catalytic chain. The vanadate transition state analogue, incorporated in chain A, is shown in bold.

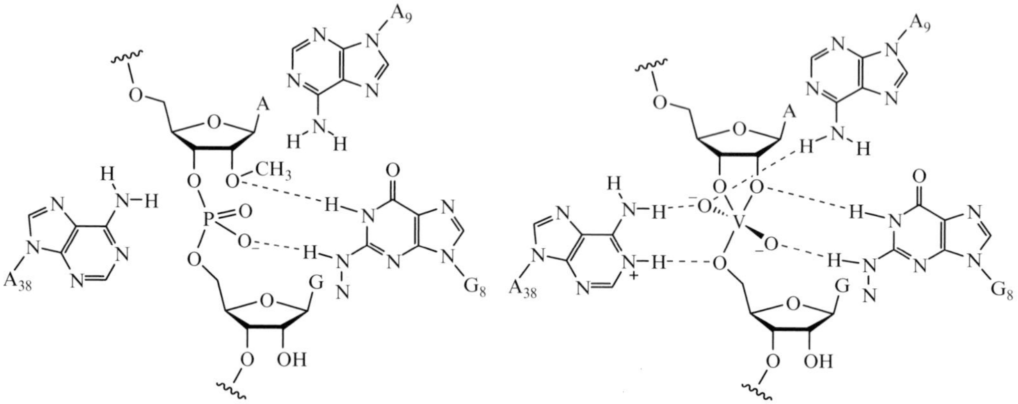

(a) Interactions in ground state   (b) Interactions in transition state

Figure 12.3 Transition state stabilisation in hairpin ribozyme.

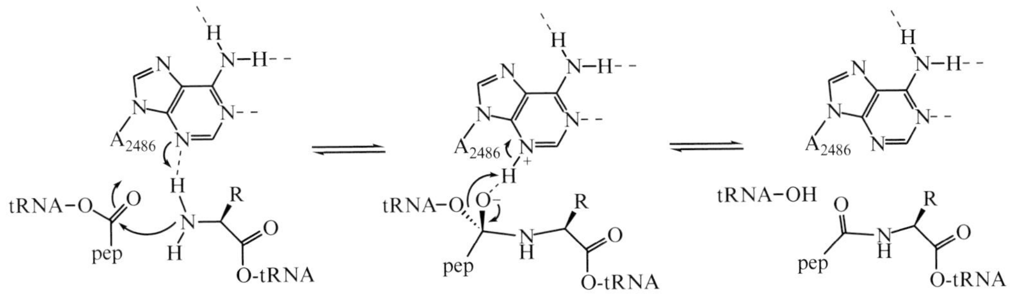

Figure 12.4 Mechanism for peptide bond formation in the catalytic site of the ribosome.

12.3   Catalytic antibodies

The immune system acts as a major line of defence against foreign substances, be they toxins, proteins, or invading micro-organisms. Upon detection of a foreign antigen the immune system generates a 'library' of up to 10^9 antibodies, some of which bind tightly and specifically to the antigen, allowing it to be targeted and destroyed by the body's 'killer' T-cells. The structure of antibodies is illustrated in Figure 12.5. They are protein molecules consisting of four polypeptide chains, two heavy (H) chains and two light (L) chains, linked together by disulphide bridges. At the two ends of each Y-shaped antibody are the 'variable' regions of the antibody in which variation in sequence is found between antibodies, and at the extreme ends of the Y are the 'hypervariable' regions which make up the antigen combining sites. The three-dimensional structure of an antibody is shown in Figure 12.6.

Antibodies bind their antigen target very tightly, typical K_d values being 10^{-9}–10^{-11} M, and very selectively. This selectivity of binding is reminiscent of the selectivity of enzyme active sites for their substrates. So they satisfy one of the criteria for enzyme catalysis: selective substrate recognition. Could antibodies also catalyse chemical reactions? This question was answered by the groups of Lerner and Schultz in 1986. They recognised that the major factor underlying enzymatic catalysis is transition state stabilisation. If an antibody could be generated that specifically recognises the transition state of a chemical reaction, and also binds less tightly the substrate and product of the reaction, then it should catalyse the reaction.

Insight into the mechanisms of chemical reactions has provided us with good models for the structures of their transition states. In many cases this has permitted the synthesis of transition state analogues. We have seen earlier how such molecules can act as potent inhibitors of enzyme-catalysed reactions, precisely because enzymes bind the transition state more tightly than either the substrate or product of the reaction. Using a synthetic transition state

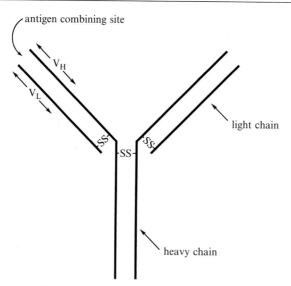

Figure 12.5 Generalised structure of antibodies.

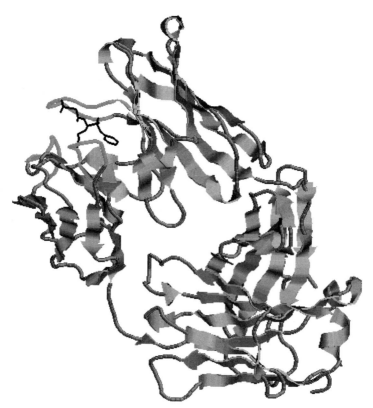

Figure 12.6 Structure of Fc fragment of a catalytic antibody (PDB file 1EAP), showing the antigen binding site, formed by three hypervariable loops (shown in red). The bound hapten ligand is shown in black.

analogue for a chemical reaction, a protocol was devised for the generation of monoclonal (i.e. single, homogeneous) catalytic antibodies, as shown in Figure 12.7.

First of all the transition state analogue is attached to a protein such as bovine serum albumen in order to generate a sizeable immune response. The 'hapten' thus formed is injected into a mouse or rabbit and the immune response triggered. The antibody-secreting cells are then isolated, and these cells are immortalised by fusing them with myeloma (cancer) cells. The immortalised antibody-secreting cells are then diluted and screened for catalytic activity, usually by a colorimetric assay. The most active cell lines are then selected, allowing the monoclonal antibodies to be purified and analysed kinetically.

The first catalytic antibodies were generated using phosphonate ester transition state analogues for the tetrahedral intermediate involved in ester hydrolysis reactions. For example, the phosphonate ester shown in Figure 12.8 acts as a transition state analogue for the hydrolysis of the corresponding ester. Haptens based on this transition state analogue elicited antibodies capable of catalysing the hydrolysis of this ester at rates 10^3–10^5-fold greater than the rate of uncatalysed ester hydrolysis.

Subsequently haptens have been designed for the more challenging amide hydrolysis reaction. Using a phosphonamidate analogue shown in Figure 12.9, a catalytic antibody 43C9 has been elicited that is capable of catalysing the amide hydrolysis reaction shown. Antibody 43C9 accelerates the hydrolysis of this amide by a factor of 10^6 at pH 9.0, and the Michaelis–Menten kinetic

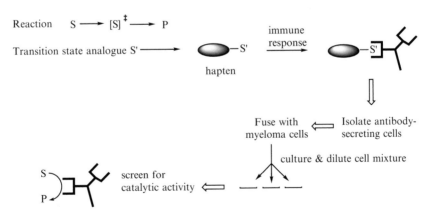

Figure 12.7 Isolation of catalytic antibodies.

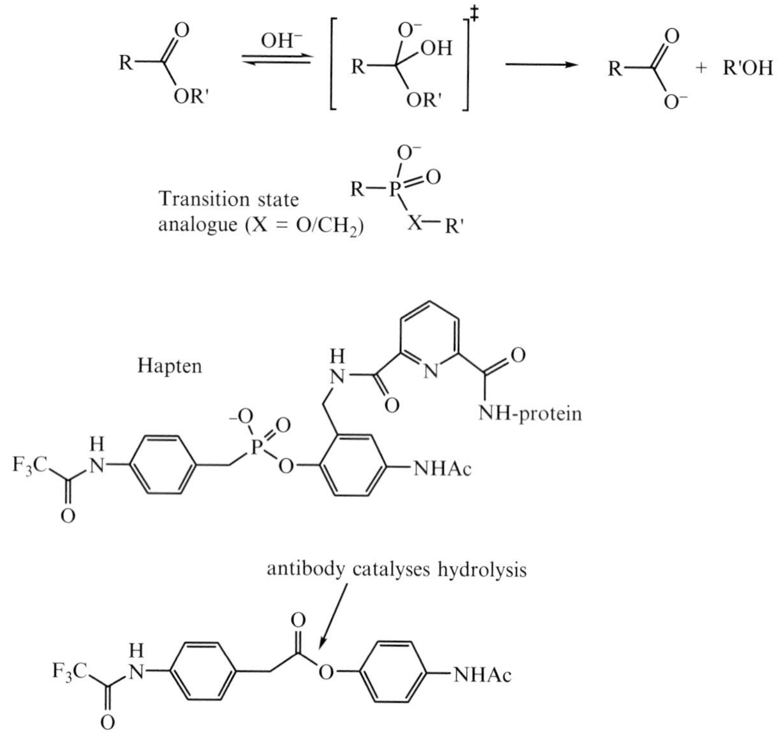

Transition state
analogue (X = O/CH$_2$)

Figure 12.8 Catalytic antibodies for ester hydrolysis.

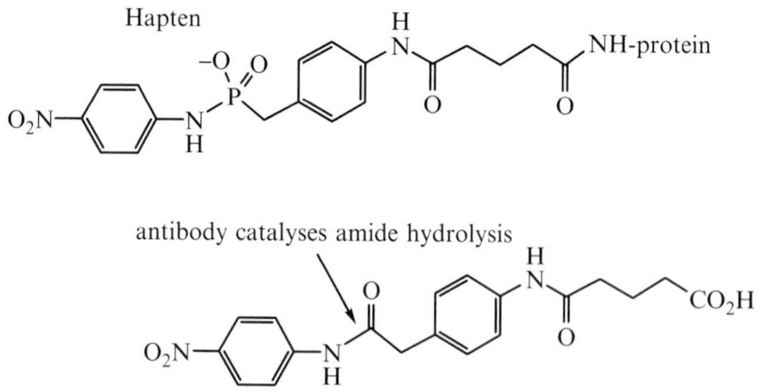

Figure 12.9 A catalytic antibody for amide hydrolysis.

parameters K_M and k_{cat} can be measured as 0.56 mM and 0.08 min^{-1}, respectively. These catalytic properties are slow by the standards of enzyme-catalysed reactions, but they validate the idea that catalytic antibodies can, in principle, be designed and generated.

How are hydrolysis reactions catalysed at the antigen combining sites of these antibodies? The amide and ester hydrolysis reactions catalysed by antibody 43C9 have been analysed in some detail, and evidence has been obtained for the existence of a covalent intermediate in the reaction formed by nucleophilic attack of a histidine side chain. Determination of the amino acid sequence of the antibody, together with molecular modelling studies, suggested that His-L91 and Arg-L96 might be involved in antibody catalysis. Mutation of either residue to glutamine gave catalytically inactive antibodies, confirming this suggestion. A mechanism proposed for the ester hydrolysis reaction is shown in Figure 12.10. In this mechanism His-L91 acts as a nucleophile and Arg-L96 stabilises the tetrahedral transition state in the reaction. Not surprisingly, this antibody is strongly inhibited by the original phosphonamidate analogue.

Catalytic antibodies have now been generated for many types of reactions, and the interested reader is referred to several reviews at the end of the chapter. The final example was mentioned in Section 10.5. Transition state analogues

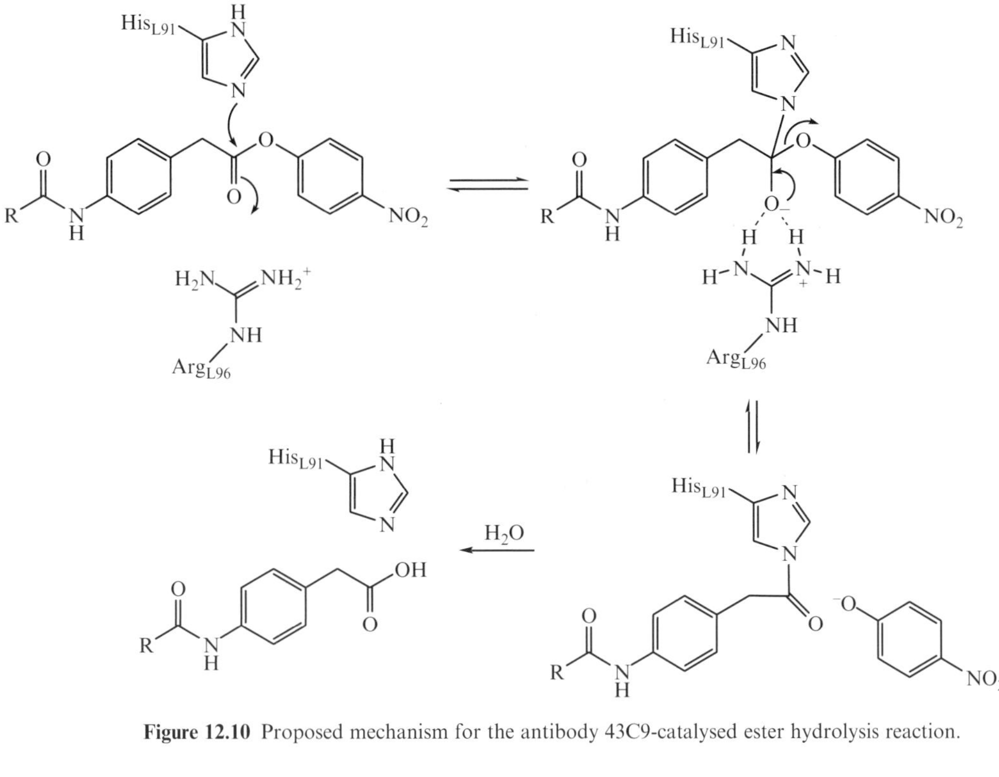

Figure 12.10 Proposed mechanism for the antibody 43C9-catalysed ester hydrolysis reaction.

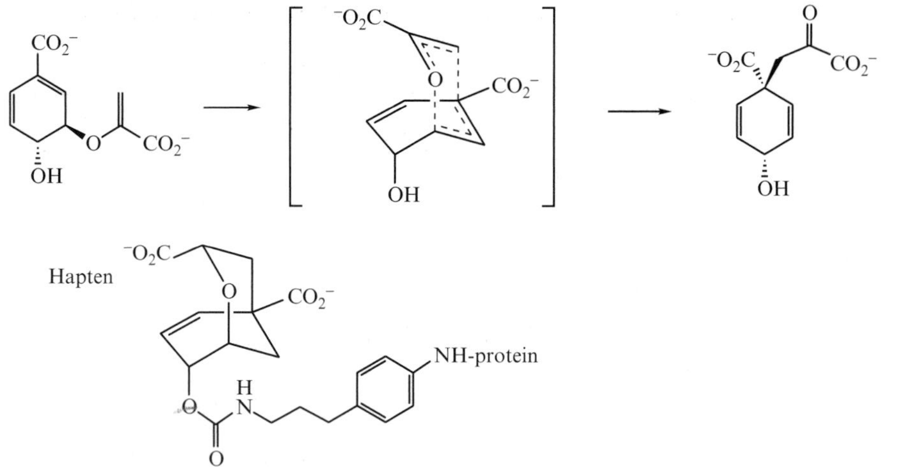

Figure 12.11 A catalytic antibody for the chorismate mutase reaction.

have been synthesised for the chorismate mutase-catalysed rearrangement of chorismate to prephenate shown in Figure 12.11. One such analogue has been used to generate catalytic antibodies capable of accelerating the rearrangement by 10^4-fold (as compared to 10^6-fold for the enzyme-catalysed reaction). The kinetic parameters for this catalytic antibody are K_M 260 μM and k_{cat} 2.7 min^{-1}.

The ability of catalytic antibodies to catalyse reactions which are difficult to achieve chemically has been demonstrated by the examples shown in Figure 12.12. Antibodies raised against a cyclic N-oxide were found to catalyse the intramolecular ring opening of a structurally related epoxide, via a disfavoured 6-endo-tet transition state, to give only the six-membered ring product. Antibodies capable of catalysing cationic cyclisation reactions have also been isolated: using cyclic N-oxide and ammonium salt haptens, antibodies capable of cyclisation of the corresponding sulphonate were found. Unexpectedly, it was found that cyclisation of the substituted alkene substrate gave a cyclopropane product, reminiscent of the monoterpene cyclase-catalysed reactions.

The fact that catalytic antibodies can be generated supports the idea that enzymes achieve much of their rate acceleration through transition state stabilisation. However, the observation that most catalytic antibodies are much slower than their enzyme counterparts raises the question: what other factors do enzymes utilise to achieve their additional rate acceleration? Perhaps the answer is that enzymes have had millions of years to perfect their catalytic abilities through such subtle ploys as use of strain, bifunctional catalysis, or protein dynamics. The other common limitation of catalytic antibodies is that they suffer from product inhibition, whereas enzymes usually bind their substrates and products relatively weakly (as discussed in Section 3.4 it is thermodynamically unfavourable for enzymes to bind their substrates strongly).

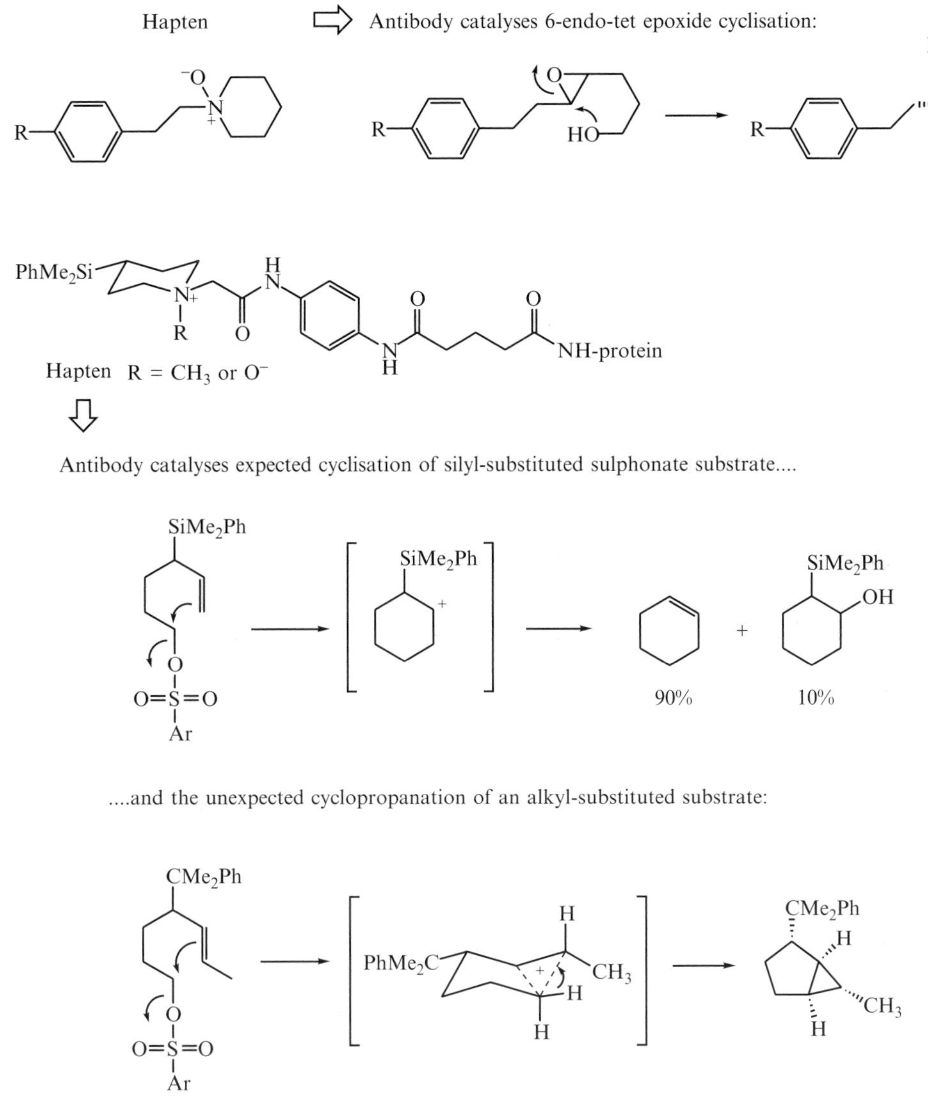
Hapten
Antibody catalyses 6-endo-tet epoxide cyclisation:
R
HO
PhMe₂Si
NH-protein
Hapten R = CH₃ or O⁻
Antibody catalyses expected cyclisation of silyl-substituted sulphonate substrate....
SiMe₂Ph
OH
O=S=O
Ar
90%
10%
....and the unexpected cyclopropanation of an alkyl-substituted substrate:
CMe₂Ph
PhMe₂C
CH₃
H

Figure 12.12 Antibody catalysis of cyclisation reactions.

12.4 Synthetic enzyme models

One of the final frontiers in biological chemistry is the design of synthetic models for biological catalysis. Can we design and synthesise small or medium-sized molecules which mimic enzymes?

First of all we need to define what requirements we need of such a model, as follows:

(1) *Selective substrate binding*. The catalyst must be able to bind its substrate effectively and specifically via a combination of electrostatic, hydrogen-bonding and hydrophobic interactions.

(2) *Pre-organisation*. The catalyst must have a rigid, well-defined three-dimensional structure, so that when it binds its substrate there is little loss of entropy in going to the transition state of the reaction.

(3) *Catalytic groups*. There must be suitably positioned catalytic groups arranged convergently (i.e. pointing inwards towards the substrate), implying that quite a large cavity is required.

(4) *Physical properties*. If the catalyst is to be useful in aqueous solution, it should be water soluble, be able to bind its substrate in water, and be active at close to pH 7.

These requirements turn out to be extremely demanding in practice, so there are only a fairly small number of synthetic enzyme-like catalysts which have been developed to date. However, this is a rapidly emerging area of research, so I will give a few examples of current models and advise the interested reader to watch this space!

Historically, the first type of enzyme models developed by Cram used cationic binding sites provided by 'crown ethers', a family of cyclic polyether molecules with a high affinity for metal ions. As well as binding metal ions, crown ethers could bind substituted ammonium cations, which was exploited in the design of the model shown in Figure 12.13. This model contains pendant thiol groups that can act as catalytic groups for ester hydrolysis, analogous to the cysteine proteases. This catalyst was found to accelerate the hydrolysis of amino acid p-nitrophenyl esters in ethanol by 10^2–10^3-fold compared with an acyclic version, showing the importance of pre-organisation of the catalyst. This system also showed some enantioselectivity, being selective for D-amino acid esters by 5–10-fold.

More recently, binding sites have been developed for anionic substrates. One example developed by Hamilton is a small molecule containing two gaunidinium side chains which is able to bind phosphodiesters via electrostatic and hydrogen-bonding interactions. This system was able to accelerate the rate of an intramolecular phosphodiester hydrolysis reaction shown in Figure 12.14 by 700-fold, presumably by a combination of transition state stabilisation and protonation of the leaving group.

The most versatile family of enzyme models in current use are the cyclodextrins developed by Breslow. These are a family of cyclic oligosaccharide molecules which form a bucket-shaped cavity capable of forming tight complexes with aromatic molecules. These systems offer the advantages that the size of the central cavity is quite large, the binding is effective, they are water soluble, and they contain pendant hydroxyl groups around the rim of the 'bucket' which can be functionalised with catalytic groups.

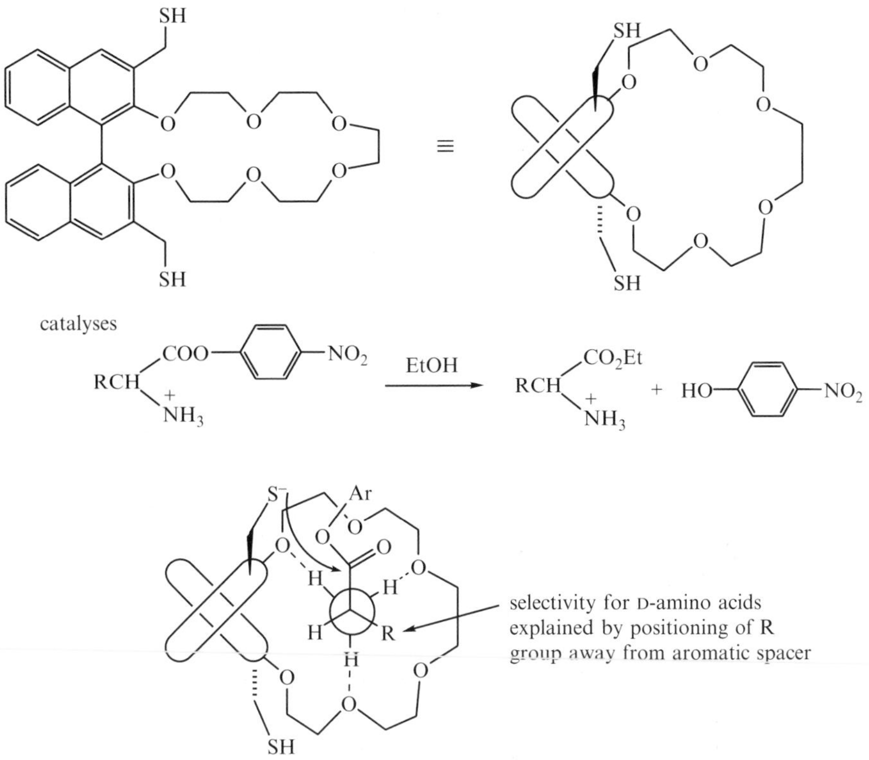

Figure 12.13 Crown ether catalytic binding site.

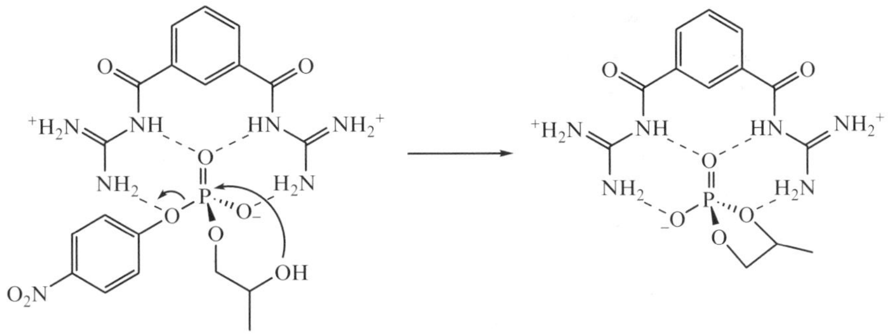

Figure 12.14 Anionic binding site.

By attaching a blocking agent to a hydroxyl group on one face of the bucket, a more hydrophobic cavity is obtained that is more effective at binding aromatic molecules. One such blocked cavity accelerated the hydrolysis of a *meta*-substituted phenyl acetate, shown in Figure 12.15, by 3300-fold over the rate of uncatalysed hydrolysis. This can be rationalised, as shown in Figure 12.15, by the participation of the free hydroxyl groups on the rim of the cavity, which come into close proximity with the *meta*-substituent.

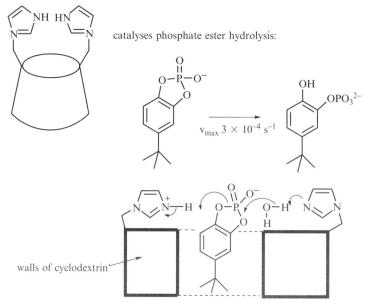

Figure 12.15 Cyclodextrin-catalysed ester hydrolysis.

Functionalisation of the free hydroxyl groups on the rim of the cavity opens up new possibilities for the design of enzyme models. One example is the cyclodextrin shown in Figure 12.16, functionalised with two imidazole side chains. This derivative was found to catalyse the regioselective hydrolysis of a cyclic phosphodiester substrate, with a rate constant of 3×10^{-4} s^{-1}. The pH/ rate profile of this catalyst showed that maximum activity was obtained at pH 6.3, at which point one of the imidazole groups is protonated and the other deprotonated. The proposed mechanism shown in Figure 12.16 involves bifunc-

catalyses phosphate ester hydrolysis:

walls of cyclodextrin

Figure 12.16 Functionalised cyclodextrin model of ribonuclease A.

tional acid/base catalysis by the two imidazole groups. The regiospecificity of phosphodiester hydrolysis can be rationalised by the orientation of the aromatic substrate in the activity, as shown. This system is a very elegant model for the mechanism of action of ribonuclease A (see Section 5.5).

Many other types of host–guest systems have been developed in recent years, for example using metal ions to co-ordinate substrate and catalyst functional groups. One alternative approach with which to finish is the use of polymers to generate chiral cavities containing catalytic groups. The idea of this approach is to use a small molecule as a template for the co-polymerisation of regular cross-linked polymer. After polymerisation the template is removed, leaving a complementary cavity inside the polymer which can bind molecules of related structure. For example, co-polymerisation of an aromatic dicarboxylic acid with an amino-acrylamide derivative gives a polymer containing pendant amino groups contained within cavities inside the polymer, as shown in Figure 12.17. After removal of the template, this polymer was found to catalyse the β-elimination reaction of a related substrate with enzyme-like Michaelis–Menten kinetics (K_M 27 mM, k_{cat} 1.1×10^{-2}min^{-1}).

In summary, chemists are using rational design and the ability to synthesise unnatural three-dimensional structures to try to mimic the remarkable catalytic properties of enzymes, and perhaps to generate new types of catalysts. Such catalysts might, for example, be able to catalyse new reactions or operate in organic solvents or at high temperatures. Our current efforts seem crude compared with the biological counterparts, but the future holds great promise.

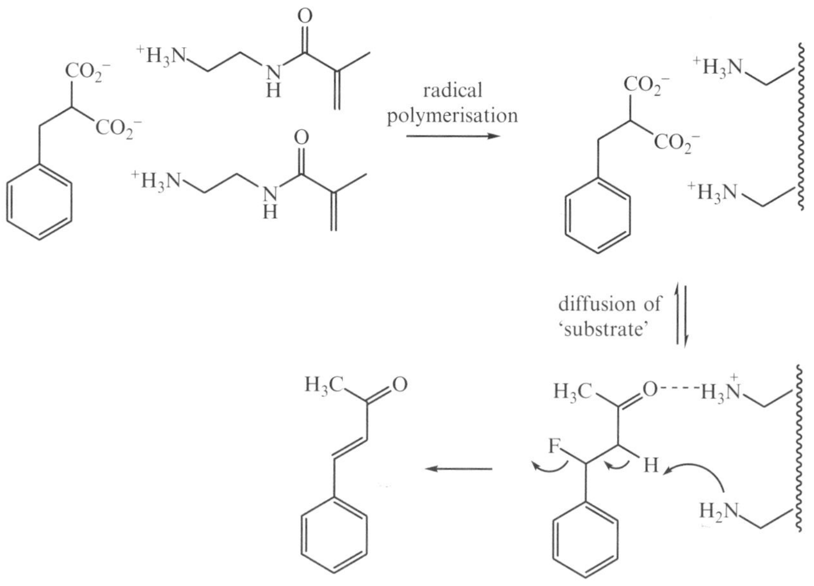

Figure 12.17 Catalytic polymers.

Problems

(1) How reversible do you think the *Tetrahymena* ribozyme-catalysed self-splicing reaction is? What is the potential significance of the reverse reaction?

(2) Suggest a hapten that could be used to induce catalytic antibodies for the lactonisation reaction below.

(3) Explain why the hapten below elicited antibodies capable of catalysing the elimination reaction shown.

(4) What type of catalytic properties would you expect from a cyclodextrin covalently modified with: (i) a thiol group; (ii) a riboflavin group (*Note*: reduced riboflavin is rapidly oxidised by dioxygen in aerobic solutions)?

Further reading

Catalytic RNA

T.R. Cech & B.L. Bass (1986) Biological catalysis by RNA. *Annu. Rev. Biochem.*, **55**, 599–630.

T.R. Cech, D. Herschlag, J.A. Piccirilli & A.M. Pyle (1992) RNA catalysis by a group I ribozyme. *J. Biol. Chem.*, **267**, 17479–82.

P. Nissen, J. Hansen, N. Ban, P.B. Moore & T.A. Steitz (2000) The structural basis of ribosome activity in peptide bond synthesis. *Science*, **289**, 920–30.

R.B. Rupert, A.P. Massey, S.T. Sigurdsson & A.R. Ferre-D'Amare (2002) Transition state stabilization by a catalytic RNA. *Science*, **298**, 1421–4.

Catalytic antibodies

S.J. Benkovic (1992) Catalytic antibodies. *Annu. Rev. Biochem.*, **61**, 29–54.

T. Li, R.A. Lerner & K.D. Janda (1997) Antibody-catalyzed cationic reactions: rerouting of chemical transformations via antibody catalysis. *Acc. Chem. Res.*, **30**, 115–21.

P.G. Schultz (1989) Catalytic antibodies. *Acc. Chem. Res.*, **22**, 287–94.

P.G. Schultz & R.A. Lerner (1993) Antibody catalysis of difficult chemical transformations. *Acc. Chem. Res.*, **26**, 391–5.

J.P. Stevenson & N.R. Thomas (2000) Catalytic antibodies and other biomimetic catalysts. *Nat. Prod. Reports*, **17**, 535–77.

J.D. Stewart, L.J. Liotta & S.J. Benkovic (1993) Reaction mechanisms displayed by catalytic antibodies. *Acc. Chem. Res.*, **26**, 396–404.

Enzyme models

R. Breslow (1986) Artificial enzymes and enzyme models. *Adv. Enzymol.*, **58**, 1–60.

R. Breslow (1995) Biomimetic chemistry and artificial enzymes: catalysis by design. *Acc. Chem. Res.*, **28**, 146–53.

H. Dugas (1996) *Bioorganic Chemistry – A Chemical Approach to Enzyme Action*, 3rd edn. Springer-Verlag, New York.

W.B. Motherwell, M.J. Bingham & Y. Six (2001) Recent progress in the design and synthesis of artificial enzymes. *Tetrahedron*, **57**, 4663–86.

Catalytic polymers

G. Wulff (2002) Enzyme-like catalysis by molecularly imprinted polymers. *Chem. Rev.*, **102**, 1–27.

Appendix 1: Cahn–Ingold–Prelog Rule for Stereochemical Nomenclature

In order to establish the configuration of a chiral centre containing four different substituents, the four substituents are ranked according to their atomic mass. The substituent with highest atomic mass is labelled '1', the next highest '2', etc. (i.e. O > N > C > H). The molecule is then drawn with substituent '4' pointing away from you, and the substituents 1, 2 and 3 connected in the direction $1 \rightarrow 2 \rightarrow 3$. If the arrows form a clockwise right-handed screw, then the centre has the R configuration. If the arrows form an anti-clockwise left-handed screw, then the centre has the S configuration. For example, the specifically deuterated sample of ethanol below has four different substituents (O > C > D > H), and the C-1 centre is designated as R. This molecule is then written as $1R$-[1-^{2}H]-ethanol.

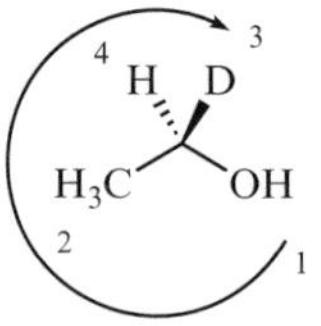

$1R$-[1-^{2}H]-ethanol

Commonly, two or more of the α-substituents have the same atomic mass. In order to prioritise two such α-substituents the atomic mass of their respective β-substituents is analysed in the same way, and the one with higher-ranking β-substituents is given the higher ranking. For example, L-alanine below has two carbon substituents, but the carboxyl group has oxygen β-substituents whereas the methyl group has hydrogen β-substituents. If an α-substituent is attached by a double bond to an atom X, then for the purposes of this analysis the α-substituent has two X β-substituents. L-Alanine is therefore written as $2S$-alanine.

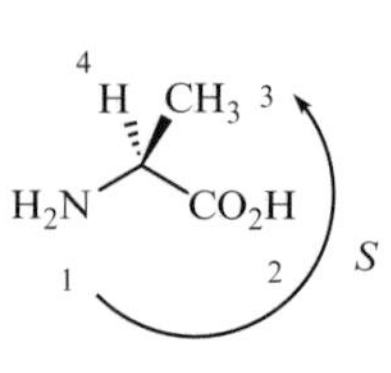

L-alanine = $2S$-alanine

Prochiral configuration is obtained by replacing the substituent of interest with the next highest available isotope. Commonly, prochiral hydrogens are of interest in enzyme-catalysed reactions, in which case the hydrogen is replaced with a deuterium atom, and the configuration of the resulting chiral centre analysed as above. Thus, in the labelled ethanol molecule above the substituent replaced with deuterium is the *proR* hydrogen.

Finally, the direction of attack onto a double bond can be defined in similar fashion, by designating the three substituents around the sp^2 carbon of interest as 1,2 and 3 as above. Look down onto one face of the double bond, and if the substituents are arranged in clockwise fashion then you are looking at the *re*-face. If they are arranged in anti-clockwise fashion then you are looking at the *si*-face. For example, the two faces of the carbonyl group of glyceraldehyde-3-phosphate are indicated below.

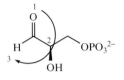

looking at the *re*-face
of the C-1 carbonyl of
glyceraldehyde-3-phosphate

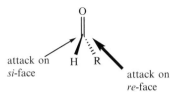

attack on   H   R
si-face

attack on
re-face

Appendix 2: Amino Acid Abbreviations

Abbreviation		Amino acid	Side chain
A	Ala	Alanine	$-CH_3$
C	Cys	Cysteine	$-CH_2SH$
D	Asp	Aspartic acid	$-CH_2CO_2H$
E	Glu	Glutamic acid	$-(CH_2)_2CO_2H$
F	Phe	Phenylalanine	$-CH_2Ph$
G	Gly	Glycine	$-H$
H	His	Histidine	$-CH_2$-imidazole
I	Ile	Isoleucine	$-CH(CH_3)CH_2CH_3$
K	Lys	Lysine	$-(CH_2)_4NH_2$
L	Leu	Leucine	$-CH_2CH(CH_3)_2$
M	Met	Methionine	$-(CH_2)_2SCH_3$
N	Asn	Asparagine	$-CH_2CONH_2$
P	Pro	Proline	$-(CH_3)_3-N_\alpha$ (cyclic)
Q	Gln	Glutamine	$-(CH_2)_2CONH_2$
R	Arg	Arginine	$-(CH_2)_3NHC(NH_2)_2^+$
S	Ser	Serine	$-CH_2OH$
T	Thr	Threonine	$-CH(CH_3)OH$
V	Val	Valine	$-CH(CH_3)_2$
W	Trp	Tryptophan	$-CH_2$-indole
Y	Tyr	Tyrosine	$-CH_2-C_6H_4-OH$

Appendix 3: A Simple Demonstration of Enzyme Catalysis

An important (and enjoyable) part of chemistry is to be able to demonstrate chemical principles by experiment. Many of the experiments described in this book are highly technical, but in principle enzymes can be easily isolated from a variety of natural sources. For readers who would like to observe enzyme catalysis at first hand, I include a brief procedure for a second year experiment that I set up at Southampton University in 1996, which involves an esterase activity easily isolated from orange peel!

The experiment involves the regioselective enzymatic hydrolysis of hydroxybenzoic acid diester derivatives to either the hydroxy-ester or the acyl-acid product (see Figure A3.1). The *para*-substituted diester methyl 4-acetoxybenzoate is commercially available (Aldrich). Alternatively, diester derivatives of *para*-, *meta*- or *ortho*-hydroxybenzoic acid can be readily synthesised by acid-catalysed esterification of the carboxyl group, followed by pyridine-catalysed acylation of the phenolic hydroxyl group (see Figure A3.2; see Vogel for standard procedures). The enzymatic hydrolysis of each diester derivative is then assayed against orange peel esterase (as a crude extract) and commercially available (Sigma) pig liver esterase (PLE) and porcine pancreatic lipase (PPL), using thin-layer chromatography to monitor the appearance of one or other hydrolysis product and to measure the rate of hydrolysis.

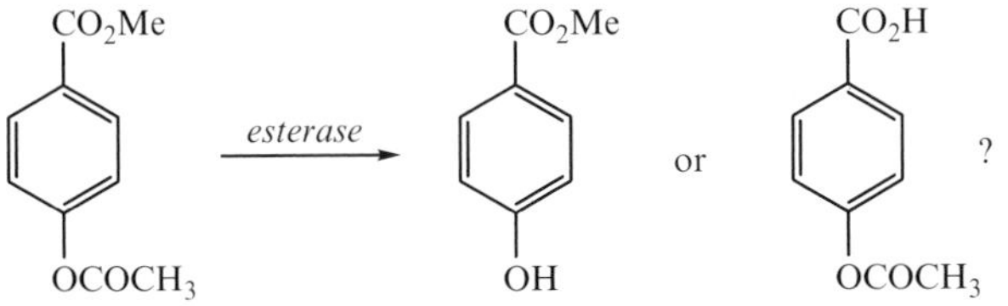

Figure A3.1

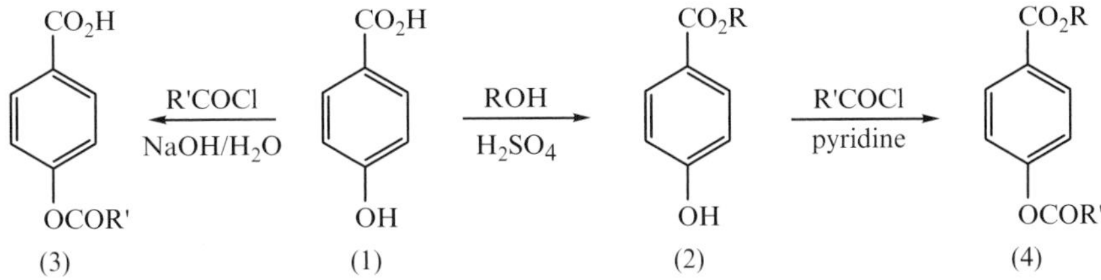

Figure A3.2

Preparation of orange peel extract

Peel the outer layer (the 'zest') of one or two oranges using a knife. Add the peel to 50–60 ml of 50 mM sodium citrate buffer (pH 5.5) containing 2.3% NaCl, and homogenise in a blender for 2 min until homogeneous. If you have access to a centrifuge, then spin at 12 000 g for 10 min, and decant the clear orange supernatant into a beaker for use in the enzyme assays. The extract is stable for at least 24 h if kept on ice.

Assays for enzyme-catalysed hydrolysis

Devise a suitable thin-layer chromatography system to separate the diester (4) from hydroxy-ester (2) and acylated acid (3). Typically 1:1 ethyl acetate/petroleum ether (60–80° fraction); dichloromethane; or dichloromethane/10% methanol are useful eluents. Dissolve 0.1 g of your diester (4) in 5 ml acetone, and set up the following incubations in screw-topped vials:

(a) 0.1 ml diester (4) solution + 0.9 ml 50 mM sodium citrate buffer (pH 5.5);
(b) 0.1 ml diester (4) solution + 0.9 ml orange peel extract;
(c) 0.1 ml diester (4) solution + 0.9 ml pig liver esterase stock (1.0 unit ml^{-1} in 50 mM potassium phosphate buffer (pH 7.0));
(d) 0.1 ml diester (4) solution + 0.9 ml porcine pancreatic lipase stock (1.0 unit ml^{-1} in 50 mM potassium phosphate buffer (pH 7.0))

Analyse the four incubations by thin-layer chromatography at 30 min, 1 h, 2 h and 24 h time points, and you should see enzymatic hydrolysis to (2) or (3). In all cases that we have examined some enzymatic hydrolysis was observed with at least one of the three enzymes, and in nearly all cases the enzymatic hydrolysis was specific for production of either hydroxy-ester (2) or acyl acid (3).

Further reading

T.D.H. Bugg, A.M. Lewin & E.R. Catlin (1997) Regiospecific ester hydrolysis by orange peel esterase – an undergraduate experiment. *J. Chem. Educ.*, **74**, 105–7.
A.I. Vogel (1989) *Vogel's Textbook of Practical Organic Chemistry*, 5th edn. Longman, Harlow.

Appendix 4: Answers to Problems

Chapter 2

(1) pH4. (a) Arginine, lysine, histidine. (b) None.

pH7. (a) Arginine, lysine, histidine (partially). (b) Aspartate, glutamate.

pH10. (a) Arginine. (b) Aspartate, glutamate, cysteine, tyrosine (partially).

(2) Donation of nitrogen lone pair into C=O bond requires nitrogen lone pair to be parallel with π bond. In fact there is some double bond character in the C−N bond, so the amide bond is planar and rigid. *Trans*-conformation more favourable due to steric repulsions between α-carbons in *cis*-conformation. Proline forms a secondary amide linkage in which there is a much smaller difference in energy between *cis*- and *trans*-conformations.

(3) 1st reading frame: Thr–Ala–Glu–Asn–Phe–Ala–Pro–Ser–Arg–Stop

2nd reading frame: Arg–Leu–Lys–Thr–Ser–His–Gln–Val–Asp–

3rd reading frame: Gly–Stop (–Lys–Leu–Arg–Thr–Lys–Ser–Ile)

Stop codon is impossible in the middle of a gene, so second reading frame appears to be the right one.

(4) 1(Met) × 4(Ala) × 6(Leu) × 6(Ser) × 2(His) × 2(Asp) × 1(Trp) × 2(Phe)× 6(*Arg*) × 4(*Val*) = 27 648.

A primer based on the His–Asp–Trp–Phe sequence would have a 1 in 8 chance of being correct.

(5) (a) This α−helix has 6 leucines on one face, forming a very hydrophobic surface. This leads to self-aggregation in water to form a four-helix bundle with hydrophobic side chains on the inside and hydrophilic side chains on the surface. There are also favourable Glu_4-Lys_8 and Glu_5-Lys_9 electrostatic interactions which stabilise the helix.

(b) This α-helix has five lysines on one face, designed to mimic the active site of acetoacetate decarboxylase, where the proximity of a second lysine residue leads to a lower pK_a value and a more nucleophilic lysine. This helix showed some catalytic activity as an oxaloacetate decarboxylase.

Chapter 3

(1) Intramolecular base catalysis, with the internal tertiary amine acting as a base to deprotonate an attacking water molecule. Tertiary amines are good bases and poor nucleophiles, so nucleophilic catalysis is not feasible.

(2) Intramolecular acid catalysis (carboxylic acid will be protonated at pH 4).

(3) (a) Phenoxide ion attacks to form five-membered ring. Effective concentration 7.3×10^4 M. Large rate acceleration due to intramolecular nucleophilic attack, five-membered ring.

(b) Either (i) general base-catalysed attack of water or (ii) nucleophilic attack by active site aspartate to give covalent ester intermediate (similar to haloalkane dehalogenase). To examine mechanism (ii) could try to detect covalent intermediate (e.g. by stopped flow methods). Rate enhancement through transition state stabilisation, bifunctional catalysis, etc.

(4) In the first catalytic cycle, the oxygen atom introduced into the product comes from the aspartate nucleophile, via hydrolysis of the ester intermediate. In subsequent cycles ^{18}O label becomes incorporated into the active site aspartate and is transferred to product.

Chapter 4

(1) 787 units mg^{-1} ÷ 28 mg μmol^{-1} = 28 μmol product min^{-1} μmol^{-1} enzyme. So $k_{cat} = 0.47$ s^{-1}.

(2) Use Lineweaver–Burk or Eadie–Hofstee plot. $v_{max} = 6.0$ nmol min^{-1}, $K_M = 0.71$ m M. $k_{cat} = 2.0$ s^{-1}, $k_{cat}/K_M = 2,800$ M$^{-1}s^{-1}$.

(4) Retention of stereochemistry. Not consistent with an S_N2-type displacement.

(5) Imine linkage formed between aldehyde group and ε-amino group of an active site lysine residue. Enzymatic reaction goes with retention of configuration at phosphorus, whereas non-enzymatic reaction goes with inversion. Suggests that non-enzymatic reaction is a single displacement, whereas enzymatic reaction is probably a double displacement reaction proceeding via a phosphoenzyme intermediate.

(6) Could try to detect phosphoenzyme intermediate using ^{32}P-labelled substrate. In D$_2$O should see ^{2}H incorporation into acetaldehyde.

Chapter 5

(1) Hydrolysis of acyl enzyme intermediate is rate-limiting step in this case. Rapid formation of acyl enzyme intermediate, releasing a stoichiometric amount of p-nitrophenol, followed by a slower hydrolysis step.

(2) Mechanism as for chymotrypsin, via acyl enzyme intermediate. Phosphorylation of active site serine by organophosphorus inhibitors gives a tetrahedral adduct resembling the tetrahedral intermediate in the mechanism, which is hydrolysed very slowly. Differences in toxicity due to: (i) presence of sulphur on parathion, which de-activates the phosphate ester (in insects this is rapidly oxidised to the phosphate ester, which then kills the

insect!); (ii) differences in active site structure between human and insect enzymes.

Antidote binds to choline site through positively-charged pyridinium group. Hydroxylamine group is a potent nucleophile which attacks the neighbouring tetrahedral phosphate ester. Thus the rate of hydrolysis of the tetrahedral phosphate adduct is rapidly accelerated.

(3) Aldehyde is attacked by active site cysteine, generating a thio-hemiacetal intermediate which mimics the tetrahedral intermediate of the normal enzymatic reaction, and is hence bound tightly by the enzyme.
(4) Acetate kinase gives acyl phosphate intermediate, acetate thiokinase involves acyl adenylate (RCO.AMP) intermediate.
(5) Glycogen phosphorylase cleaves glucose units successively from end of chain. Reaction proceeds with retention of configuration at the anomeric position, so a covalent intermediate is probably formed (cf. lysozyme) by attack of an active site carboxylate. Displacement by phosphate gives α-D-glucose-1-phosphate. Phosphoglucose isomerase contains phosphorylated enzyme species which transfers phosphate to C-6 to give 1,6-dipho-spho-glucose. Dephosphorylation at C-1 regenerates phosphoenzyme species. Glucose-6-phosphatase is straightforward phosphate monoester hydrolysis. Defect in glycogen phosphorylase leads to inability to utilise glycogen, so unable to maintain periods of physical exercise.
(6) Processed by enzyme to give covalent 2'-fluoro glycosyl enzyme intermediate. 2'-Fluoro substituent is electron-withdrawing, destabilises oxonium ion, therefore slows down the hydrolysis of the covalent intermediate.

Chapter 6

(1) Alcohol dehydrogenase: $(-0.16\,(CH_3CHO))-(-0.32(\,NAD^+))=+0.16V$.
Enoyl reductase: $(+0.19\,(enoyl\,CoA))-(-0.32\,(NAD^+))= +0.49V$.
Acyl CoA dehydrogenase: $(+0.25\,(cytc_{ox}))-(+0.19\,(enoyl\,CoA)) = +0.06V$.

In acyl CoA dehydrogenase the redox potential for the intermediate FAD must be close to $+0.19V$ if electron transfer is to be thermodynamically favourable. This is right at the top end of the redox potential range for flavin.

(2) Enzyme transfers *proR* hydrogen of NADPH onto C-3 position of substrate. Overall *syn*-addition of hydrogens from NADPH and water.
(3) Transfer of H^* onto enzyme-bound NAD^+. Resulting C-4 ketone assists the E1cb elimination of C-6 hydroxyl group to give unsaturated ketone intermediate. Transfer of H^* from cofactor to C-6 gives product.
(4) Transfer of *proS* hydrogen of NADPH to FAD. Then reverse of acyl CoA dehydrogenase mechanism: transfer of $H^\bullet$ onto β-position giving α-radical; electron transfer from flavin semiquinone to give α-carbanion; protonation

from water at α-position. Note that the same hydrogen transferred from NADPH to FAD is then transferred to substrate.

(5) Formation of flavin hydroperoxide intermediate from $FADH_2$ and O_2. Attack on flavin hydroperoxide *para* to phenolic hydroxyl group, followed by elimination of nitrite to give quinone. Quinone then reduced to hydroquinone by second equivalent of NADH.

(6) Mechanism of hydroxylation as for general mechanism, via iron(IV)-oxo species. Triple bond of inhibitor is epoxidised to give reactive alkene epoxide intermediate, which rearranges with 1,2-shift of H^* to give a ketene intermediate. This is attacked either by water, giving the by-product, or by an active site nucleophile, leading to covalent modification.

Chapter 7

(1) Opening of monosaccharide at C-1 reveals aldehyde substrate. Since enzyme requires no cofactors it presumably proceeds through imine linkage at C-2 of pyruvate, followed by deprotonation at C-3 to give enamine intermediate. Carbon–carbon bond formation between enamine and aldehyde, followed by hydrolysis of resulting imine linkage.

(2) Sequential addition of three malonyl CoA units as for fatty acid synthase gives a tetraketide intermediate. Formation of carbanion between first and second ketone groups, followed by reaction with thioester terminus, leads to formation of chalcone. Formation of carbanion adjacent to thioester terminus, followed by reaction with first ketone group, leads after decarboxylation (β-keto-acid) to resveratrol. Very similar reactions, so similar active sites, but differences in position of carbanion formation and carbon–carbon bond formation.

(3) Attack of bicarbonate onto phosphate monoester gives enol intermediate and carboxyphosphate. Attack of enol at C-3 onto carboxyphosphate gives carboxylated product.

(4) Reverse of normal biotin mechanism. Attack of N-1 of biotin cofactor onto oxaloacetate gives pyruvate and carboxy-biotin intermediate, which carboxylates propionyl CoA to give methylmalonyl CoA.

(5) Attack of TPP anion onto keto group gives tetrahedral adduct. Cleavage of α,β-bond using TPP as electron sink gives enamine intermediate. This reacts with aldehyde of second substrate to give another tetrahedral adduct. Detachment from cofactor regenerates TPP anion.

(6) Attack of TPP anion onto ketone gives tetrahedral adduct. Decarboxylation gives enamine intermediate as in normal mechanism. At this point the enamine intermediate is oxidised by FAD to give acetyl adduct (probably via H$^\bullet$ transfer followed by single electron transfer). Hydrolysis by attack of water gives acetate product, and regenerates TPP anion. FADH$_2$ re-oxidised to FAD by O$_2$.

(7) 1,3-migration of pyrophosphate to give linalyl PP. Attack at C-1 to form six-membered ring gives tertiary carbonium ion. Formation of second ring with concerted attack of pyrophosphate gives product. No positional isotope exchange in pyrophosphate means that pyrophosphate is bound very tightly (to Mg^{2+}) by enzyme, and is not even free to rotate.

(8) Reaction to copalyl PP commenced by protonation of terminal alkene to give tertiary carbonium ion. Two ring closures followed by loss of proton to form C=C. Loss of pyrophosphate followed by formation of six-membered ring. Closure to form final five-membered ring followed by 1,2-alkyl shift and elimination.

(9) Aromatic precursor made from tetraketide intermediate which is methylated by SAM: carbanion formation between first and second ketones is followed by attack on terminal thioester carbonyl. Formation of phenoxy radical *para* to methyl group. Carbon–carbon bond formation via radical coupling. Re-aromatisation of left-hand ring followed by attack of phenoxide onto $\alpha\beta$-unsaturated ketone of right-hand ring.

Chapter 8

(1) Treat with substrate and NaB^3H_4 (or ^{14}C-substrate + $NaBH_4$), degrade labelled enzyme by protease digestion, purify labelled peptide and sequence.

(2) Overall *syn*-addition. Attack of carboxylate onto $\alpha\beta$-unsaturated carboxylic acid to give enolate intermediate, which protonates on same face.

(3) Possibilities are: (i) elimination of water to give enol intermediate, which protonates in β-position; (ii) 1,2-hydride shift from α- to β-position (cf. Pinacol re-arrangement). In mechanism (ii) an α-^2H substituent would be transferred to β-position, in mechanism (i) probably not (unless single base responsible for proton transfer). No intramolecular proton transfer observed in practice, so probably mechanism (i).

(4) Oxidation of C-3' hydroxyl group by enzyme-bound NAD^+. Formation of C-3' ketone assists elimination of methionine by E1cb mechanism, giving $\alpha\beta$-unsaturated ketone intermediate. Addition of water at C-5', followed by reduction of C-3' ketone, regenerating enzyme-bound NAD^+.

(5) Isochorismate synthase could be 1,5-addition of water, but all three reactions can be rationalised by attack of an enzyme active site nucleophile at C-2 and allylic displacement of water, giving a covalent intermediate. Attack of water at C-6 gives isochorismate. Attack of ammonia at C-6 or C-4, followed by elimination of pyruvate, gives anthranilate or *p*-aminobenzoate, respectively.

Chapter 9

(1) Use of PLP as a four-electron sink. Formation of threonine-PLP adduct, followed by deprotonation in α-position, gives ketimine intermediate. Deprotonation at β-position possible using imine as electron sink. Either one-base or two-base mechanisms possible for epimerisation process.

(2) Formation of aldimine adduct followed by α-deprotonation gives $\alpha\beta$-unsaturated ketimine intermediate. This can be attacked by an active site nucleophile at γ-position to covalently modify enzyme.

(3) Formation of aldimine adduct with inhibitor followed by α-deprotonation gives a ketimine intermediate which is a tautomeric form of an aromatic amine. Rapid aromatisation gives a covalently modified PLP adduct.

(4) Use of PLP as a four-electron sink. Formation of amino acid-PLP adduct followed by α-deprotonation. Removal of C-3 *proS* hydrogen using imine as electron sink followed by elimination of phosphate gives βγ-unsaturated intermediate. Protonation at γ-position, followed by attack of water at β-position, and reprotonation at α-position.

(5) Attachment of PLP onto α-amino group followed by α-deprotonation gives ketiminine intermediate. Hydration of γ-ketone group is followed by cleavage of βγ-bond, using imine as an electron sink. C–C cleavage and reprotonation proceeds with overall retention of stereochemistry.

Chapter 10

(1) One *S* centre is epimerised by the epimerase enzyme with introduction of an α-^{2}H, but if left to equilibrate both *S* centres would undergo enzyme-catalysed exchange with ^{2}H$_2$O. *R* centre is decarboxylated, with replacement by ^{2}H, so in principle three atoms of ^{2}H would be found in the L-lysine product.

(2) Deprotonation at γ-position to form dienol intermediate is followed by reprotonation at α-position. 2-Chlorophenol is processed to give δ-chloro intermediate. Upon deprotonation at the γ-position loss of Cl$^-$ gives a γδ-unsaturated lactone. This is processed by opening of the lactone, and reduction of the αβ-bouble bond by an NADH-dependent reductase.

(3) Protonation of the dienol at C-5 is followed by attack of water (or an active site nucleophile) at the C-6 ketone. Cleavage of the C-5,C-6 bond is then facilitated by the presence of an αβ-unsaturated ketone group which can act as an electron sink.

(4) Possibilities are: (i) reversible attack of an active site nucleophile at the amide carbonyl, allowing free rotation of the tetrahedral intermediate; (ii) a 'strain' mechanism in which the enzyme binds the substrate in a strained conformation close to the transition state for rotation of the amide bond. Available evidence points to mechanism (ii), and it is thought that cyclosporin A acts as a mimic of this strained intermediate. (Note: the immunosuppressant activity is due to the complex formed between cyclosporin A and this protein, which is called cyclophilin.)

(5) Concerted four-electron pericyclic reaction is a disfavoured process. Stepwise reaction possible by transfer of phosphate group to active site group, followed by reaction of enol intermediate with phosphoenzyme species.

Chapter 11

(1) (a) Abstraction of hydrogen atom by Ado$-$CH$_2$ radical at C-2, followed by cyclisation onto C-4 to give a cyclopropyl radical; fragmentation of other C$-$C bond in cyclopropane ring to give product radical.

(b) Formation of radical at C-2, followed by fragmentation of C-3–C-4 bond to give alkenyl radical, which re-attacks acrylic acid at C-2 to give product radical.

(2) (a) Abstraction of H* to give radical intermediate; cyclisation onto ester carbonyl to give cyclopropyl intermediate; fragmentation to give product radical.

(b) Possibly SAM-dependent radical reaction, or generation by protein radical or haem Fe=O.

(3) Abstraction of hydrogen atom from methyl group of toluene; attack of radical on C=C of succinate to give product radical; abstraction of hydrogen atom from glycine α-CH_2 to give product, and regenerate glycyl radical.

(4) Generation of tyrosyl radical and cysteine radical (one-electron oxidations by copper centre); radical coupling to form C–S bond, followed by re-aromatisation.

(5) Attack of methanol at quinone C=O to form hemiketal intermediate, followed by abstraction of adjacent hydrogen by hemiketal O$^-$ and elimination of formaldehyde (using the other C=O as electron sink), to give reduced PQQ.

Chapter 12

(1) In principle the reaction is reversible, which would allow the ribozyme to integrate itself into a piece of RNA (reminiscent of the life cycle of some viruses).

(2) A cyclic phosphonate ester would be a good transition state analogue. Such an analogue was used to elicit antibodies capable of catalysing this lactonisation reaction with an enantiomeric excess of 94%.

(3) At neutral pH the tertiary amine will be protonated. When exposed to the immune response, this will elicit a complementary antibody containing a negatively-charged carboxylate group in close proximity. This group functions as a base for the elimination reaction. This 'bait and switch' trick has been used in other cases also.

(4) (a) Might expect to mimic cysteine proteases, hydrolysing aromatic ester (and possibly amide) substrates.

(b) Might expect to mimic FAD-containing oxidases, oxidising aromatic amines and aromatic thioethers (to sulphoxides).

Index